T0211507

Handbook of Experimental Pharmacology

Volume 257

The *Handbook of Experimental Pharmacology* is one of the most authoritative and influential book series in pharmacology. It provides critical and comprehensive discussions of the most significant areas of pharmacological research, written by leading international authorities. Each volume in the series represents the most informative and contemporary account of its subject available, making it an unrivalled reference source.

HEP is indexed in PubMed and Scopus.

More information about this series at http://www.springer.com/series/164

Anton Bespalov • Martin C. Michel •
Thomas Steckler
Editors

Good Research Practice in Non-Clinical Pharmacology and Biomedicine

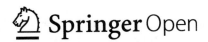

Editors
Anton Bespalov
Partnership for Assessment &
Accreditation of Scientific Practice
Heidelberg, Baden-Württemberg, Germany

Martin C. Michel
Department of Pharmacology
Johannes Gutenberg University
Mainz, Rheinland-Pfalz, Germany

Thomas Steckler
Janssen Pharmaceutica N.V.
Beerse, Belgium

ISSN 0171-2004   ISSN 1865-0325   (electronic)
Handbook of Experimental Pharmacology
ISBN 978-3-030-33658-5   ISBN 978-3-030-33656-1   (eBook)
https://doi.org/10.1007/978-3-030-33656-1

This book is an open access publication.

This Springer imprint is published by the registered company Springer Nature Switzerland AG.
The registered company address is: Gewerbestrasse 11, 6330 Cham, Switzerland

Preface

Pharmacologists and other experimental life scientists study samples to infer conclusions about how a molecule, cell, organ, and/or organism work in health and disease and how this can be altered by drugs. The concept of inference implies that whatever is reported based on the sample under investigation is representative for the molecule, cell, organ, or organism under investigation in general. However, this generalizability requires that two fundamental conditions are met: First, what is being reported must be a true representation of what has been found. This sounds trivial, but if data are selected, e.g., by unexplained removal of outliers or reporting is biased by focusing on the findings in support of a hypothesis, reported data become a biased rather than a true representation of what has been found. Second, what has been found must be robust, i.e., other investigators doing a very similar experiment should come up with similar findings. This requires a complete reporting of what exactly has been done. It also requires that biases at the level of sampling, measuring, and analyzing are reduced as much as feasible. These are scientific principles that have been known for a long time. Nonetheless, scientific practice apparently often ignores them—in part based on felt pressure to generate as many articles in high-profile journals as possible. While the behavior of each participant (investigators, institutions, editors, publishers, and funders) follows an understandable logic, the result is counterproductive for the greater aims of scientific investigation. Against this background, this volume in the series Handbook of Experimental Pharmacology discusses various aspects related to the generation and reporting of robust data. It is published 100 years after the first volume of the series, and this anniversary is a fitting occasion to reflect on current practice and to discuss how the robustness of experimental pharmacology can be enhanced.

It has become clear in the past decade, that many if not most preclinical study findings are not reproducible. Governmental and nongovernmental funding bodies have realized that and demand that the scientific community improves its standards of scientific rigor. Particularly, the use of experimental animals in biomedical research creates the ethical imperative that they are utilized in robustly designed studies only.

The challenge to generate robust results of high quality is not a unique to experimental biology, nor is it fundamentally different in academia and pharmaceutical industry. In some areas subject to regulatory oversight, such as manufacturing or in clinical trials, clear rules and regulations have been established that are internationally agreed upon, such as those of Good Manufacturing Practice, Good Laboratory Practice, or Good Clinical Practice. The exploratory nature, complexity, and often unexpectedness of experimental pharmacology may in many ways be unfit for such formal rules. Nonetheless, we as a community need to improve our standards also in the nonregulated areas of biomedical research. Indeed, various groups of individuals and organizations have developed guidelines for doing robust science and its reporting, which are summarized and discussed by Kabitzke et al. in the chapter "Guidelines & Initiatives for Good Research Practice". Many of the conclusions reached in this book will not sound new to those who are engaged in clinical research because most of them have been established long ago in the context of evidence-based medicine as discussed by Lefevre and Balice-Gordon in the chapter "Learning from Principles of Evidence-Based Medicine to Optimize Non-clinical Research Practices". Thus, limited robustness of many published studies is mainly not a problem of not knowing better but of implementation of established best practice. Bongiovanni et al. discuss means of quality assurance in a non-regulated environment in the chapter "Quality in Non-GxP Research Environment".

The second part of this volume consists of chapters discussing quality aspects in the design, execution, analysis, and performance of studies in experimental pharmacology. Four chapters discuss aspects of study design, including general principles (Huang et al., in the chapter "General Principles of Preclinical Study Design"), the role and characteristics of exploratory vs. confirmatory studies (Dirnagl, in the chapter "Resolving the Tension Between Exploration and Confirmation in Preclinical Biomedical Research"), the role and challenges of randomization and blinding (Bespalov et al., in the chapter "Blinding and Randomization"), and the need for appropriate positive and negative controls (Moser, in the chapter "Out of Control? Managing Baseline Variability in Experimental Studies with Control Groups"). Most of the arguments being made in these chapters will sound familiar to those engaged in clinical trials. More specific for experimental research are the next three chapters dealing with the quality of research tools (Doller and Wes, in the chapter "Quality of Research Tools"), genetic background and sex of experimental animals (Sukoff Rizzo et al., in the chapter "Genetic Background and Sex: Impact on Generalizability of Research Findings in Pharmacology Studies"), and aspects of quality in translational studies (Erdogan and Michel, in the chapter "Building Robustness Intro Translational Research").

Even if study design and execution have followed quality principles, findings may not be robust if their analysis and reporting fall short of those principles. If methods being used and results being obtained are not described with enough granularity, they become irreproducible on technical grounds: you can only confirm something if you know what has been done and found. Emmerich and Harris (in the chapter "Minimum Information and Quality Standards for Conducting, Reporting, and Organizing In Vitro Research") and Voehringer and Nicholson (in the chapter "Minimum Information in In Vivo Research") propose minimum information to be provided in the reporting of findings from in vitro and in vivo research, respectively. Also, lack of understanding of the principles of statistical analysis and misinterpretation of *P*-values are widely seen as a major contributing factor to poor robustness. Lew (in the chapter "A Reckless Guide to *P*-Values: Local Evidence, Global Errors") summarizes key principles of statistical analysis—in a surprisingly readable and entertaining manner given the inherent dryness of the subject.

As most research is carried out by teams, it becomes essential that each member documents what has been done and found in a manner accessible to other team members. Gerlach et al. (in the chapter "Electronic Lab Notebooks and Experimental Design Assistants") discuss the advantages and challenges of using electronic lab notebooks and electronic study design assistants as well as the ALCOA (Attributable, Legible, Contemporaneous, Original, and Accurate) and FAIR (Findable, Accessible, Interoperable, Reusable) principles and Hahnel (in the chapter "Data Storage") issues of data storage. Even when data have been generated, analyzed, and reported to the highest standards, some degree of disagreement is expected in the scientific literature. Systematic reviews and meta-analysis have long been powerful tools of clinical medicine to aggregate the available evidence. Macleod et al. (in the chapter "Design of Meta-Analysis Studies") discuss how this experience can be leveraged for experimental research. When all of this is said and done, findings should be published to become available to the scientific community. Based on his perspective as editor of a high-profile journal, Hrynaszkiewicz (in the chapter "Publishers' Responsibilities in Promoting Data Quality and Reproducibility") discusses how publishers can contribute to the unbiased reporting of robust research.

The question arises whether increased overall robustness of published data in the experimental life sciences can be achieved based on individual investigators doing the right thing or whether structural elements are required to support this. This book highlights three specific aspects why efforts to introduce and maintain high research quality standards cannot be reduced to isolated guidelines, recommendations, and policies. First, such efforts will not be successful without viewing them in the broad context of research environment and infrastructure and without providing a possibility for the changes to trigger feedback. Gilis (in the chapter "Quality Governance in Biomedical Research") discusses the importance of fit-for-purpose quality governance. Second, many areas of modern research environment are already a subject to existing legal and institutional rules and policies. Good research practice does not come into conflict but can effectively learn from how other changes were introduced and are maintained. Guillen and Steckler (in the chapter "Good Research Practice: Lessons from Animal Care & Use") focus on the care and use of experimental animals that cannot be separated from issues related to study design, execution, and analysis. Third, most biomedical research today is conducted by teams of scientists, often across different countries. Vaudano (in the chapter "Research Collaborations and Quality in Research: Foes or Friends?") stresses the importance of the role of transparency and data sharing in scientific collaborations. However, all of this would be incomplete by ignoring the elephant in the room: what will it cost in terms of financial, time, and other resources to implement quality in research? Grondin and coworkers (in the chapter "Costs of Implementing Quality in Research Practice") explore how quality can be achieved with limited impact on often already limited resources.

This volume has largely drawn on authors who participate as investigators in the European Quality In Preclinical Data (EQIPD; https://quality-preclinical-data.eu) consortium. EQIPD is part of the Innovative Medicine Initiatives (https://imi. europa.eu), a public–private partnership between the European Commission and the European Federation of Pharmaceutical Industries and Associations. EQIPD was launched in October 2017 to develop simple, sustainable solutions that facilitate improvements in data quality without adversely impacting innovation or freedom of research. EQIPD brings together researchers from 29 institutions in academia, small businesses, and major pharmaceutical companies from eight countries. Moreover, it involves stakeholders from various nonprofit organizations not only in Europe but also in Israel and the USA and a range of international advisors and collaborators including the U.S. National Institutes of Health. Some of the contributors to this volume come from this expanded group of EQIPD participants.

The efforts of authors and editors of this volume would have limited impact if they were not accessible to the wider community of experimental pharmacologists and biologists in general. Therefore, we are happy that not only each chapter will be listed individually on PubMed for easy retrieval but also available as open access. Therefore, we would like to thank the sponsors making open access publication of this volume possible: AbbVie Inc., Boehringer Ingelheim Pharma GmbH & Co. KG, Janssen Pharmaceutica NV, Pfizer Inc., Sanofi-Aventis Recherche et Dévelopement, and UCB Biopharma SPRL, all of whom are members of EQIPD. Finally, we would like to thank our families and specifically spouses, Inna, Martina, and Regine, who tolerated the time we spent on preparing this volume.

<table>
<tr><td>Heidelberg, Germany</td><td>Anton Bespalov</td></tr>
<tr><td>Mainz, Germany</td><td>Martin C. Michel</td></tr>
<tr><td>Beerse, Belgium</td><td>Thomas Steckler</td></tr>
</table>

Contents

Quality in Non-GxP Research Environment

Sandrine Bongiovanni, Robert Purdue, Oleg Kornienko, and René Bernard

Contents

S. Bongiovanni (✉)
Quality Assurance, Novartis Institutes for BioMedical Research (NIBR), Novartis Pharma AG, Basel, Switzerland
e-mail: sandrine.bongiovanni@novartis.com

R. Purdue
Information and Compliance Management, Novartis Institutes for BioMedical Research Inc., Cambridge, MA, USA
e-mail: robert.purdue@novartis.com

O. Kornienko
External Services Quality Assurance, Novartis Institutes for Biomedical Research, Cambridge, MA, USA
e-mail: oleg.kornienko@novartis.com

R. Bernard
Department of Experimental Neurology, Clinic of Neurology, Charité Universitätsmedizin, Berlin, Germany
e-mail: rene.bernard@charite.de

© The Author(s) 2019   1
A. Bespalov et al. (eds.), *Good Research Practice in Non-Clinical Pharmacology and Biomedicine*, Handbook of Experimental Pharmacology 257,
https://doi.org/10.1007/164_2019_274

Abstract

There has been increasing evidence in recent years that research in life sciences is lacking in reproducibility and data quality. This raises the need for effective systems to improve data integrity in the evolving non-GxP research environment. This chapter describes the critical elements that need to be considered to ensure a successful implementation of research quality standards in both industry and academia. The quality standard proposed is founded on data integrity principles and good research practices and contains basic quality system elements, which are common to most laboratories. Here, we propose a pragmatic and risk-based quality system and associated assessment process to ensure reproducibility and data quality of experimental results while making best use of the resources.

Keywords

ALCOA+ principles · Data integrity · Data quality · EQIPD · European Quality in Preclinical Data · European Union's Innovative Medicines Initiative · Experimental results · Good research practice · IMI · Non-GxP research environment · Quality culture · Reproducibility · Research quality standard · Research quality system · Risk-based quality system assessment · Transparency

1   Why Do We Need a Quality Standard in Research?

Over the past decades, numerous novel technologies and scientific innovation initiated a shift in drug discovery and development models. Progress in genomics and genetics technologies opened the door for personalized medicine. Gene and targeted therapies could give the chance of a normal life for genetically diseased patients. For example, adeno-associated viruses, such as AAV9, are currently used to create new treatments for newborns diagnosed with spinal muscular atrophy (SMA) (Mendell et al. 2017; Al-Zaidy et al. 2019). Similarly, the use of clustered regularly interspaced short palindromic repeats (CRISPR) (Liu et al. 2019) or proteolysis targeting chimeras (PROTACs) (Caruso 2018) is leading to novel cancer therapy developments. The broader use of digitalization, machine learning and artificial intelligence (AI) (Hassanzadeh et al. 2019) in combination with these technologies will revolutionize the drug discovery and clinical study design and accelerate drug development (Pangalos et al. 2019).

Regulators all over the world are closely monitoring these breakthrough scientific advances and drug development revolution. While they evaluate the great promise of innovative medicines, they also raise questions about potential safety risks, ethics and environment. Consequently, new ethical laws and regulations are emerging to mitigate the risks without slowing down innovation. For example, the UK Human Tissue Act became effective in 2006, followed by the Swiss Human Research Act in January 2014 (Swiss-Federal-Government, Effective 1 January 2014); the EU General Data Protection Regulation (No.679/2016, the GDPR) came into effect on May 25, 2018 (EMA 2018a); and the guideline on good pharmacogenomics practice has been in effect since September 2018 (EMA 2018b).

This is exemplified by the EMA Network Strategy to 2020 (EMA 2015), which aims both to promote innovation and to better understand associated risks, in order to provide patients with safe and novel drugs or treatments on the market more rapidly.

This evolving research and regulatory environment, along with many other new challenges, such as aggressive patent litigation cases, increasing burden for approval and reimbursement of new molecular entities (NMEs), challenging market dynamics and high societal pressure enforce radical changes in the research and drug development models of the pharmaceutical industry (Gautam and Pan 2016). In response, most of the pharmaceutical companies have refocused on portfolio management, acquired promising biotechnology companies and developed research collaborations with academia (Palmer and Chaguturu 2017). The goal is to speed up drug development in order to deliver new drugs and new treatments to their patients and customers. Thus, transition from research to drug development should be more efficient. To do so, robust data quality, integrity and reproducibility became essential, and the development of a quality culture across the entire value chain emerged to be critical. Indeed, while many drug development areas already applied the various good practice (GxP) standards and guidances, no recognized quality standard governed discovery and early development. Conversely, discovery activities had to comply with many regulations, such as biosafety, controlled substances and data privacy; thus, there was a real risk of exposure in non-GXP research.

In order to mitigate these newly emerging risks and speed up drug development, some pharmaceutical companies decided to develop their own internal research quality standard (RQS), based on good scientific practices and data integrity, to promote robust science and data quality. The foundations of RQS were the WHO: "Quality Practices in Basic Biomedical Research" (WHO 2005), first published in 2001, and the "Quality in Research Guideline for working in non-regulated research", published by the British Research Quality Association RQA, in 2006 and revised in 2008 and 2014 (RQA-Working-Party-on-Quality-in-Non-Regulated-Research 2014).

Academic research institutions and laboratories are as committed as their pharmaceutical counterparts to good scientific practices but are largely operating without defined standards. Many universities hold their scientists accountable for good scientific practices, which are mainly focused on preventing misconduct and promoting a collaborative environment. Academic output is measured by the amount of publications, often in prestigious journals. Peer review of manuscripts is seen by academics as the main quality control element. During the last decade, the replication and reproducibility crisis in biomedical sciences has exposed severe quality problems in the planning and conduct of research studies in both academia and pharmaceutical industry. Academic crisis response elements include public transparency measures such as preregistration, open-access publication and open data (Kupferschmidt 2018; Levin et al. 2016).

As a result of the replication crisis, which hinges on poor quality of experimental design and resulting data, quality management now has a historic chance to be introduced in the academic biomedical world. Such a system incorporates openness and transparency as key elements for quality assurance (Dirnagl et al. 2018).

2 Critical Points to Consider Before Implementing a Quality Standard in Research

2.1 GxP or Non-GxP Standard Implementation in Research?

Many activities performed in discovery phase and early development are not conducted under GxP standard but need to comply with a number of regulations. Thus, the implementation of an early phase quality standard could help to mitigate the gap and reduce risk exposure. A simple solution could be to apply good laboratory practice (GLP) standards to all research activities in order to mitigate the gap of quality standard.

The classical GxP standards were often born reactively, out of disaster and severe malpractices, which compromised human health. The GLP, for example, originate from the early 1970s, when the Food and Drug Administration (FDA) highlighted several compliance findings in preclinical studies in the USA, such as mis-identification of control and treated animals, suppressed scientific findings, data inventions, dead animal replacements and mis-dosing of test animals. These cases emphasized the need for better control of safety data to minimize risk, in study planning and conduct, in order to both improve the data reliability and protect study participant life. As a result, the FDA created the GLP regulations, which became effective on June 20, 1979. The FDA also launched their Bioresearch Monitoring Program (BIMO), which aimed to conduct routine inspection and data reviews of nonclinical laboratories, in order to evaluate their compliance with the FDA GLP regulation requirements (FDA 1979). Thereafter, the Organisation for Economic Co-operation and Development (OECD) launched their GLP regulation in Europe. Each country, which adopted GLP into their law, tended to add some specificities to their application of GLPs.

Regulated research, which delivers data directly supporting patient safety, is one research area, where GLP were mostly implemented successfully to ensure data integrity and reliability for regulatory approval. Accredited regulatory research laboratories employ continuously trained personnel to perform mainly routine analysis, following defined standard operating procedures (SOPs). Regulatory activities are systematically reviewed/audited by quality assurance groups and inspected by regulators. Thus, developing and maintaining GLP standards needs resources from both research laboratories and regulatory bodies.

In contrast, early discovery research rarely delivers results, which directly impact human health. Therefore the implementation of GxP standards might not be required by the scope of discovery activities (Hickman et al. 2018). However, discovery science would benefit from the use of best scientific practices and quality standards, in order to enhance research robustness and effectiveness and proactively achieve compliance. Many discovery laboratories, hosted either in academia, small biotechs or industries, use cutting-edge technologies, constantly develop novel methods and need the flexibility that GxP standards do not offer. Furthermore, when resources are limited, as often in academia, the implementation of GxP standards is often unbearable. In addition, governmental oversight would increase the burden on the part of the regulatory agencies to come up with specific regulations, check documentation and perform additional inspections.

Therefore the main argument for not extending GxP regulation to non-GxP research is that it would stifle the creativity of researchers, slow down innovation and seriously limit early discovery research. Pragmatic, risk-based and science-driven research quality standards could fit with the discovery activities' scope and requirement of this research activity and ensure data integrity while saving resources.

2.1.1 Diverse Quality Mind-Set

The success of the development and implementation of a research quality standard relies first on understanding the mind-set of GxP group associates and non-GxP researchers.

Experienced GxP scientists, working in conventional science performing routine well-developed and validated assays, generally apply standards consistently and straightforwardly. Risks in such GxP areas are pretty well understood and predicate rules apply. GxP researchers are used to audits and regulatory inspections. Quality assurance departments usually have these activities under strict scrutiny and help to ensure that study documentation is ready for inspection.

In early discovery, the oversight of quality professionals might be lighter. The scientists might be less familiar with audit or inspections. Thus, many pharma companies have implemented clear internal research guidelines, and a number of universities have dedicated teams both to ensure data integrity and to conduct scientists training.

Academic researchers operate under laboratory conditions similar to those in industrial non-GxP research and are united in their commitment to produce high-quality data. There are academic institutional and funder requirements to preserve research data for at least 10 years after a research project ended, many of which support scientific publications. However, there are varying levels of requirements for documentation, aside from laboratory notebooks, which are still in paper format at most universities, despite the fact that most data are nowadays created and preserved in digital format. But the documentation practices are slowly adapting in academic research laboratories: electronic laboratory notebooks are gaining popularity (Dirnagl and Przesdzing 2016), and more and more institutions are willing to cover licensing costs for their researchers (Kwok 2018). Another group of academic stakeholders are funders, who have tightened the requirements in the application phase. Grant application should include data management plans describing processes to collect, preserve data and ensure their public access. These promising developments might mark the beginning of documentation quality standards in academic biomedical research.

2.2   Resource Constraints

The development of phase-appropriate standards, which provide enough flexibility for innovation and creativity while using best practices ensuring documentation quality and data integrity, is complex and requires time and resources. Thus, both a consistent senior management support and a strong partnership between quality professionals and research groups are mandatory to succeed in both the implementation and the maintenance of the research quality standard.

Research groups, which have the right quality culture/mind-set, could require less inputs from a quality organization.

While these requirements are relatively easy to implement in a pharmaceutical setting, the current academic research environment presents a number of hindrances: usually, academic institutions transfer the responsibilities for data integrity to the principal investigators. While many universities have quality assurance offices, their scope might be limited to quality of teaching and not academic research. Internal and external funding sources do not always support a maintainable quality assurance structure needed to achieve research quality characteristics including robustness, reproducibility and data integrity (Begley et al. 2015). However, more and more academia are increasing their efforts to address research quality.

3   Non-GxP Research Standard Basics

The foundation of any quality standards in regulated and non-regulated environments are good documentation practices, based on data integrity principles, named ALCOA+. Thus, a non-GxP Research standard should focuses on data integrity and research reproducibility. The rigor and frequency of its application need to be adapted to the research phase to which it is applied: in early discovery, focus is laid on innovation, protection of intellectual property and data integrity. In contrast, many other elements have to be consistently implemented, such as robust method validation, equipment qualification in nonclinical confirmatory activities or clinical samples analysis under exploratory objectives of clinical protocols and early development.

3.1 Data Integrity Principles: ALCOA+

Essential principles ensuring data integrity throughout the lifecycle are commonly known by the acronym "ALCOA". Stan Woollen first introduced this acronym in the early 1990s when he worked at the Office of Enforcement, in the USA. He used it to memorize the five key elements of data quality when he presented the GLP and FDA's overall BIMO program (Woollen 2010). Since then, QA professionals used commonly the acronym ALCOA to discuss data integrity. Later on, four additional elements, extracted from the Good Automated Manufacturing Practice (GAMP) guide "A Risk-Based Approach to GxP Complaint Laboratory Computerized Systems" (Good Automated Manufacturing Practice Forum 2012), completed the set of integrity principles (ALCOA+). The ALCOA+ consists of a set of principles, which underpins any quality standards:

Principle	Meaning
Attributable	The source of data is identified: who/when created a record and who/when/why changed a record
Legible	Information is clear and readable. In other words, documentation is comprehensive and understandable without need for specific software or knowledge

(continued)

Contemporaneous	Information is recorded at the time of data generation and/or event observation
Original	Source information is available and preserved in its original form
Accurate	There are no errors or editing without documented amendments

Additional elements:

Principle	Meaning
Complete	All data is recorded, including repeat or reanalysis performed
Available	Data is available and accessible at any time for review or audit and for the lifetime of the record
Consistent	Harmonized documentation process is constantly applied
Enduring	Data is preserved and retrievable during its entire lifetime

In order to ensure data integrity and compliance with ALCOA+ principles, all scientific and business practices should underpin the RQS. This standard needs to contain a set of essential quality system elements that can be applied to all types of research, in a risk-based and flexible manner. At a minimum, the following elements should be contained.

3.2   Research Quality System Core Elements

3.2.1   Management and Governance

Management support is critical to ensure that resources are allocated to implement, maintain and continuously improve processes to ensure sustained compliance with RQS. Roles and responsibilities should be well defined, and scientists should be trained accordingly. Routine quality system assessments, conducted by QA and/or scientists themselves, should be also implemented.

3.2.2   Secure Research Documentation and Data Management

Scientists should document their research activities by following the ALCOA+ principles, in a manner to allow reproducibility and straightforward data reconstruction of all activities. Data management processes should ensure long-term data security and straightforward data retrieval.

3.2.3   Method and Assay Qualification

Methods and key research processes should be consistently documented and available for researchers conducting the activity. Assay acceptance/rejection criteria should be predefined. Studies should be well designed to allow statistical relevance. Routine QC and documented peer reviews of research activities and results should be conducted to ensure good scientific quality and reliability. Any change to the method should be documented.

3.2.4 Material, Reagents and Samples Management

Research materials, reagents and samples should be fit for purpose and documented in a manner to permit reproducibility of the research using equivalent items with identical characteristics. Their integrity should be preserved through their entire life cycle until their disposal, which should be consistent with defined regulation or guidance. Research specimens should be labelled to facilitate traceability and storage conditions.

3.2.5 Facility, Equipment and Computerized System Management

Research facilities should be fit for their research activity purpose and provide safe and secure work environments. Research equipment and computerized system, used in the laboratory, should be suitable for the task at hand and function properly. Ideally, their access should be restricted to trained users only, and an activity log should be maintained to increase data traceability.

3.2.6 Personnel and Training Records Management

Research personnel should be competent, trained to perform their research functions in an effective and safe manner. Ideally, in industry environment, personnel and training records should be maintained and available for review.

3.2.7 Outsourcing/External Collaborations

The RQS should be applied to both internal and external activities (conducted by other internal groups, external research centres, academic laboratories or service providers). Agreement to comply with requirements of RQS should be signed off before starting any research work with research groups outside of the organization. Assessment and qualification of an external partner's quality system are recommended and should be conducted in a risk-based manner (Volsen et al. 2014).

3.3 Risk- and Principle-Based Quality System Assessment Approach

The risk-based and principle-based approaches are the standard biopharma industry quality practice to balance resources, business needs and process burden in order to maximize the impact of an assessment. The risk-based approach is essentially an informed and intelligent way to prioritize frequency and type of assessment (remote, on-site) across a large group of service providers.

The principle-based trend reflects the fact that it may not be possible to anticipate and prescriptively address a myriad of emerging nuances and challenges in a rapidly evolving field. Cell and gene therapy (e.g. CAR-NK and CAR-T), digital medicine, complex drug/device interfaces and new categories of biomarkers are just some of the recent examples demanding a flexible and innovative quality mind-set:

- CAR-NK and CAR-T Immuno-oncology therapy is an example where patient is treated with his own or donor's modified cells. Multiple standards and regulations apply. Researchers perform experiments under a combination of sections of good clinical practice (GCP) and good tissue practice (GTP) in a hospital setting (Tang et al. 2018a, b).
- Digital therapeutics are another emerging biopharmaceutical field (Pharmaceuticalcommerce.com 2019). Developers utilize knowledge of wearable medical devices, artificial intelligence and cloud computing to boost the effectiveness of traditional chemical or biological drugs or create standalone therapies. As software becomes a part of treatment, it brings a host of nontraditional quality challenges such as health authority pre-certification, management of software updates and patient privacy when using their own devices.

For the above examples, it is important to adhere to ALCOA+ principles as no single quality standard can cover all the needs.

As quality is by design a support function to serve the needs of researchers, business and traditional quality risk factors need to come together when calculating an overall score.

A simple 3X4 Failure Mode and Effects Analysis (FMEA) – like risk matrix – can be constructed using the following example:

Suppose that:

- A pharmaceutical company wants to use an external service provider and works on coded human tissue, which is a regulated activity by law, in several countries, such as Switzerland and the UK:
  – Quality risk factor 1. Severity is medium.
- This laboratory was already audited by the quality assurance of the pharmaceutical company, and gaps were observed in data security and integrity. Remediation actions were conducted by this laboratory to close these gaps:
  – Quality risk factor 2. Severity is high.
- The planned activity will be using a well-established method that the pharma company needs to transfer to the Swiss laboratory. Since the method need to be handoff, the risk is medium:
  – Business risk factor 1. Severity is medium.
- The data generated by the laboratory may be used later in an Investigational New Drug (IND) Application. This is a submission critical, and it will be filed to Health Authorities.
  – Business risk factor 2. Severity is high.

Risk factor	Severity		
	Low	Medium	High
Quality 1	1	3	9
Quality 2	1	3	9
Business 1	1	3	9
Business 2	1	3	9

The risk matrix is balanced for quality and business components. Final business risk is calculated as a product of two business component severity scores such as medium $\times$ high $= 3 \times 9 = 27$. Quality risk is calculated in the same fashion.

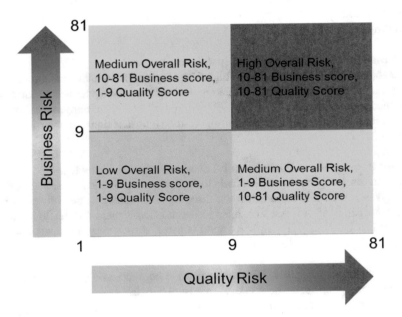

4   How Can the Community Move Forward?

The improvement of research reproducibility is not only about process implementation but also about promoting quality culture. The research community needs to join force to build a harmonized and recognized quality culture in research, providing tools, guidelines and policies to ensure data quality and research reproducibility.

4.1   Promoting Quality Culture

A process might be far easier for building systems than building a culture of quality. Very often goals are set around cost, speed and productivity. But what is the cost of working on poor processes and with low quality?

In the Oxford dictionary, culture is defined as "The ideas, customs and social behaviour of a particular people or society" and quality as "The standard of something as measured against other things of a similar kind; the degree of excellence of something" (Oxford-Dictionary 2019). So what are the building blocks, which could allow the research community to build a strong quality culture and which elements could influence scientist's behaviours to strive for research excellence?

4.1.1 Raising Scientist Awareness, Training and Mentoring

In order to embark on the quality journey, researchers should understand the benefits of embracing robust quality:

First Benefit: Help Ensure Their Sustained Success

Great science can lead to patents, publications, key portfolio management decisions, scientific advances and drug submissions. Robust processes position researcher for sustained success, preserving their scientific credibility and enabling, for example, to defend their patent against litigation, make the right decisions, answer regulator's questions.

Second Benefit: Serve Patients and Advance Scientific Knowledge

The main researcher focus, which fuels their motivation to innovate and go forward, is to advance scientific knowledge and discover new pathways, new drugs and new treatment. Efficient processes enhance research effectiveness and lead to scientific discoveries. Data integrity supports good science, drug safety, products and treatment development for patients and customers.

Once awareness is raised, researchers need to be trained on basic documentation processes and good scientific practices to ensure data integrity and quality. Targeted training should be added on new guidelines, processes and regulations applied to their specific activities (e.g. human tissue use, natural products, pharmacogenomics activities).

4.1.2 Empowering of Associates

The best way to engage researchers is to empower them to perform some changes in order to improve processes and systems. These changes need to be documented, fit for purpose and organized within the quality framework, managed and governed by the senior management. Managers should lead by example, embrace the change in quality culture and interact more with their staff during study planning or laboratory meetings. They should also encourage people to speak up when they observe inaccuracies in the results or potential fraud.

4.1.3 Incentives for Behaviours Which Support Research Quality

A culture that emphasizes research quality can be fostered by providing appropriate incentives for certain behaviours that are aligned with the quality objectives. Such incentives can come in form of promotions, monetary rewards or public recognition. Awards for best practices to ensure data integrity could be a start. Not all incentives must be endured. Some are only necessary to introduce or change a certain practice. Incentives permit an uptake to be measured and the more visible incentives within an institution improve the reach. There is a great variability in effectiveness of a certain incentive. Questionnaires are a useful instrument to find out which incentives are effective for a certain target research population. Any incentives that do not promote quality need to be critically evaluated by the management (Lesmeister 2018; Finkel 2019).

4.1.4 Promoting a Positive Error Culture

"Error is human" and errors will happen in any research laboratory environment, no matter what precautions are taken. However, errors can be prevented from reoccurring and serve as teaching examples for quality assurance and risk management. For this to happen, a positive error culture needs to be created by leaders that embrace learning and do not punish reported errors. The possibility of anonymous reporting is a crucial element as a seed for community trust, so error reporting is not used for blaming and shaming. Next, a guided discussion of reported errors with the laboratory personnel needs to take place, and potential consequences can be discussed. Such a community effort empowers laboratory workers and makes them part of the solution.

An example of a system to manage errors is the free "Laboratory Critical Incident and Error Reporting System" (LabCIRS) software which permits to record all incidents anonymously and to analyse, discuss and communicate them (Dirnagl and Bernard 2018).

4.2 Creating a Recognized Quality Standard in Research: IMI Initiative – EQIPD

Large pharmaceutical companies, service providers and academia are facing the same challenges. They need to manage budget and portfolio, keep credibility and serve customers and patients. Research reproducibility, accuracy and integrity are a benefit to all. For the first time, an Innovative Medicines Initiative (IMI 2008) project on quality was launched in October 2017, named European Quality In Preclinical Data (EQIPD 2017). EQIPD is a 3-year project co-funded by the EU's Innovative Medicines Initiative (IMI 2008) and the European Federation of Pharmaceutical Industries and Associations (EFPIA).

Pharmaceutical companies and academia joined forces to foster a quality culture and develop a "unified non-GxP research quality standard", which is expect to be released in 2020 (Steckler et al. 2018; Macleod and Steckler 2019).

The aim of this project is to establish best practices, primarily in the preclinical neuroscience field but also applicable to the overall non-GxP research, that are harmonized across the pharmaceutical industry to improve data quality and reproducibility in discovery and exploratory research. The EQIPD members are working together to develop simple and sustainable solutions to facilitate implementation of robust research quality systems and expansion of knowledge on principles necessary to address robustness and quality.

4.3   Funders Plan to Enhance Reproducibility and Transparency

The NIH proposed first to implement a mandatory training regarding result reproducibility and transparency and good experimental design. Starting in 2019, the NIH research grant applications now have to include components that address reproducibility, rigor and transparency. Applications must include measures to ensure robust and unbiased experimental design, methodology, analysis, interpretation and reporting of results. More relevant biological models should be considered, and the rigor of prior research that the application is based on should be reviewed. NIH asked publishers to get more involved, promote peer-review and data disclosure. In addition, the whole research community is encouraged to work together in order to improve research reproducibility (National-Institutes-of-Health-NIH 2019).

European funders as well aim to enhance reproducibility, mainly by increased transparency and public data availability of research results. The most prominent EU project with that goal is the European Open Science Cloud (EOSC 2018). A key feature of the EOSC is that the shared data conforms to the FAIR criteria: findable, accessible, interoperable and reusable (Wilkinson et al. 2016, 2019). Also at the national funder level, more calls of applications emerge that specifically address scientific rigor and robustness in non-GLP research (German-Federal-Ministry-of-Education-and-Research 2018).

5   Conclusion

In conclusion, the strategic collaboration between pharmaceutical companies, service providers and academia is critical to help develop both quality culture and standards in research, which could help enhance research reproducibility and data integrity. As resources are often limited, a pragmatic quality system combined with a risk-based approach could mitigate the gaps and proactively address the ever-changing regulatory environment, which continuously expands quality expectations.

References

Al-Zaidy S, Pickard AS, Kotha K, Alfano LN, Lowes L, Paul G, Church K, Lehman K, Sproule DM, Dabbous O, Maru B, Berry K, Arnold WD, Kissel JT, Mendell JR, Shell R (2019) Health outcomes in spinal muscular atrophy type 1 following AVXS-101 gene replacement therapy. Pediatr Pulmonol 54(2):179–185

Begley CG, Buchan AM, Dirnagl U (2015) Robust research: institutions must do their part for reproducibility. Nature 525(7567):25–27

Caruso C (2018) Arvinas, pfizer team up on PROTACs. Cancer Discov 8(4):377–378

Dirnagl U, Bernard R (2018) Errors and error management in biomedical research. In: Hagen JU (ed) How could this happen? Managing errors in organizations. Springer, Cham, pp 149–160

Dirnagl U, Przesdzing I (2016) A pocket guide to electronic laboratory notebooks in the academic life sciences. F1000Res 5:2

Dirnagl U, Kurreck C, Castanos-Velez E, Bernard R (2018) Quality management for academic laboratories: burden or boon? Professional quality management could be very beneficial for academic research but needs to overcome specific caveats. EMBO Rep 19(11):e47143

EMA (2015) Eu-medicines-agencies-network-strategy-2020-working-together-improve-health. https://www.ema.europa.eu/documents/other/eu-medicines-agencies-network-strategy-2020-working-together-improve-health_en.pdf. Accessed 24 May 2019

EMA (2018a) General data protection regulation. Directive 95/46/EC. https://eur-lex.europa.eu/legal-content/EN/TXT/PDF/?uri=CELEX:32016R0679. Accessed 24 May 2019

EMA (2018b) Guideline on good pharmacogenomic practice (EMA/CHMP/718998/2016). http://www.ema.europa.eu/docs/en_GB/document_library/Scientific_guideline/2018/03/WC500245944.pdf. Accessed 24 May 2019

EOSC (2018) European Open Science Cloud (EOSC). https://www.eosc-portal.eu/. Accessed 24 May 2019

EQIPD (2017) European quality in preclinical data. https://quality-preclinical-data.eu/. Accessed 24 May 2019

FDA (1979) FDA issues good laboratory practice (GLP) regulations. https://www.fda.gov/inspections-compliance-enforcement-and-criminal-investigations/compliance-program-guidance-manual-cpgm/bioresearch-monitoring-program-bimo-compliance-programs. Accessed 24 May 2019

Finkel A (2019) To move research from quantity to quality, go beyond good intentions. Nature 566(7744):297

Gautam A, Pan X (2016) The changing model of big pharma: impact of key trends. Drug Discov Today 21(3):379–384

German-Federal-Ministry-of-Education-and-Research (2018) Richtlinie zur Förderung von konfirmatorischen präklinischen Studien. https://www.gesundheitsforschung-bmbf.de/de/8344.php. Accessed 24 May 2019

Good Automated Manufacturing Practice Forum I. S. f. P. E.-I (2012) ISPE GAMP® good practice guide: a risk-based approach to GxP compliant laboratory computerized systems, 2nd ed.

Hassanzadeh P, Atyabi F, Dinarvand R (2019) The significance of artificial intelligence in drug delivery system design. Adv Drug Deliv Rev. https://doi.org/10.1016/j.addr.2019.05.001

Hickman S, Ogilvie B, Patterson T (2018) Good laboratory practice – to GLP or not to GLP? Drug Discov World 19:61–66

IMI (2008) Innovative medicines initiative. http://www.imi.europa.eu/content/mission. Accessed 24 May 2019

Kupferschmidt K (2018) More and more scientists are preregistering their studies. Should you? https://www.sciencemag.org/news/2018/09/more-and-more-scientists-are-preregistering-their-studies-should-you. Accessed 24 May 2019

Kwok R (2018) How to pick an electronic laboratory notebook. Nature 560:269–270

Lesmeister F (2018) How to create a culture of quality improvement. https://www.bcg.com/en-ch/publications/2018/how-to-create-culture-quality-improvement.aspx. Accessed 24 May 2019

Levin N, Leonelli S, Weckowska D, Castle D, Dupre J (2016) How do scientists define openness? Exploring the relationship between open science policies and research practice. Bull Sci Technol Soc 36(2):128–141

Liu B, Saber A, Haisma HJ (2019) CRISPR/Cas9: a powerful tool for identification of new targets for cancer treatment. Drug Discov Today 24(4):955–970

Macleod M, Steckler T (2019) A European initiative to unclog pipeline for new medicines. Nature 568(7753):458

Mendell JR, Al-Zaidy S, Shell R, Arnold WD, Rodino-Klapac LR, Prior TW, Lowes L, Alfano L, Berry K, Church K, Kissel JT, Nagendran S, L'Italien J, Sproule DM, Wells C, Cardenas JA, Heitzer MD, Kaspar A, Corcoran S, Braun L, Likhite S, Miranda C, Meyer K, Foust KD, Burghes AHM, Kaspar BK (2017) Single-dose gene-replacement therapy for spinal muscular atrophy. N Engl J Med 377(18):1713–1722

National-Institutes-of-Health-NIH (2019) Guidance: rigor and reproducibility in grant applications. https://grants.nih.gov/policy/reproducibility/guidance.htm. Accessed 24 May 2019

Oxford-Dictionary (2019) LEXICO dictionary powered by Oxford. https://www.lexico.com/en.

Palmer M, Chaguturu R (2017) Academia–pharma partnerships for novel drug discovery: essential or nice to have? https://doi.org/10.1080/17460441.2017.1318124

Pangalos M, Rees S et al (2019) From genotype to phenotype: leveraging functional genomics in oncology. Drug Discov World Spring 2019

Pharmaceuticalcommerce.com (2019) Prescription digital therapeutics are entering the healthcare mainstream. https://pharmaceuticalcommerce.com/brand-marketing-communications/prescrip tion-digital-therapeutics-are-entering-the-healthcare-mainstream/. Accessed 07 June 2019

RQA-Working-Party-on-Quality-in-Non-Regulated-Research (2014) Guidelines for quality in non-regulated research. https://www.therqa.com/resources/publications/booklets/guidelines-for-quality-in-non-regulated-scientific-research-booklet/. Accessed 24 May 2019

Steckler T, Macleod M, Kas MJH, Gilis A, Wever KE (2018) European quality in preclinical data (EQIPD): een breed consortium voor het verbeteren van de kwaliteit van proefdieronderzoek. Biotech 57(2):18–23

Swiss-Federal-Government (Effective 1 January 2014) Human Research Act HRA. https://www.admin.ch/opc/en/classified-compilation/20061313/201401010000/810.30.pdf. Accessed 24 May 2019

Tang X, Yang L, Li Z, Nalin AP, Dai H, Xu T, Yin J, You F, Zhu M, Shen W, Chen G, Zhu X, Wu D, Yu J (2018a) Erratum: first-in-man clinical trial of CAR NK-92 cells: safety test of CD33-CAR NK-92 cells in patients with relapsed and refractory acute myeloid leukemia. Am J Cancer Res 8(9):1899

Tang X, Yang L, Li Z, Nalin AP, Dai H, Xu T, Yin J, You F, Zhu M, Shen W, Chen G, Zhu X, Wu D, Yu J (2018b) First-in-man clinical trial of CAR NK-92 cells: safety test of CD33-CAR NK-92 cells in patients with relapsed and refractory acute myeloid leukemia. Am J Cancer Res 8(6):1083–1089

Volsen SG, Masson MM, Bongiovanni S (2014) A diligent process to evaluate the quality of outsourced, non GxP. Quasar 128:16–20

WHO (2005) Quality practices in basic biomedical research (QPBR). https://www.who.int/tdr/publications/training-guideline-publications/handbook-quality-practices-biomedical-research/en/. Accessed 24 May 2019

Wilkinson MD, Dumontier M, Aalbersberg IJ, Appleton G, Axton M, Baak A, Blomberg N, Boiten JW, da Silva Santos LB, Bourne PE, Bouwman J, Brookes AJ, Clark T, Crosas M, Dillo I, Dumon O, Edmunds S, Evelo CT, Finkers R, Gonzalez-Beltran A, Gray AJ, Groth P, Goble C, Grethe JS, Heringa J, Hoen PA, Hooft R, Kuhn T, Kok R, Kok J, Lusher SJ, Martone ME, Mons A, Packer AL, Persson B, Rocca-Serra P, Roos M, van Schaik R, Sansone SA, Schultes E, Sengstag T, Slater T, Strawn G, Swertz MA, Thompson M, van der Lei J, van Mulligen E, Velterop J, Waagmeester A, Wittenburg P, Wolstencroft K, Zhao J, Mons B (2016) The FAIR guiding principles for scientific data management and stewardship. Sci Data 3:160018

Wilkinson MD, Dumontier M, Jan Aalbersberg I, Appleton G, Axton M, Baak A, Blomberg N, Boiten JW, da Silva Santos LB, Bourne PE, Bouwman J, Brookes AJ, Clark T, Crosas M, Dillo I, Dumon O, Edmunds S, Evelo CT, Finkers R, Gonzalez-Beltran A, Gray AJG, Groth P, Goble C, Grethe JS, Heringa J, Hoen PAC, Hooft R, Kuhn T, Kok R, Kok J, Lusher SJ, Martone ME, Mons A, Packer AL, Persson B, Rocca-Serra P, Roos M, van Schaik R, Sansone SA, Schultes E, Sengstag T, Slater T, Strawn G, Swertz MA, Thompson M, van der Lei J, van Mulligen E, Jan V, Waagmeester A, Wittenburg P, Wolstencroft K, Zhao J, Mons B (2019) Addendum: the FAIR Guiding Principles for scientific data management and stewardship. Sci Data 6(1):6
Woollen SW (2010) Data quality and the origin of ALCOA. http://www.southernsqa.org/newsletters/Summer10.DataQuality.pdf. Accessed 24 May 2019

Guidelines and Initiatives for Good Research Practice

Patricia Kabitzke, Kristin M. Cheng, and Bruce Altevogt

Contents

Abstract

This chapter explores existing data reproducibility and robustness initiatives from a cross-section of large funding organizations, granting agencies, policy makers, journals, and publishers with the goal of understanding areas of overlap and potential gaps in recommendations and requirements. Indeed, vigorous stakeholder efforts to identify and address irreproducibility have resulted in the development of a multitude of guidelines but with little harmonization.

P. Kabitzke (✉)
Cohen Veterans Bioscience, Cambridge, MA, USA

The Stanley Center for Psychiatric Research, Broad Institute of MIT & Harvard, Cambridge, MA, USA
e-mail: pkabitzk@broadinstitute.org

K. M. Cheng
Pfizer Inc., New York, NY, USA
e-mail: Kristin.Cheng@Pfizer.com

B. Altevogt
Pfizer Inc., Silver Spring, MD, USA
e-mail: Bruce.Altevogt@Pfizer.com

© The Author(s) 2019
A. Bespalov et al. (eds.), *Good Research Practice in Non-Clinical Pharmacology and Biomedicine*, Handbook of Experimental Pharmacology 257,
https://doi.org/10.1007/164_2019_275

19

This likely results in confusion for the scientific community and may pose a barrier to strengthening quality standards instead of being used as a resource that can be meaningfully implemented. Guidelines are also often framed by funding bodies and publishers as recommendations instead of requirements in order to accommodate scientific freedom, creativity, and innovation. However, without enforcement, this may contribute to uneven implementation. The text concludes with an analysis to provide recommendations for future guidelines and policies to enhance reproducibility and to align on a consistent strategy moving forward.

Keywords
Data quality · Good research practice · Guidelines · Preclinical · Reproducibility

1   Introduction

The foundation of many health care innovations is preclinical biomedical research, a stage of research that precedes testing in humans to assess feasibility and safety and which relies on the reproducibility of published discoveries to translate research findings into therapeutic applications.

However, researchers are facing challenges while attempting to use or validate the data generated through preclinical studies. Independent attempts to reproduce studies related to drug development have identified inconsistencies between published data and the validation studies. For example, in 2011, Bayer HealthCare was unable to validate the results of 43 out of 67 studies (Prinz et al. 2011), while Amgen reported its inability to validate 47 out of 53 seminal publications that claimed a new drug discovery in oncology (Begley and Ellis 2012).

Researchers attribute this inability to validate study results to issues of robustness and reproducibility. Although defined with some nuanced variation across research groups, reproducibility refers to achieving similar results when repeated under similar conditions, while robustness of a study ensures that similar results can be obtained from an experiment even when there are slight variations in test conditions or reagents (CURE Consortium 2017).

Several essential factors could account for a lack of reproducibility and robustness such as incomplete reporting of basic elements of experimental design, including blinding, randomization, replication, sample size calculation, and the effect of sex differences. Inadequate reporting may be due to poor training of the researchers to highlight and present technical details, insufficient reporting requirements, or page limitations imposed by the publications/journals. This results in the inability to replicate or further use study results since the necessary information to do so is lacking.

The limited presence of opportunities and platforms to contradict previously published work is also a contributing factor. Only a limited number of platforms allow researchers to publish scientific papers that point out any shortcomings of previously published work or highlight a negative impact of any of the components found during the study. Such data is equally essential and informative as any positive data/findings from a study, and limited availability of such data can result in irreproducibility.

Difficulty in accessing unpublished data is also a contributing factor. Negative or validation data are rarely welcomed by high-impact journals, and unpublished dark data related to published results (such as health records or performance on tasks which did not result in a significant finding) may comprise essential details that may help to reproduce the results of the study or build on its results.

For the past decade, stakeholders, such as researchers, journals, funders, and industry leaders, have been aggressively involved in identifying and taking steps to address the issue of reproducibility and robustness of preclinical research findings. These efforts include maintaining and, in some cases, strengthening scientific quality standards including examining and developing policies that guide research, increasing requirements for reagent and data sharing, and issuing new guidelines for publication.

One important step that stakeholders in the scientific community have taken is to support the development and implementation of guidelines. However, the realm of influence for a given type of stakeholder has been limited. For example, journals usually issue guidelines related to reporting of methods and data, whereas funders may issue guidelines pertaining primarily to study design and, increasingly, data management and availability. In addition, the enthusiasm with which stakeholders have tried to address the "reproducibility crisis" has led to the generation of a multitude of guidelines. This has resulted in a littered landscape where there is overlap without harmonization, gaps in recommendations or requirements that may enhance reproducibility, and slow updating of guidelines to meet the needs of promising, rapidly-evolving computational approaches. Worse yet, the perceived increased burden to meet requirements and lack of clarity around what guidelines to follow reduce compliance as it may leave researchers, publishers, and funding organizations confused and overwhelmed. The goal of this chapter is to compile and review the current state of existing guidelines to understand the overlaps, perform a gap analysis on what may still be missing, and to make recommendations for the future of guidelines to enhance reproducibility in preclinical research.

2 Guidelines and Resources Aimed at Improving Reproducibility and Robustness in Preclinical Data

2.1 Funders/Granting Agencies/Policy Makers

Many funders and policy makers have acknowledged the issue of irreproducibility and are developing new guidelines and initiatives to support the generation of data that are robust and reproducible. This section highlights guidelines, policies, and resources directly related to this issue in preclinical research by the major international granting institutions and is not intended to be an exhaustive review of all available guidelines, policies, and resources. Instead, the organizations reviewed represent a cross-section of many of the top funding organizations and publishers in granting volume and visibility. Included also is a network focused specifically on robustness, reproducibility, translatability, and reporting transparency of preclinical data with membership spanning across academia, industry, and publishing. Requirements pertaining to clinical research are included when guidance documents are also used for preclinical research. The funders, granting agencies, and policy makers surveyed included:

- National Institutes of Health (NIH) (Collins and Tabak 2014; LI-COR 2018; Krester et al. 2017; NIH 2015, 2018a, b)
- Medical Research Council (MRC) (Medical Research Council 2012a, b, 2016a, b, 2019a, b, c)
- The World Health Organization (WHO) (World Health Organization 2006, 2010a, b, 2019)
- Wellcome Trust (Wellcome Trust 2015, 2016a, b, 2018a, b, 2019a, b; The Academy of Medical Sciences 2015, 2016a, b; Universities UK 2012)
- Canadian Institute of Health Research (CIHR) (Canadian Institutes of Health Research 2017a, b)
- Deutsche Forschungsgemeinschaft (DFG)/German Research Foundation (Deutsche Forschungsgemeinschaft 2015, 2017a, b)
- European Commission (EC) (European Commission 2018a, b; Orion Open Science 2019)
- Institut National de la Santé et de la Recherche Médicale (INSERM) (French Institute of Health and Medical Research 2017; Brizzi and Dupre 2017)
- US Department of Defense (DoD) (Department of Defense 2017a, b; National Institutes of Health Center for Information Technology 2019)
- Cancer Research UK (CRUK) (Cancer Research UK 2018a, b, c)
- National Health and Medical Research Council (NHMRC) (National Health and Medical Research Council 2018a, b, 2019; Boon and Leves 2015)
- Center for Open Science (COS) (Open Science Foundation 2019a, b, c, d; Aalbersberg 2017)
- Howard Hughes Medical Institute (HHMI) (ASAPbio 2018)
- Bill & Melinda Gates Foundation (Gates Open Research 2019a, b, c, d, e)
- Innovative Medicines Initiative (IMI) (Innovative Medicines Initiative 2017, 2018; Community Research and Development Information Service 2017; European Commission 2017)
- Preclinical Data Forum Network (European College of Neuropsychopharmacology 2019a, b, c, d)

2.2   Publishers/Journal Groups

Journal publishers and groups have been revising author instructions and publication policies and guidelines, with an emphasis on detailed reporting of study design, replicates, statistical analyses, reagent identification, and validation. Such revisions are expected to encourage researchers to publish robust and reproducible data (National Institutes of Health 2017). Those publishers and groups considered in the analysis were:

- NIH Publication Guidelines Endorsed by Journal Groups (Open Science Foundation 2019d)
- Transparency and Openness Promotion (TOP) Guidelines for Journals (Open Science Foundation 2019d; Nature 2013)
- Nature Journal (Nature 2017, 2019; Pattinson 2012)
- PLOS ONE Journal (The Science Exchange Network 2019a, b; Fulmer 2012; Baker 2012; Powers 2019; PLOS ONE 2017a, b, 2019a, b, c; Bloom et al. 2014; Denker et al. 2017; Denker 2016)
- Journal of Cell Biology (JCB) (Yamada and Hall 2015)
- Elsevier (Cousijn and Fennell 2017; Elsevier 2018, 2019a, b, c, d, e, f, g; Scholarly Link eXchange 2019; Australian National Data Service 2018)

2.3   Summary of Overarching Themes

Guidelines implemented by funding bodies and publishers/journals to attain data reproducibility can take on many forms. Many agencies prefer to frame their guidelines as recommendations in order to accommodate scientific freedom, creativity, and innovation. Therefore, typical guidelines that support good research practices differ from principles set forth by good laboratory practices, which are based on a more formal framework and tend to be more prescriptive.

In reviewing current guidelines and initiatives around reproducibility and robustness, key areas that can lead to robust and reproducible research were revealed and are discussed below.

Research Design and Analysis Providing a well-defined research framework and statistical plan before initiating the research reduces bias and thus helps to increase the robustness and reproducibility of the study.

Funders have under taken various initiatives to support robust research design and analysis, including developing guidance on granting applications. These require researchers to address a set of objectives in the grant proposal including the strengths and weakness of the research, details on the experimental design and methods of the study, planned statistical analyses, and sample sizes. In addition, researchers are often required to abide by existing reporting guidelines such as ARRIVE and asked to provide associated metadata.

Some funders, including NIH, DFG, NHMRC, and HHMI, have developed well-defined guidance documents focusing on robustness and reproducibility for applicants, while others, including Wellcome Trust and USDA, have started taking additional approaches to implement such guidelines. For instance, a symposium was held by Wellcome Trust, while USDA held an internal meeting to identify approaches and discuss solutions to include strong study designs and develop rigorous study plans.

As another example, a dedicated annexure, "Reproducibility and statistical design annex," is required from the researchers in MRC-funded research projects to provide information on methodology and experimental design.

Apart from funders, journals are also working to improve study design quality and reporting, such as requiring that authors complete an editorial checklist before submitting their research in order to enhance the transparency of reporting and thus the reproducibility of published results. Nearly all journals, including *Nature Journal of Cell Biology*, and *PLOS ONE* and the major journal publisher Elsevier have introduced this requirement.

Some journals are also prototyping alternate review models such as early publication to help verify study design. For instance, in Elsevier's *Registered Reports* initiative, the experimental methods and proposed analyses are preregistered and reviewed before study data is collected. The article gets published on the basis of its study protocol and thus prevents authors from modifying their experiments or excluding essential information on null or negative results in order to get their articles published. However, this has been implemented in a limited number of journals in the Elsevier portfolio. *PLOS ONE* permits researchers to submit their articles before a peer review process is conducted. This allows researchers/authors to seek feedback on draft manuscripts before or in parallel to formal review or submission to the journal.

Training and Support Providing adequate training to researchers on the importance of robust study design and experimental methods can help to capture relevant information crucial to attaining reproducibility.

Funders such as MRC have deployed training programs to train both researchers and new panel members on the importance of experimental design and statistics and on the importance of having robust and reproducible research results.

In addition to a detailed guidance handbook for biomedical research, WHO has produced separate, comprehensive training manuals for both trainers and trainees to learn how to implement their guidelines. Also, of note, the Preclinical Data Forum Network, sponsored by the European College of Neuropsychopharmacology (European College of Neuropsychopharmacology 2019e) in Europe and Cohen Veterans Bioscience (Cohen Veterans Bioscience 2019) in the United States, organizes yearly training workshops to enhance awareness and to help junior scientists further develop their experimental skills, with prime focus on experimental design to generate high-quality, robust, reproducible, and relevant data.

Reagents and Reference Material Developing standards for laboratory reagents are essential to maintain reproducibility.

Funders such as HHMI require researchers to make all tangible research materials including organisms, cell lines, plasmids, or similar materials integral to a publication through a repository or by sending them out directly to requestors.

Laboratory Protocols Providing detailed laboratory protocols is required to reproduce a study. Otherwise, researchers may introduce process variability when attempting to reproduce the protocol in their own laboratories. These protocols can also be used by reviewers and editors during the peer review process or by researchers to compare methodological details between laboratories pursuing similar approaches.

Funders such as INSERM took the initiative to introduce an *electronic lab book*. This platform provides better research services by digitizing the experimental work. This enables researchers to better trace and track the data and procedures used in experiments.

Journals such as *PLOS ONE* have taken an initiative wherein authors can deposit their laboratory protocols on repositories such as *protocols.io*. A unique digital object identifier (DOI) is assigned to each study and linked to the Methods section of the original article, allowing researchers to access the published work of these authors along with the detailed protocols used to obtain the results.

Reporting and Review Providing open and transparent access to the research findings and study methods and publishing null or negative results associated with a study facilitate data reproducibility.

Funders require authors to report, cite, and store study data in its entirety, and have developed various initiatives to facilitate data sharing. For instance, CIHR and NHMRC have implemented an *open access policy*, which requires researchers to store their data in specific repositories to improve discovery and facilitate interaction among researchers, gain Creative Commons Attribution license (CC BY) for their research to allow other researchers to access and use the data in parts or as a whole, and link their research activities via identifiers such as digital object identifiers (DOIs) and ORCID to allow appropriate citation of datasets and provide recognition to data generators and sharers.

Wellcome Trust and Bill & Melinda Gates Foundation have launched their own publishing platforms – *Wellcome Open Research* and *Gates Open Research*, respectively – to allow researchers to publish and share their results rapidly.

Other efforts focused on data include the European Commission, which aims to build an open research platform "European Open Science Cloud" that can act as a virtual repository of research data of publicly funded studies and allow European researchers to store, process, and access research data.

In addition, the Preclinical Data Forum Network has been working toward building a data exchange and information repository and incentivizing the publication of negative data by issuing the world's first price for published "negative" scientific results.

Journals have also taken various initiatives to allow open access of their publications. Some journals such as *Nature* and *PLOS ONE* require data availability statements to be submitted by researchers to help in locating the data, and accessing details for primary large-scale data, through details of repositories and digital object identifiers or accession numbers.

Journals also advise authors to upload their raw and metadata in appropriate repositories. Some journals have created their separate cloud-based repository, in addition to those publicly available. For instance, Elsevier has created *Mendeley Data* to help researchers manage, share, and showcase their research data. And, JCB has established JCB DataViewer, a cross-platform repository for storing large amounts of raw imaging and gel data, for its published manuscripts. Elsevier has also partnered with platforms such as Scholix and FORCE11, which allows data citation, encouraging reuse of research data, and enabling reproducibility of published research.

3   Gaps and Looking to the Future

A gap analysis of existing guidelines and resources was performed, addressing such critical factors as *study design, transparency, data management, availability of resources and information, linking relevant research, publication opportunities, consideration of refutations,* and *initiatives to grow.* It should be noted that these categories were not defined de novo but based on a comprehensive review of the high-impact organizations considered.

We considered the following observed factors within each category to understand where organizations are supporting good research practices with explicit guidelines and/or initiatives and to identify potential gaps:

- Study Design
  - Scientific premise of proposed research: Guidelines to support current or proposed research that is formed on a strong foundation of prior work.
  - Robust methodology to address hypothesis: Guidelines to design robust studies that address the scientific question. This includes justification and reporting of the experimental technique, statistical analysis, and animal model.
  - Animal use guidelines and legal permissions: Guidelines regarding animal use, clinical trial reportings, or legal permissions.
  - Validation of materials: Guidelines to ensure validity of experimental protocol, reagent, or equipment.
- Transparency
  - Comprehensive description of methodology: Guidelines to ensure comprehensive reporting of method and analysis to ensure reproducibility by other researchers. For example, publishers may include additional space for

researchers to detail their methodology. Similar to "robust methodology to address hypothesis" but more focused on post-collection reporting rather than initial design.

- Appropriate acknowledgments: Guidelines for authors to appropriately acknowledge contributors, such as co-authors or references.
- Reporting of positive and negative data: Guidelines to promote release of negative data, which reinforces unbiased reporting.

• Data Management
- Early design of data management: Guidelines to promote early design of data management.
- Storage and preservation of data: Guidelines to ensure safe and long-term storage and preservation of data.
- Additional tools for data collection and management: Miscellaneous data management tools developed (e.g., electronic lab notebook).

• Availability of Resources and Information
- Data availability statements: A statement committing researchers to sharing data (usually upon submission to a journal or funding organization).
- Access to raw or structured data: Guidelines to share data in publicly available or institutional repositories to allow for outside researchers to reanalyze or reuse data.
- Open or public access publications: Guidelines to encourage open or public access publications, which allows for unrestricted use of research.
- Shared access to resources, reagents, and protocols: Guidelines to encourage shared access to resources, reagents, and protocols. This may include requirements for researchers to independently share and ship resources or nonprofit reagent repositories.

• Linking Relevant Research
- Indexing data, reagents, and protocols: Guidelines to index research components, such as data, reagent, or protocols. Indexing using a digital object identifier (DOI) allows researchers to digitally track use of research components.
- Two-way linking of relevant datasets and publications: Guidelines to encourage linkage between publications. This is particularly important in clinical research when multiple datasets are compiled to increase analytical power.

• Publication Opportunities
- Effective review: Guidelines to expedite or strengthen the review process, such as a checklist for authors or reviewers to complete or additional responsibilities of the reviewer.
- Additional peer review and public release processes: Opportunities to release research conclusions independent from the typical journal process.
- Preregistration: Guidelines to encourage preregistration, a process where researchers commit to their study design prior to collecting data. This reduces bias and increases clarity of the results.

• Consideration of Refutations

- Attempts to resolve failures to reproduce: Guidelines for authors and organizations to address any discrepancies in results or conclusions
- Initiatives to Grow
  - Develop resources: Additional resources developed to increase reproducibility and rigor in research. This includes training workshops.
  - Work to develop responsible standards: Commitments and overarching goals made by organization to increase reproducibility and rigor in research.

As part of the *study design*, it appeared that there is a dearth of guidelines to ensure validity of experimental protocols, reagents, or equipment. Variability and incomplete reporting of reagents used is a known and oft-cited source of irreproducibility.

The most notable omission regarding *transparency* were guidelines to promote the release of report negative data to reinforce unbiased reporting. This also results in poor study reproducibility since, overwhelmingly, only positive data are reported for preclinical studies.

Most funding agencies have seriously begun initiatives addressing *data management* to ensure safe and long-term storage and preservation of data and are developing, making available, or promoting data management tools (e.g., electronic lab notebook). However, these ongoing activities do not often include guidelines to promote the early design of data management, which may reduce errors and ease researcher burden by optimizing and streamlining the process from study design to data upload.

To that point, a massive shift can be seen as both funders and publishers intensely engage in guidelines around the *availability of resources and information*. Most of this effort is in the ongoing development of guidelines to share data in publicly available or institutional repositories to allow for outside researchers to reanalyze or reuse data. This is to create a long-term framework for new strategies to research that will allow for "big data" computational modeling, deep-learning artificial intelligence, and mega-analyses across species and measures. However, not many guidelines were found that encourage shared access to resources, reagents, and protocols. This may include requirements for researchers to independently share and ship resources or nonprofit reagent repositories.

Related are guidelines for *linking relevant research*. This includes guidelines to index research components, such as data, reagents, or protocols with digital object identifiers (DOIs) that allow researchers to digitally track the use of research components and guidelines to encourage two-way linking of relevant datasets and publications. This is historically a common requirement for clinical studies and is currently being developed for preclinical research, but not consistently across the organizations surveyed.

On the reporting side, the most notable exclusion to *publication opportunities* guidelines were those that encourage preregistration, a process whereby researchers commit to their study design prior to collecting data and publishers agree to publish results whether they be positive or negative. These would serve to reduce both experimental and publication biases and increase clarity of the results.

In the category *consideration of refutations*, which, broadly, are attempts to resolve failures to reproduce a study, few guidelines exist. However, there is ongoing work to develop guidelines for authors and organizations to address discrepancies in results or conclusions and a commitment from publishers that they will consider publications that do not confirm previously published research in their journal.

Lastly, although many organizations cite a number of *initiatives to grow*, there appear to be notable gaps both in the development of additional resources and work to develop responsible standards. One initiative that aims to develop solutions to address the issue of data reproducibility in preclinical neuroscience research is the EQIPD (European Quality in Preclinical Data) project, launched in October 2017 with support from the Innovative Medicines Initiative (IMI). The project recognizes poor data quality as the main concern resulting in the non-replication of studies/ experiments and aims to look for simple, sustainable solutions to improve data quality without impacting innovation. It is expected that this initiative will lead to a cultural change in data quality approaches in the medical research and drug development field with the final intent to establish guidelines that will strengthen robustness, rigor, and validity of research data to enable a smoother and safer transition from preclinical to clinical testing and drug approval in neuroscience (National Institutes of Health 2017; Nature 2013, 2017; Vollert et al. 2018).

In terms of providing additional resources, although some organizations emphasize training and workshops for researchers to enhance rigor and reproducibility, it is unclear if and how organizations themselves assess the effectiveness and actual implementation of their guidelines and policies. An exception may be WHO's training program, which provides manuals for both trainer and trainee to support the implementation of their guidelines.

More must also be done to accelerate work to develop consensus, responsible standards. As funders, publishers, and preclinical researchers alike begin recognizing the promise of computational approaches and attempt to meet the demands for these kinds of analyses, equal resources and energy must be devoted to the required underlying standards and tools. To be able to harmonize data across labs and species, ontologies and CDEs must be developed and researchers must be trained and incentivized to use them. Not only may data that have already been generated offer profound validation opportunities but also the ability to follow novel lines of research agnostically based on an unbiased foundation of data. In acquiring new data, guidelines urging preclinical scientists to collect and upload all experimental factors, including associated dark data in a usable format may bring the field closer to understanding if predictive multivariate signatures exist, embrace deviations in study design, and may be more reflective of clinical trials.

Overall, the best path forward may be for influential organizations to develop a comprehensive plan to enhance reproducibility and align on a standard set of policies. A coherent road map or strategy would ensure that all known factors related to this issue are addressed and reduce complications for investigators.

References

Aalbersberg IJ (2017) Elsevier supports TOP guidelines in ongoing efforts to ensure research quality and transparency. https://www.elsevier.com/connect/elsevier-supports-top-guidelines-in-ongoing-efforts-to-ensure-research-quality-and-transparency. Accessed 1 May 2018

ASAPbio (2018) Peer review meeting summary. https://asapbio.org/peer-review/summary. Accessed 1 May 2018

Australian National Data Service (2018) Linking data with Scholix. https://www.ands.org.au/working-with-data/publishing-and-reusing-data/linking-data-with-scholix. Accessed 1 May 2018

Baker M (2012) Independent labs to verify high-profile papers. https://www.nature.com/news/independent-labs-to-verify-high-profile-papers-1.11176. Accessed 1 May 2018

Begley CG, Ellis LM (2012) Raise standards for preclinical cancer research. Nature 483(7391):531

Bloom T, Ganley E, Winker M (2014) Data access for the open access literature: PLOS's data policy. https://journals.plos.org/plosbiology/article?id=10.1371/journal.pbio.1001797. Accessed 1 May 2018

Boon W-E, Leves F (2015) NHMRC initiatives to improve access to research outputs and findings. Med J Aust 202(11):558

Brizzi F, Dupre P-G (2017) Inserm Labguru Pilot Program digitizes experimental data across institute. https://www.labguru.com/case-study/Inserm_labguru-F.pdf. Accessed 1 May 2018

Canadian Institutes of Health Research (2017a) Reproducibility in pre-clinical health research. Connections 17(9)

Canadian Institutes of Health Research (2017b) Tri-agency open access policy on publications. http://www.cihr-irsc.gc.ca/e/32005.html. Accessed 1 May 2018

Cancer Research UK (2018a) Develop the cancer leaders of tomorrow: our research strategy. https://www.cancerresearchuk.org/funding-for-researchers/our-research-strategy/develop-the-cancer-leaders-of-tomorrow. Accessed 1 May 2018

Cancer Research UK (2018b) Data sharing guidelines. https://www.cancerresearchuk.org/funding-for-researchers/applying-for-funding/policies-that-affect-your-grant/submission-of-a-data-sharing-and-preservation-strategy/data-sharing-guidelines. Accessed 1 May 2018

Cancer Research UK (2018c) Policy on the use of animals in research. https://www.cancerresearchuk.org/funding-for-researchers/applying-for-funding/policies-that-affect-your-grant/policy-on-the-use-of-animals-in-research#detail1. Accessed 1 May 2018

Cohen Veterans Bioscience (2019) Homepage. https://www.cohenveteransbioscience.org. Accessed 15 Apr 2019

Collins FS, Tabak LA (2014) NIH plans to enhance reproducibility. Nature 505(7485):612–613

Community Research and Development Information Service (2017) European quality in preclinical data. https://cordis.europa.eu/project/rcn/211612_en.html. Accessed 1 May 2018

Cousijn H, Fennell C (2017) Supporting data openness, transparency & sharing. https://www.elsevier.com/connect/editors-update/supporting-openness,-transparency-and-sharing. Accessed 1 May 2018

CURE Consortium (2017) Curating for reproducibility defining "reproducibility." http://cure.web.unc.edu/defining-reproducibility/. Accessed 1 May 2018

Denker SP (2016) The best of both worlds: preprints and journals. https://blogs.plos.org/plos/2016/10/the-best-of-both-worlds-preprints-and-journals/?utm_source=plos&utm_medium=web&utm_campaign=plos-1702-annualupdate. Accessed 1 May 2018

Denker SP, Byrne M, Heber J (2017) Open access, data and methods considerations for publishing in precision medicine. https://www.plos.org/files/PLOS_Biobanking%20and%20Open%20Data%20Poster_2017.pdf. Accessed 1 May 2018

Department of Defense (2017a) Guide to FY2018 research funding at the Department of Defense (DOD). August 2017. https://research.usc.edu/files/2011/05/Guide-to-FY2018-DOD-Research-Funding.pdf. Accessed 1 May 2018

Department of Defense (2017b) Instruction for the use of animals in DoD programs. https://www. esd.whs.mil/Portals/54/Documents/DD/issuances/dodi/321601p.pdf. Accessed 1 May 2018

Deutsche Forschungsgemeinschaft (2015) DFG guidelines on the handling of research data. https:// www.dfg.de/download/pdf/foerderung/antragstellung/forschungsdaten/guidelines_research_ data.pdf. Accessed 1 May 2018

Deutsche Forschungsgemeinschaft (2017a) DFG statement on the replicability of research results. https://www.dfg.de/en/research_funding/announcements_proposals/2017/info_wissenschaft_ 17_18/index.html. Accessed 1 May 2018

Deutsche Forschungsgemeinschaft (2017b) Replicability of research results. A statement by the German Research Foundation. https://www.dfg.de/download/pdf/dfg_im_profil/reden_ stellungnahmen/2017/170425_stellungnahme_replizierbarkeit_forschungsergebnisse_en.pdf. Accessed 1 May 2018

Elsevier (2018) Elsevier launches Mendeley Data to manage entire lifecycle of research data. https://www.elsevier.com/about/press-releases/science-and-technology/elsevier-launches- mendeley-data-to-manage-entire-lifecycle-of-research-data. Accessed 15 Apr 2019

Elsevier (2019a) Research data. https://www.elsevier.com/about/policies/research-data#Principles. Accessed 15 Apr 2019

Elsevier (2019b) Research data guidelines. https://www.elsevier.com/authors/author-resources/ research-data/data-guidelines. Accessed 15 Apr 2019

Elsevier (2019c) STAR methods. https://www.elsevier.com/authors/author-resources/research- data/data-guidelines. Accessed 15 Apr 2019

Elsevier (2019d) Research elements. https://www.elsevier.com/authors/author-resources/research- elements. Accessed 15 Apr 2019

Elsevier (2019e) Research integrity. https://www.elsevier.com/about/open-science/science-and- society/research-integrity. Accessed 15 Apr 2019

Elsevier (2019f) Open data. https://www.elsevier.com/authors/author-resources/research-data/ open-data. Accessed 15 Apr 2019

Elsevier (2019g) Mendeley data for journals. https://www.elsevier.com/authors/author-resources/ research-data/mendeley-data-for-journals. Accessed 15 Apr 2019

European College of Neuropsychopharmacology (2019a) Preclinical Data Forum Network. https:// www.ecnp.eu/research-innovation/ECNP-networks/List-ECNP-Networks/Preclinical-Data- Forum.aspx. Accessed 15 Apr 2019

European College of Neuropsychopharmacology (2019b) Output. https://www.ecnp.eu/research- innovation/ECNP-networks/List-ECNP-Networks/Preclinical-Data-Forum/Output.aspx. Accessed 15 Apr 2019

European College of Neuropsychopharmacology (2019c) Members. https://www.ecnp.eu/research- innovation/ECNP-networks/List-ECNP-Networks/Preclinical-Data-Forum/Members. Accessed 15 Apr 2019

European College of Neuropsychopharmacology (2019d) ECNP Preclinical Network Data Prize. https://www.ecnp.eu/research-innovation/ECNP-Preclinical-Network-Data-Prize.aspx. Accessed 15 Apr 2019

European College of Neuropsychopharmacology (2019e) Homepage. https://www.ecnp.eu. Accessed 15 Apr 2019

European Commission (2017) European quality in preclinical data. https://quality-preclinical-data. eu/wp-content/uploads/2018/01/20180306_EQIPD_Folder_Web.pdf. Accessed 1 May 2018

European Commission (2018a) The European cloud initiative. https://ec.europa.eu/digital-single- market/en/european-cloud-initiative. Accessed 1 May 2018

European Commission (2018b) European Open Science cloud. https://ec.europa.eu/digital-single- market/en/european-open-science-cloud. Accessed 1 May 2018

French Institute of Health and Medical Research (2017) INSERM 2020 strategic plan. https://www. inserm.fr/sites/default/files/2017-11/Inserm_PlanStrategique_2016-2020_EN.pdf. Accessed 1 May 2018

Fulmer T (2012) The cost of reproducibility. SciBX 5:34

Gates Open Research (2019a) Introduction. https://gatesopenresearch.org/about. Accessed 15 Apr 2019

Gates Open Research (2019b) Open access policy frequently asked questions. https://www.gatesfoundation.org/How-We-Work/General-Information/Open-Access-Policy/Page-2. Accessed 15 Apr 2019

Gates Open Research (2019c) How to publish. https://gatesopenresearch.org/for-authors/article-guidelines/research-articles. Accessed 15 Apr 2019

Gates Open Research (2019d) Data guidelines. https://gatesopenresearch.org/for-authors/data-guidelines. Accessed 15 Apr 2019

Gates Open Research (2019e) Policies. https://gatesopenresearch.org/about/policies. Accessed 15 Apr 2019

Innovative Medicines Initiative (2017) European quality in preclinical data. https://www.imi.europa.eu/projects-results/project-factsheets/eqipd. Accessed 1 May 2018

Innovative Medicines Initiative (2018) 78th edition. https://www.imi.europa.eu/news-events/news letter/78th-edition-april-2018. Accessed 1 May 2018

Krester A, Murphy D, Dwyer J (2017) Scientific integrity resource guide: efforts by federal agencies, foundations, nonprofit organizations, professional societies, and academia in the United States. Crit Rev Food Sci Nutr 57(1):163–180

LI-COR (2018) Tracing the footsteps of the data reproducibility crisis. http://www.licor.com/bio/blog/reproducibility/tracing-the-footsteps-of-the-data-reproducibility-crisis/. Accessed 1 May 2018

Medical Research Council (2012a) MRC ethics series. Good research practice: principles and guidelines. https://mrc.ukri.org/publications/browse/good-research-practice-principles-and-guidelines/. Accessed 1 May 2018

Medical Research Council (2012b) UK Research and Innovation. Methodology and experimental design in applications: guidance for reviewers and applicants. https://mrc.ukri.org/documents/pdf/methodology-and-experimental-design-in-applications-guidance-for-reviewers-and-applicants/. Accessed 1 May 2018

Medical Research Council (2016a) Improving research reproducibility and reliability: progress update from symposium sponsors. https://mrc.ukri.org/documents/pdf/reproducibility-update-from-sponsors/. Accessed 1 May 2018

Medical Research Council (2016b) Funding reproducible, robust research. https://mrc.ukri.org/news/browse/funding-reproducible-robust-research/. Accessed 1 May 2018

Medical Research Council (2019a) Proposals involving animal use. https://mrc.ukri.org/funding/guidance-for-applicants/4-proposals-involving-animal-use/. Accessed 15 Apr 2019

Medical Research Council (2019b) Grant application. https://mrc.ukri.org/funding/guidance-for-applicants/2-the-application/#2.2.3. Accessed 15 Apr 2019

Medical Research Council (2019c) Open research data: clinical trials and public health interventions. https://mrc.ukri.org/research/policies-and-guidance-for-researchers/open-research-data-clinical-trials-and-public-health-interventions/. Accessed 15 Apr 2019

National Health and Medical Research Council (2018a) National Health and Medical Research Council open access policy. https://www.nhmrc.gov.au/about-us/resources/open-access-policy. Accessed 15 Apr 2019

National Health and Medical Research Council (2018b) 2018 NHMRC symposium on research translation. https://www.nhmrc.gov.au/event/2018-nhmrc-symposium-research-translation. Accessed 15 Apr 2019

National Health and Medical Research Council (2019) Funding. https://www.nhmrc.gov.au/funding. Accessed 15 Apr 2019

National Institutes of Health (2017) Principles and guidelines for reporting preclinical research. https://www.nih.gov/research-training/rigor-reproducibility/principles-guidelines-reporting-pre clinical-research. Accessed 1 May 2018

National Institutes of Health Center for Information Technology (2019) Federal Interagency Traumatic Brain Injury Research Informatics System. https://fitbir.nih.gov/. Accessed 1 May 2018

Nature (2013) Enhancing reproducibility. Nat Methods 10:367

Nature (2017) On data availability, reproducibility and reuse. 19:259

Nature (2019) Availability of data, material and methods. http://www.nature.com/authors/policies/availability.html. Accessed 15 Apr 2019

NIH (2015) Enhancing reproducibility through rigor and transparency. https://grants.nih.gov/grants/guide/notice-files/NOT-OD-15-103.html. Accessed 1 May 2018

NIH (2018a) Reproducibility grant guidelines. https://grants.nih.gov/reproducibility/documents/grant-guideline.pdf. Accessed 1 May 2018

NIH (2018b) Rigor and reproducibility in NIH applications: resource chart. https://grants.nih.gov/grants/RigorandReproducibilityChart508.pdf. Accessed 1 May 2018

Open Science Foundation (2019a) Transparency and openness guidelines. Table summary funders. https://osf.io/kzxby/. Accessed 15 Apr 2019

Open Science Foundation (2019b) Transparency and openness guidelines. https://osf.io/4kdbm/. Accessed 15 Apr 2019

Open Science Foundation (2019c) Transparency and openness guidelines for funders. https://osf.io/bcj53/. Accessed 15 Apr 2019

Open Science Foundation (2019d) Open Science Framework, guidelines for Transparency and Openness Promotion (TOP) in journal policies and practices. https://osf.io/9f6gx/wiki/Guidelines/?_ga=2.194585006.2089478002.1527591680-566095882.1527591680. Accessed 15 Apr 2019

Orion Open Science (2019) What is Open Science? https://www.orion-openscience.eu/resources/open-science. Accessed 15 Apr 2019

Pattinson D (2012) PLOS ONE launches reproducibility initiative. https://blogs.plos.org/everyone/2012/08/14/plos-one-launches-reproducibility-initiative/. Accessed 1 May 2018

PLOS ONE (2017a) Making progress toward open data: reflections on data sharing at PLOS ONE. https://blogs.plos.org/everyone/2017/05/08/making-progress-toward-open-data/. Accessed 1 May 2018

PLOS ONE (2017b) Protocols.io tools for PLOS authors: reproducibility and recognition. https://blogs.plos.org/plos/2017/04/protocols-io-tools-for-reproducibility/. Accessed 1 May 2018

PLOS ONE (2019a) Data availability. https://journals.plos.org/plosone/s/data-availability. Accessed 15 Apr 2019

PLOS ONE (2019b) Submission guidelines. https://journals.plos.org/plosone/s/submission-guidelines. Accessed 15 Apr 2019

PLOS ONE (2019c) Criteria for publication. https://journals.plos.org/plosone/s/criteria-for-publication. Accessed 15 Apr 2019

Powers M (2019) New initiative aims to improve translational medicine hit rate. http://www.bioworld.com/content/new-initiative-aims-improve-translational-medicine-hit-rate-0. Accessed 15 Apr 2019

Prinz F, Schlange T, Asadullah K (2011) Believe it or not: how much can we rely on published data on potential drug targets? Nat Rev Drug Discov 10(9):712

Scholarly Link eXchange (2019) Scholix: a framework for scholarly link eXchange. http://www.scholix.org/home. Accessed 15 Apr 2019

The Academy of Medical Sciences (2015) Reproducibility and reliability of biomedical research: improving research practice. In: Symposium report, United Kingdom, 1–2 April 2015

The Academy of Medical Sciences (2016a) Improving research reproducibility and reliability: progress update from symposium sponsors. https://acmedsci.ac.uk/reproducibility-update/. Accessed 1 May 2018

The Academy of Medical Sciences (2016b) Improving research reproducibility and reliability: progress update from symposium sponsors. https://acmedsci.ac.uk/file-download/38208-5631f0052511d.pdf. Accessed 1 May 2018

The Science Exchange Network (2019a) Validating key experimental results via independent replication. http://validation.scienceexchange.com/. Accessed 15 Apr 2019

The Science Exchange Network (2019b) Reproducibility initiative. http://validation.scienceexchange.com/#/reproducibility-initiative. Accessed 15 Apr 2019

Universities UK (2012) The concordat to support research integrity. https://www.universitiesuk.ac.uk/policy-and-analysis/reports/Pages/research-concordat.aspx. Accessed 1 May 2018

Vollert J, Schenker E, Macleod M et al (2018) A systematic review of guidelines for rigour in the design, conduct and analysis of biomedical experiments involving laboratory animals. Br Med J Open Sci 2:e000004

Wellcome Trust (2015) Research practice. https://wellcome.ac.uk/what-we-do/our-work/research-practice. Accessed 1 May 2018

Wellcome Trust (2016a) Why publish on Wellcome Open Research? https://wellcome.ac.uk/news/why-publish-wellcome-open-research. Accessed 1 May 2018

Wellcome Trust (2016b) Wellcome to launch bold publishing initiative. https://wellcome.ac.uk/press-release/wellcome-launch-bold-publishing-initiative. Accessed 1 May 2018

Wellcome Trust (2018a) Grant conditions. https://wellcome.ac.uk/sites/default/files/grant-conditions-2018-may.pdf. Accessed 1 May 2018

Wellcome Trust (2018b) Guidelines on good research practice. https://wellcome.ac.uk/funding/guidance/guidelines-good-research-practice. Accessed 1 May 2018

Wellcome Trust (2019a) Wellcome Open Research. https://wellcomeopenresearch.org/. Accessed 15 Apr 2019

Wellcome Trust (2019b) Wellcome Open Research. How it works. https://wellcomeopenresearch.org/about. Accessed 15 Apr 2019

World Health Organization (2006) Handbook: quality practices in basic biomedical research. https://www.who.int/tdr/publications/training-guideline-publications/handbook-quality-practices-biomedical-research/en/. Accessed 1 May 2018

World Health Organization (2010a) Quality Practices in Basic Biomedical Research (QPBR) training manual: trainee. https://www.who.int/tdr/publications/training-guideline-publications/qpbr-trainee-manual-2010/en/. Accessed 1 May 2018

World Health Organization (2010b) Quality Practices in Basic Biomedical Research (QPBR) training manual: trainer. https://www.who.int/tdr/publications/training-guideline-publications/qpbr-trainer-manual-2010/en/. Accessed 1 May 2018

World Health Organization (2019) Quality practices in basic biomedical research. https://www.who.int/tdr/publications/quality_practice/en/. Accessed 15 Apr 2019

Yamada KM, Hall A (2015) Reproducibility and cell biology. J Cell Biol 209(2):191

Learning from Principles of Evidence-Based Medicine to Optimize Nonclinical Research Practices

Isabel A. Lefevre ⓘ and Rita J. Balice-Gordon ⓘ

Contents

Abstract

Thousands of pharmacology experiments are performed each day, generating hundreds of drug discovery programs, scientific publications, grant submissions, and other efforts. Discussions of the low reproducibility and robustness of some of this research have led to myriad efforts to increase data quality and thus

I. A. Lefevre (✉)
Rare and Neurologic Diseases Research, Sanofi, Chilly-Mazarin, France
e-mail: Isabel.Lefevre@sanofi.com

R. J. Balice-Gordon
Rare and Neurologic Diseases Research, Sanofi, Framingham, MA, USA

© The Author(s) 2019   35
A. Bespalov et al. (eds.), *Good Research Practice in Non-Clinical Pharmacology and Biomedicine*, Handbook of Experimental Pharmacology 257,
https://doi.org/10.1007/164_2019_276

reliability. Across the scientific ecosystem, regardless of the extent of concerns, debate about solutions, and differences among goals and practices, scientists strive to provide reliable data to advance frontiers of knowledge. Here we share our experience of current practices in nonclinical neuroscience research across biopharma and academia, examining context-related factors and behaviors that influence ways of working and decision-making. Drawing parallels with the principles of evidence-based medicine, we discuss ways of improving transparency and consider how to better implement best research practices. We anticipate that a shared framework of scientific rigor, facilitated by training, enabling tools, and enhanced data sharing, will draw the conversation away from data unreliability or lack of reproducibility toward the more important discussion of how to generate data that advances knowledge and propels innovation.

Keywords

Data reliability · Decision-making · Evidence-based medicine · Nonclinical pharmacology · Research methodology

1   Introduction

Over the last 10 years, debate has raged about the quality of scientific evidence, expanding from a conversation among experts, amplified by systematic reviews and meta-analyses published in peer-reviewed journals, into a heated discussion splashed across mainstream press and social media. What is widely perceived as a "reproducibility crisis" is the subject of countless, and sometimes inaccurate, statements on the poor "reproducibility," "replicability," insufficient "rigor," "robustness," or "validity" of data and conclusions. In the context of nonclinical pharmacological data, these are cited as foundational for later clinical trial failure. The decision to advance a compound to human testing is based on a substantial body of evidence supporting the efficacy and safety of a therapeutic concept. Nonclinical studies that support, for example, an investigational new drug (IND) filing or a clinical trial application (CTA), which gate studies in humans, are reviewed under quality control procedures; most safety studies must comply with regulations laid out by health authorities, whereas nonclinical efficacy studies are usually performed in a nonregulated environment (see chapter "Quality in Non-GxP Research Environment"). If clinical trial results support both efficacy and safety of the intervention, health authorities review *all* of the evidence, to determine whether or not to approve a new therapeutic.

Once a new therapeutic is made available to patients and their physicians, clinical trial findings and real-world observations contribute to forming a larger body of evidence that can be used for decision-making by a physician considering which treatment option would best benefit a patient. In many countries, medical students are taught to critically appraise all the accessible information in order to choose the "best possible option," based upon the "best possible evidence"; this process is part of evidence-based medicine (EBM), also known as "medicine built on proof." In EBM, clinical evidence is ranked according to the risk of underlying bias, using the available sources of evidence, from case studies through randomized, controlled clinical trials (RCTs) to clinical trial meta-analyses. Well-designed randomized trial results are generally viewed to be of higher reliability, or at least less influenced by internal bias, than observational studies or case reports. Since meta-analysis aims to provide a more trustworthy estimate of the effect and its magnitude (effect size), meta-analyses of RCTs are regarded as the most reliable source for recommending a given treatment, although this can be confounded if the individual RCTs themselves are of low quality.

A well-established framework for rating quality of evidence is the Grading of Recommendations, Assessment, Development, and Evaluation (GRADE) system (http://www.gradeworkinggroup.org). GRADE takes the EBM process a step further, rating a body of evidence, and considering internal risk of bias, imprecision, inconsistency, indirectness, and publication bias of individual studies as reasons for rating the quality of evidence down, whereas a large effect, or a dose-response relationship, can justify rating it up (Balshem et al. 2011). The Cochrane Collaboration, which produces systematic reviews of health interventions, now requires authors to use GRADE (https://training.cochrane.org/grade-approach). The British Medical Journal has developed a suite of online tools (https://bestpractice.bmj.com/info/us/toolkit) with a section on how to use GRADE, and various electronic databases and journals that summarize evidence are also available to clinicians. In a recent development, the GRADE Working Group has begun to explore how to rate evidence from nonclinical animal studies, and the first attempt to implement GRADE in the nonclinical space has successfully been performed on a sample of systematic reviews and examples, with further efforts planned (Hooijmans et al. 2018). In contrast, with the exception of those who also have medical training or clinical research experience, most scientists are unaware of the guiding principles of EBM and are unfamiliar with formal decision-enabling algorithms. At least in part due to the diversity of nonclinical experiments, systematic reviews and meta-analyses are far less common in nonclinical phases than in clinical ones, and there are very few broadly accepted tools with which to assess nonclinical data quality (Hooijmans et al. 2018; Sena et al. 2014). Pioneering work in this area came from the stroke field, with nonclinical research guidelines and an assessment tool elaborated by STAIR, the Stroke Therapy Academic Industry Roundtable (Hooijmans et al. 2014) (https://www.thestair.org). The CAMARADES collaboration (originally the "Collaborative Approach to Meta-Analysis and Review of Animal Data from Experimental Stroke") has now extended its scope to support groups wishing to undertake systematic reviews and meta-analyses of animal studies in research on neurological diseases (http://www.dcn.ed.ac.uk/camarades). The Systematic Review Centre for Laboratory Animal Experimentation (SYRCLE) has designed a comprehensive method to systematically review evidence from animal studies (Hooijmans et al. 2014), based on the Cochrane risk of bias tool. SYRCLE's tool covers different forms of bias and several domains of study design, many of which are common to both clinical and nonclinical research (Table 2 in Hooijmans et al. 2014). As a consequence, measures known to reduce bias in clinical settings, such as randomization and blinding, are recommended for implementation in nonclinical research. Although the tool was primarily developed to guide systematic reviewers, it can also be used to assess the quality of any in vivo experimental pharmacology study. However, these structured approaches have had limited uptake in other fields.

The attention to sources of bias that can influence study conduct, outcomes, and interpretation is an essential element of EBM. A catalog of bias is being collaboratively constructed, to map all the biases that affect health evidence (https://catalogofbias.org). In the nonclinical space, despite a number of publications and material from training courses and webinars (e.g., http://neuronline.sfn.org/Collections/Promoting-Awareness-and-Knowledge-to-Enhance-Scientific-Rigor-in-Neuroscience), an equivalent, generalizable framework, or a common standard for rating evidence quality is lacking, as is a unified concept of what constitutes the best possible material for decision-making. Discussions are also limited by confusing and varied terminology, although attempts have been made to clarify and harmonize terms and definitions (Goodman et al. 2016). Here we will use the word "reliability," in its generally accepted sense of accuracy and dependability, of something one expects to be able to rely on to make a decision. As a consequence, reliability also reflects the extent to which something can consistently be repeated. Both meanings apply to experimental pharmacology studies in all parts of the biomedical ecosystem.

The EBM framework is used to find reliable answers to medical questions. Here we will describe the purposes, current practices, and factors that contribute to bias in addressing scientific questions. We will consider which EBM principles can apply to nonclinical pharmacology work and how to strengthen our ability to implement best research practices, without limiting innovation that is urgently needed.

2 Current Context of Nonclinical, Nonregulated Experimental Pharmacology Study Conduct: Purposes and Processes Across Sectors

2.1 Outcomes and Deliverables of Nonclinical Pharmacology Studies in Industry and Academia

Experimental pharmacology studies in biopharma companies and nonclinical contract research organizations (CROs) can have various purposes, such as furthering the understanding of a disease mechanism, developing a model or assay, or characterizing the effects of a novel compound. Such studies can also document a patent application and/or generate data on the efficacy or safety of a compound that is to enter clinical development, in which case the study report may ultimately be part of a regulatory submission to a health authority. In academia, the primary goal is to provide experimental evidence to answer scientific questions and disseminate new knowledge by publishing the findings; academic scientists also occasionally file patents and, in collaboration with biopharma companies, perform studies that may in turn become part of regulatory submission dossiers. Academic drug discovery platforms, which have sprouted in recent years, mainly aim to provide nonclinical data that will be further leveraged in biopharma drug discovery programs, although it is increasingly common that these data are used to advance clinical studies as well.

Different business models and end goals across academia and industry, and different outcomes of nonclinical research, imply different processes and deliverables, which can be associated with a step or feature in EBM, as described in Table 1.

Investigating a scientific hypothesis is often done in a stepwise manner; from an initial idea, several questions can be asked in parallel, and answers are generated in both an incremental and iterative manner, by performing additional experiments and repeating cycles. Choices and decisions are made at each step, based on data; if these data are under- or overestimated, their interpretation will be biased, affecting subsequent steps. For example, in a drug discovery project, inaccurate estimates of in vitro potency or in vivo efficacy can skew the doses tested in nonclinical safety experiments, and bias the estimate of the dosage range in which *only* the desired response is observed in nonclinical species, and, most importantly, affect the subsequent determination of the corresponding dosage range to be tested in humans. As sponsors of clinical trials, among other responsibilities, biopharma companies have an ethical duty to conduct human trials *only* if there is a solid foundation for a potential clinical benefit with limited safety risks. In academic research, individuals and institutions are accountable to funders and to the community for contributing to the body of scientific knowledge. In all fields and sectors, biased interpretations of experimental data can result in wasted experiments; scientists are therefore responsible for the quality of the evidence generated.

2.2   Scientific Integrity: Responsible Conduct of Research and Awareness of Cognitive Bias

Over the last two decades, many governments and agencies involved in funding and conducting research have taken a strong stance on scientific integrity, issuing policies and charters at international, national, and institutional levels. Compliance with these policies is mandatory for employees and scientists applying for funding (examples: MRC, https://mrc.ukri.org/publications/browse/good-research-practice-principles-and-guidelines; NIH, https://grants.nih.gov/policy/research_integrity/what-is.htm; CNRS, http://www.cnrs.fr/comets/IMG/pdf/guide_2017-en.pdf). Scientific integrity means absolute honesty, transparency, and accountability in the conduct and reporting of research. Responsible research practices encompass the adherence to these principles and the systematic use of measures aiming to reduce cognitive and experimental bias.

Training on responsible scientific conduct is now mandatory at masters or PhD level in many universities; at any stage of their career, scientists can access training resources on scientific integrity and responsible research practices (see list made by EMBO, http://www.embo.org/science-policy/research-integrity/resources-on-research-integrity; NIH, Responsible Conduct of Research Training, https://oir.nih.gov/sourcebook/ethical-conduct/responsible-conduct-research-training; Mooc, https://www.fun-mooc.fr/courses/course-v1:Ubordeaux+28007EN+session01/about#). The US Department

Table 1 Parallel between EBM and nonclinical research purposes and processes across organizations

	In private sector nonclinical research	In academic nonclinical research	In EBM
Outcomes and deliverables	Patents (intellectual property) Decision to move a compound to clinical development Nonregulated study reports CRO study reports, data for customers Additions to catalog	Publications Patents (intellectual property) Study reports and data provided to public or private funders	Recommendations Guidelines Treatment decisions
Process	**Purpose in biopharma companies and CROs**	**Purpose in academia**	**Relevant EBM feature**
Initiating a research project	Driven by company strategy Triggered by prior data, exploratory studies, literature CROs: mainly triggered by requests from customers and market opportunities	Driven by science and funding opportunities Triggered by prior data, exploratory studies, literature, serendipity	Framing a question, collecting all available data, and ranking quality of evidence
Existence and use of guidelines	Company nonclinical quality and compliance rules, best research practice guidelines, patent department rules	Variable; rules of institutions, funding agencies, grant applications, journals, built into collaborations	Guidelines on use of EBM and EBM guidelines
Use of experimental bias reduction measures in study design and execution	Variable; field-dependent Detailed study plans usually mandatory for compounds selected to enter clinical development (less so for early test compounds) and systematically used by CROs	Variable; field-dependent; funding or grant-dependent; increasing due to pressure from funders, journals, peers; awareness that credibility is suffering	Core feature of EBM: studies with lowest risk of bias assumed to be most reliable
Biostatistics: access and use	Company biostatisticians and software (mostly proprietary); mandatory review of statistical analyses for compounds entering clinical development CROs: variable	Variable, somewhat "do-it-yourself": depending on statistical literacy or access to relevant expertise, widespread use of commercially available suites, free online tools	Adequate study power Meta-analyses
Data: integrity, access, and sharing	Electronic lab notebooks, electronic data storage, dedicated budgets Mandatory archive of all data and metadata for clinical stage compounds Restricted company-only access	Variable, depending on institution and resources, in particular to fund long-term safekeeping of data Ability to access data highly variable	Access to all data in real time

of Health and Human Services Office of Research Integrity has developed responsible conduct of research training courses that incorporate case studies from an academic research context (https://ori.hhs.gov/rcr-casebook-stories-about-researchers-worth-discussing). Several companies have adopted a similar case-based approach from a biopharma context.

Inaccuracy and biased interpretations are not necessarily due to purposeful scientific misconduct; in fact, most of the time, they are inadvertent, as the consequence of poor decision-making, training, or other circumstances. Mistakes can be made and can remain undetected when there is no formal process to critically review study design in advance of execution, an essential step when study outcomes gate decisions with long-term consequences, in particular for human subjects and patients. One aspect of review is to examine the multiple forms of bias that compromise data reliability, confounding evidence, and its analysis and interpretation. Experimental protocols can be biased, as can be experimenters, based on individual perceptions and behaviors: this is known as cognitive bias, i.e., the human tendency to make systematic errors, sometimes without even realizing it. Particularly problematic is confirmation bias, the tendency to seek and find confirmatory evidence for one's beliefs, and to ignore contradictory findings. Scientists can work to develop evidence to support a hypothesis, rather than evidence to contradict one. Beyond designing and performing experiments to support a hypothesis, confirmation bias can extend to reporting only those experiments that support a particular expectation or conclusion. While confirmation bias is generally subconscious, competition – for resources, publications, and other recognitions – can obscure good scientific practice. Confirmation bias can be both a cause and a consequence of publication or reporting bias, i.e., omissions and errors in the way results are described in the literature or in reports; it includes "positive" results bias, selective outcome reporting bias, "Hot stuff" bias, "All is well literature" bias, and one-sided reference bias (see definitions in https://catalogofbias.org).

In industry and academia, there are both common and specific risk factors conducive to cognitive bias, and awareness of this bias can be raised with various countermeasures, including those listed in Table 2.

2.3   Initiating a Research Project and Documenting Prior Evidence

Scientists running nonclinical pharmacology studies may have different goals, depending on where they work, but initiating a research project or study is driven by questions arising from prior findings in all parts of the biomedical ecosystem. When deciding to test a new hypothesis from emergent science, or when setting up a novel experimental model or assay, scientists generally read a handful of articles or reviews, focusing on the most recent findings. Many scientists methodically formulate an answerable question, weighing the strength of the available evidence and feasibility as primary drivers. Published findings can be weighed heavily as "truth," or disregarded, based on individual scientific judgment and many other factors. When subjective factors, such as journal impact factor, author prominence, or

Table 2 Factors that contribute to manifestations of bias and potential countermeasures

Contributing factors	In biopharma companies and CROs	In academia
Awareness and knowledge of risks of bias or misconduct	Multiple levels of review and quality control can highlight unconscious biases In-house training programs on responsible conduct of research increasingly common	Growing number of online material and training programs (see examples in Sect. 2.2)
Risk factors conducive to bias or misconduct	"Pace of business": compensation linked to performance/timelines, competitive landscape, career aspirations, customer deadlines	"Publish or perish": priority given to novel findings due to academic competition, career aspirations, funding mechanisms, and durations
Measures and incentives to increase responsible conduct	Occasional individual performance metrics CROs: responsible conduct linked to credibility, a key factor of company success	Recognition, publication, citation in leading journals with strict reporting guidelines, awards for reproducibility attempts (e.g., https://www.ecnp.eu/research-innovation/ECNP-Preclinical-Network-Data-Prize.aspx)

other subjective reasons, are weighed more heavily than the strength of the evidence, a form of bias is embedded from the conception of a research project. Similarly to the flowchart approach used in EBM, where the first step is to frame the clinical question and retrieve all the related evidence, explicitly defining a question and systematically reviewing the literature should be a common practice in nonclinical pharmacology. When deciding to work on a target, biopharma scientists also have to consider whether modulating it could potentially result in adverse effects, so the background evidence to be weighed may have other aspects than for an academic research project. An obstacle to a comprehensive assessment of prior data is that data can be published, unpublished, undisclosed, or inaccessible behind a paywall or another company's firewall or simply out of reach due to past archival practices (see Sect. 2.7). Publication and selective outcome reporting biases will therefore be present in most attempts to review and weigh prior evidence. Thus, in practice, the data a scientist will evaluate at the start of a research project is often incomplete, raising the possibility of flawed experimental design, execution and interpretation, as well as the risk of confirmation and related biases.

2.4   Existence and Use of Guidelines

Recommendations on how to design and conduct nonclinical, nonregulated research studies can be found in scientific publications, in scientific society or institution guidelines, and in grant application guidelines. Although recommended "best research practices" have been around for at least a decade, there are no consensus, universal nonclinical pharmacology quality guidelines, but instead a collection of constantly evolving, context, and type-of-experiment-specific suggestions.

Biopharma companies and nonclinical CROs generally have internal guidelines. Scientists are expected to record results in real time in laboratory notebooks, should an organization or individual need to document data and timelines to establish inventorship. Guidelines produced by research quality departments therefore focus on how scientists should record the results of their research, and deviations from standard operating procedures, in order to fulfill legal and regulatory requirements, more than on study design or the use of measures to reduce experimental bias. In the private sector, research quality guidelines and best practice recommendations are generally confidential documents. In publications, research quality guidelines and implementation are rarely mentioned. While indirect, study reporting guidelines (see Sect. 2.7) are slightly more cited, but determining to what extent these were followed is far from trivial.

2.5   Use of Experimental Bias Reduction Measures in Study Design and Execution

The core principle of EBM is that the most reliable evidence comes from clinical studies with the lowest risk of bias and typically those that are designed with adequate power, randomization, blinding, and a pre-specified endpoint, in a clinically relevant patient population. There are many resources to help investigators plan human studies, such as the SPIRIT statement (http://www.spirit-statement.org), an evidence-based guideline for designing clinical trial protocols, which is being developed into a web-based protocol building tool. There are fewer resources to assist scientists in designing nonclinical studies; an example is the NC3Rs' Experimental Design Assistant (EDA, https://www.nc3rs.org.uk/experimental-design-assistant-eda) for in vivo animal studies. Experimental protocols can be found in publications or online, but they are primarily written to provide details on technical aspects, and do not always explicitly address the different sources of experimental bias.

In biopharma research, study plans which describe the study design and experimental methods in full detail, including the planned statistical methods and analyses, and any deviations to these plans as the study progresses, are usually mandatory for studies that are critical for decision-making. Study plans are more rarely written for exploratory, pilot studies. Nonclinical CROs use study plan templates that include statistical analysis methodologies, which are generally shared with customers. In our experience, CROs and academic drug discovery centers are very willing to discuss and adapt study designs to suit customer needs. Collaboratively building a study plan is a good opportunity to share knowledge, ensure that a study is conducted and reported according to expectations, and work to identify and reduce conscious and unconscious biases. Across all sectors, planning ahead for in vivo pharmacology studies is more elaborate than for in vitro experiments, due to animal ethics requirements and the logistics of animal care and welfare. However, nonclinical study plans are not normally published, whereas clinical trial protocols are available in online databases such as the EU (https://www.clinicaltrialsregister.eu) and US (https://clinicaltrials.gov/) registers. A few initiatives, such as OSF's "preregistration challenge" (Open Science Foundation, Preregistration Challenge, Plan, Test, Discover, https://osf.io/x5w7h), have begun to promote formal preregistration of nonclinical study protocols, as a means to improve research quality (Nosek et al. 2018). However, preregistering every single nonclinical pharmacological study protocol in a public register would be difficult in practice, for confidentiality considerations, but also due to a perceived incompatibility with the pace of research in all sectors.

Overall, our experience in the field of neuroscience is that the implementation of experimental bias reduction measures is highly variable, within and across sectors, and meta-analyses of scientific publications have shown that there is clearly room for improvement, at least in the reporting of these measures (van der Worp et al. 2010; Egan et al. 2016).

Different field- and sector-related practices and weights on bias reduction measures, such as blinding and randomization (see chapter "Blinding and Randomization"), can be expected. In the clinical setting, blinding is a means to reduce observer bias, which, along with randomization to reduce selection bias, underlies the higher ranking of RCTs over, for example, open-label trials. Both blinding and randomization are relevant to nonclinical studies because the awareness of treatment or condition allocation can produce observer bias in study conduct and data analysis. Neurobehavioral measures are among the most incriminated for their susceptibility to observer bias. But even automated data capture can be biased if there are no standards for threshold and cutoff values. Observer bias is also a risk, for example, when visually counting immunolabeled cells, selecting areas for analysis in brain imaging data, and choosing recording sites or cells in manual electrophysiology experiments. Blinding has its limitations; blinding integrity may lost, such as when using transgenic mice (which are often noticeably different in appearance or behavior compared to wild-type littermates) or in pathological settings that induce visible body changes, and the experimenter's unawareness of group allocation will not be sufficient to limit the effect observing animals can have on their behavior (analogous to the Hawthorne effect in social sciences, see https://catalogofbias.org/biases/hawthorne-effect/).

Differences in resource availability will influence practices, since training experimenters, standardizing animal handling and husbandry, and earmarking suitable lab space and equipment, among other considerations, are contingent upon funding. Nonclinical CROs are most likely to have strong guidelines, or at least evidence-based standard operating procedures, and to follow them, since credibility, transparency, and customer satisfaction are business-critical. The systematic use of inclusion/exclusion criteria and blinding should be implemented as standard practice in all sectors of the biomedical ecosystem. However, while in the industry there is a tendency to optimize workflows through standardization, and similarly in academia, strong lab "traditions," one size does not necessarily fit all. Specific technical constraints may apply, in particular for randomization. For instance, in some in vitro experiments, features such as "edge effect" or "plate effect" need to be factored into the randomization procedure (https://paasp.net/simple-randomisation); liquid chromatography-coupled mass spectrometry experiments require additional caution, since randomizing the order in which samples from different groups or conditions are tested may be counterproductive if the risk of potential cross-contamination is not addressed. Randomizing the order of procedures, while often a sound measure to prevent procedural bias, may actually increase the risk of bias, if animals behave differently depending on prior procedures or paradigms. While randomization and blinding will generally be effective in reducing risks of selection and observer bias, they have no effect on non-contemporaneous bias, when control groups or samples are tested or analyzed at a separate time from treated ones.

Thus, both in EBM and in nonclinical research, high-quality designs aim to take into account all of the known sources of bias and employ the best available countermeasures. Among these, there are two universally critical items, a pre-specified endpoint with an estimate of the predicted effect size and the corresponding adequate statistical power to detect the predicted effect, given the sample size, all of which require a prior statistical plan.

2.6   Biostatistics: Access and Use to Enable Appropriate Design of Nonclinical Pharmacology Studies

Establishing an a priori statistical plan, as part of the study design, remains far from customary in nonclinical pharmacology, mainly because scientists can lack the adequate awareness and knowledge to do so. The latest Research Integrity report by the Science and Technology Committee in the UK (https://publications.parlia ment.uk/pa/cm201719/cmselect/cmsctech/350/350.pdf) emphasized that scientists need to learn and understand the principles of statistics, rather than simply being told of a list of statistical tests and software that does the analyses. In our experience, biologists' statistical proficiency appears to mostly be based on local custom and varies widely even in the same field of biology. This is illustrated by misleading phrases in methods sections of publications, such as "the number of animals used was the minimum required for statistical analysis," or "post hoc comparisons were carried out between means as appropriate," or "animals were randomly assigned to 4 groups," or "the experiments were appropriately randomized" (sic). A side effect of this phenomenon is that it hampers critical assessments of published papers; biologists confronted with unfamiliar terms may struggle to capture which study designs and analyses were actually conducted.

In practice, more attention is paid to statistics once the data have been generated. In nonclinical CROs the statistical analyses are provided to the customer in the full study reports. In biopharma companies, for clinical development candidate compounds, it is generally mandatory that the proposed statistical analyses are developed and/or validated by a statistician. Many companies have developed robust proprietary statistics software, with specific wording and a selection of internally approved tests and analysis tools. Although in-house applications are validated and updated, they are not ideal for sharing results and analyses with external partners. Overall, and despite a call for cultural change in the interactions between scientists and nonclinical statisticians (Peers et al. 2012), it seems that the nonclinical pharmacology community remains under-resourced in this area. Insight gained through discussions on data quality among partners of several European initiatives suggests that there are too few research biostatisticians in all biomedical arenas.

When a thorough process is established beforehand, choosing a pre-specified endpoint to test a hypothesis and estimating an effect size for this endpoint are essential. While both are required in EBM, these are less common in nonclinical research. Clinical studies aim to detect a predetermined effect size, or a clinically relevant direction and effect magnitude, based on prior knowledge. In contrast, scientists generally have a rough idea of values that would be negligible, due to biological variation or to inaccuracy or imprecision, but considering which values are biologically meaningful tends to be done after, rather than before, running an experiment. When generating a hypothesis, i.e., in exploratory or pilot studies, it may be possible to choose an endpoint of interest, without necessarily defining its direction and amplitude. In contrast, prior estimates of effect size are essential when the aim is to demonstrate a pharmacological effect in confirmatory studies upon which decisions about next steps are based. This distinction between exploratory and confirmatory studies (Kimmelman et al. 2014 and chapter "Resolving the Tension Between Exploration and Confirmation in Preclinical Biomedical Research") is a determining factor in study design, but remains an underused concept in nonclinical work.

Arguably the most serious consequence of insufficient planning is that nonclinical studies are too often underpowered (Table 2 in Button et al. 2013) or are of unknown power, when publications fail to reveal how sample sizes were chosen (Carter et al. 2017). Despite its central role in the null hypothesis significance testing framework, which remains the most used in nonclinical pharmacology, for many scientists, statistical power is one of the least well-understood aspects of statistics. This may be because it is generally explained using abstract mathematical terms, and its role more extensively discussed in clinical research, or in psychology, than in biology. However, recognizing that inadequately powered studies can lead to unreliable conclusions on the direction and magnitude of an effect in a sample of the whole population is just as important in nonclinical pharmacology as it is in EBM. Assay development is by definition exploratory in initial attempts; but when the assay is going to be used routinely, sample sizes to achieve a desired statistical power need to be determined. Unfortunately, this is not yet the norm in nonclinical pharmacology, where decisions are often made on so-called converging evidence from several underpowered studies with different endpoints or on a single published study of unknown power, offering little confidence that the same effect(s) would be seen in the whole population from which the sample was taken.

As discussed above (see Sect. 2.5), randomization is essential to prevent selection bias across all sectors of research. Randomization can be achieved even with limited resources and applied in many nonclinical pharmacology studies regardless of their purpose and type, without necessarily involving statistical expertise. The randomization procedure must however be part of the study design, and statistical evaluation before a study is conducted can help determine which procedure is best suited.

2.7   Data Integrity, Reporting, and Sharing

Notwithstanding the existence of vast amounts of electronic storage space and sophisticated software to ensure file integrity, retaining, and potentially sharing, original datasets and protocols is not yet straightforward. Barriers to widespread data sharing are slowly being overcome, but there remains a need for long-term funding, and the ability to browse data long after the software used to generate or store them has become obsolete.

In biopharma companies and CROs, it is customary to retain all original individual and transformed data, with information on how a study was performed, in laboratory notebooks and annexes. Scientists working in industry are all aware that the company owns the data; one does not lose or inadvertently misplace or destroy the company's property, and in audits, quality control procedures, preparation for regulatory filings, or patent litigation cases, to name a few, original data must often be produced. This also applies to studies conducted by external collaborators. For compounds that are tested in human trials (including compounds that reach the market), all data and metadata must be safely stored and retrievable for 30 years after the last administration in humans. It is thus common practice to keep the records decades after they were generated (see item GRS023 in https://india-pharma.gsk.com/media/733695/records-retention-policy-and-schedule.pdf). Such durations exceed by far the life of the software used to generate or store the data and require machine-readable formats. Paper laboratory notebooks are also stored for the duration; their contents are notoriously difficult to retrieve as time passes, and teams or companies disperse. Electronic source data in FDA-regulated clinical investigations are expected to be attributable, legible, contemporaneous, original, and accurate (ALCOA). This expectation is also applied to nonregulated nonclinical data in many biopharma companies and in nonclinical CROs. The recent FAIR (findable, accessible, interoperable, reusable) guiding principles for scientific data management and stewardship (Wilkinson et al. 2016) are intended to facilitate data access and sharing while maintaining confidentiality if needed. To this date, broadly sharing raw data and protocols from biopharma research remains rare (but see Sect. 3.1).

Generally speaking, data generated in academia destined for publication are not as strictly managed. Institutional policies (see examples of data retention policies: Harvard,   https://vpr.harvard.edu/files/ovpr-test/files/research_records_and_data_retention_and_maintenance_guidance_rev_2017.pdf;   MRC,   https://mrc.ukri.org/documents/pdf/retention-framework-for-research-data-and-records/)   may   state that data should be retained for a minimum of 3 years after the end of a research project, a period of 7–10 years or more, or as long as specified by research funder, patent law, legislative, and other regulatory requirements. Effective record-keeping and retention is limited by funding and by the rapid turnover of the scientists performing most of the experiments; a classic problem is the struggle to find the data generated by the now long-gone postdoctoral associate. Access to original, individual data can be requested by other scientists or required by journals and funding agencies or, on rare occasions, for investigations of scientific misconduct. Although academic data and metadata sharing is improving (Wallach et al. 2018), with extended supplementary materials and checklists, preprint servers, data repositories (Figshare, https://figshare.com; OSF, https://osf.io; PRIDE, https://www.ebi.ac.uk/pride/archive), and protocol sharing platforms (https://experiments.springernature.com; https://www.protocols.io), universal open access to data is yet to be achieved.

In biopharma companies, there is an enormous amount of early discovery studies, including but not limited to assay development and screening campaigns, with both "positive" and "negative" data, that are not intended per se for publication, even though many could be considered precompetitive. A relatively small proportion of conducted studies is eventually published. However, for each compound entering clinical development, all the results that are considered relevant are documented in the nonclinical pharmacology study reports that support IND and CTA filings. A summary of the data is included in the nonclinical overview of the application dossiers and in the Investigator's Brochure. From these documents it is often difficult to assess the quality of the evidence, since they contain relatively little experimental or study information (Wieschowski et al. 2018); study design features are more likely to be found in the study reports, although there are no explicit guidelines for these (Langhof et al. 2018). The study reports themselves are confidential documents that are usually only disclosed to health authorities; they are intended to be factual and include study plans and results, statistical plans and analyses, and individual data.

In academia, publishing is the primary goal; publication standards and content are set by guidelines from funders, institutions, partners, peer reviewers, and most importantly by journals and editorial policies. In recent years, journal guidelines to authors have increasingly focused on good reporting practices, implementing recommendations from landmark publications and work shepherded by institutions such as the NC3Rs with the ARRIVE guidelines (Kilkenny et al. 2010), and the NIH (Landis et al. 2012), mirroring coordinated initiatives to improve clinical trial reporting guidelines, such as the EQUATOR network (https://www.equator-net work.org). Yet despite the impressive list of journals and institutions that have officially endorsed the ARRIVE guidelines, meta-research shows that there is much to be improved in terms of compliance (Jin et al. 2018; Hair et al. 2019). Moreover, there is no obligation to publish every single study performed or to report all experiments of a study in peer-reviewed journals; an important amount, possibly as much as 50%, remain unpublished (ter Riet et al. 2012).

3   Overcoming Obstacles and Further Learning from Principles of Evidence-Based Medicine

3.1   Working Together to Improve Nonclinical Data Reliability

Many conversations among researchers, basic and clinical, resemble the one between Professor Benchie and Doctor Athena (Macleod 2015), in which Athena concludes that they should be able to improve reliability and translatability, at least a little, by learning from the strengths and weaknesses of their respective backgrounds.

Strictly following the EBM and GRADE rules would require that scientists appraise *all* the available nonclinical evidence with relevance to the question being asked. This should be the case when deciding whether to take a compound to the clinic, but is unlikely to happen for other purposes. Scientists would nevertheless benefit from a basic understanding of the methodology, strengths and weaknesses of systematic review and meta-analysis. Meta-analyses are often performed in collaborations, and a recent feasibility study using crowd-sourcing for clinical study quality assessment suggests that this could be a way forward, since experts and novices obtained the same results (Pianta et al. 2018). Combined with recently developed and highly promising machine learning algorithms (Bannach-Brown et al. 2019), collaborative efforts could increase the pace and reduce human error in systematic reviews and meta-analysis.

In recent years, private sector organizations, academic institutions, disease foundations, patient associations, and government bodies have formed consortia to tackle a wide variety of complex questions, in a precompetitive manner. Many of these partnerships bring together basic and clinical researchers and also aim to share experimental procedures and unpublished findings. Collective efforts have produced consensus recommendations, based on the critical appraisal of published and unpublished data, in fields such as stroke (Macleod et al. 2009) and pain (Knopp et al. 2015; Andrews et al. 2016). In IMI Europain (home page: www.imieuropain.org), the group of scientists and clinicians working on improving and refining animal models of chronic pain, addressing the clinical relevance of endpoints used in animal models and methodologies to reduce experimental bias, held teleconference meetings roughly 10 times a year over 5 years, which represents a substantial amount of shared data and expertise. Leveraging this combined expertise and aiming to develop a novel, non-evoked outcome measure of pain-related behavior in rodents, IMI Europain partners from both academia and industry accomplished a multicenter nonclinical study (Wodarski et al. 2016), in the spirit of a phase 3 multicenter clinical trial. One of the important lessons learned during this study was that absolute standardization should not be the goal, since circumstantial differences such as site location cannot be erased, leading to pragmatic accommodations for local variations in laboratory practice and procedures. An effort to uncover evidence-based drivers of reliability in other subfields of neuroscience is ongoing in IMI EQIPD (home page: https://quality-preclinical-data.eu), with the overarching goal of building broadly applicable tools for managing nonclinical data quality. Discussions on emerging pathways of neurodegenerative disease within the IMI neurodegeneration strategic governance group led to a single, collectively written article describing independent attempts that failed to reproduce or extend the findings of a prominent publication (Latta-Mahieu et al. 2018). A culture of collaboration is thus growing, and not only in large consortia. Co-designing nonclinical studies is now the preferred practice in bilateral partnerships or when studies are outsourced by biopharma companies to nonclinical CROS or academic drug discovery centers.

3.2   Enhancing Capabilities, from Training to Open Access to Data

Research quality training should aim to provide the ability to recognize the different forms of bias and how to minimize risks, covering the full scope of data reliability, rather than solely focusing on compliance or on scientific integrity. In the private sector, laboratory notebook compliance audits are routinely performed; checklists are used to assess whether scientists have correctly entered information in laboratory notebooks. When releasing individual audit results to scientists, these compliance checklists or, in all sectors, the Nature Reporting Summary (https://www.nature.com/documents/nr-reporting-summary.pdf) checklist can also be used as tools for continuing training.

Initial and continuing training in statistics should be an absolute priority for all biologists. Those who are privileged to work closely with biostatisticians should aim to establish a common language, and a meaningful engagement of both parties from the start, to be able to translate the scientific question to a statistical one and co-build study designs, with the most stringent criteria for confirmatory studies.

Learning to read a paper and to critically appraise evidence and keeping in mind that low-quality reporting can confound the appraisal and that even high-profile publications may have shortcomings should also be part of training, continued in journal clubs, and carried over to post-publication peer review (e.g., PubPeer, https://pubpeer.com). Paying particular attention to the methods section and any supplementary methods information, searching for sample size considerations, randomization, and blinding, before interpreting data presented in figures, is an effective way to remember that independent evaluation of the data, with its strengths and limitations, is the core responsibility of scientists in all research endeavors.

The fact that many clinical trial findings remain unpublished is still a major roadblock for EBM, which various organizations have been tackling in recent years (see links in https://www.eupati.eu/clinical-development-and-trials/clinical-study-results-publication-and-application). In biopharma companies, proprietary nonclinical data include a considerable amount of study replicates, sometimes spread over several years. Many attempts are also made to reproduce data reported in the literature (Begley and Ellis 2012; Prinz et al. 2011; Djulbegovic and Guyatt 2017), but most of these remain undisclosed. In recent years, several independent groups have been instrumental in coordinating and publishing reproducibility studies, such as the Reproducibility Initiative collaboration between Science Exchange, PLOS, figshare, and Mendeley (http://validation.scienceexchange.com/#/reproducibility-initiative), the Center for Open Science (The Reproducibility Project, a collaborative effort by the Center for Open Science: https://cos.io), and a unique nonprofit-driven initiative in amyotrophic lateral sclerosis (Scott et al. 2008). In sectors of the biomedical ecosystem where the focus is more on exploring new ideas, generating and testing hypotheses, or confirming and extending a team's own work rather than replicating that of others, a substantial amount of work, possibly as much as 50% (ter Riet et al. 2012), remains unpublished. Thus, in the nonclinical space, the obstacles to widespread open access to data have yet to be overcome.

4   Conclusion and Perspectives

Although the term evidence-based medicine was first introduced almost 30 years ago, building upon efforts over several decades to strengthen a data-driven practice of medicine, there are still misconceptions and resistance to the approach, as well as challenges to its practical implementation, despite a number of striking illustrations of its impact (Djulbegovic and Guyatt 2017). Adapting the conceptual toolbox of EBM and using it to optimize nonclinical research practices and decision-making will likely also require time, and most importantly, strong commitment and well-targeted, well-focused advocacy from all stakeholders. Several lessons from EBM particularly deserve the attention of nonclinical scientists, such as the importance of framing a question, critically appraising prior evidence, carefully designing a study that addresses that question, and assessing the quality of the data before moving to the next step (Fig. 1).

In medicine, reviewing the evidence aims to inform the decision about how to treat a patient; in science, the decision can be about whether or not to pursue a project, about which experiment to do next, which assay to develop, whether the work is sufficient for publication, or whether the aggregated evidence supports testing a compound in humans. In all sectors, a universal framework, with customizable tools, such as those available in the clinical setting, higher standards in data, and metadata management practices and sharing would help scientists assess and generate more reliable data.

Adapting EBM principles to nonclinical research need not undermine the freedom to explore. Assessing the quality of prior work should not paralyze scientists or prevent them from thinking out of the box, and the effective implementation of measures, such as blinding and randomization, to reduce bias should not produce a bias against novelty. Exploratory studies aiming to generate new hypotheses may follow less strict designs and statistical approaches, but when they are followed by confirmatory studies, a novel body of evidence and knowledge is formed, which can propel innovation through significance and impact. Indeed, "Innovative research projects are expected to generate data that is reproducible and provides a foundation for future studies" (http://grants.nih.gov/reproducibility/faqs.htm#4831). In other words, to be truly innovative, novel findings should endure beyond the initial excitement they create. If publications were collaboratively appraised using an adaptation of GRADE ratings, journals could develop novel impact metrics to reflect these ratings and the endurance of the findings.

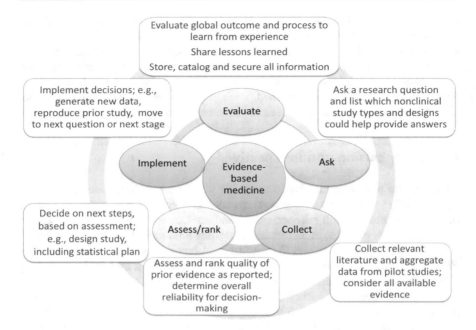

Fig. 1 Adapting the five evidence-based medicine steps to nonclinical pharmacology research

In drug discovery and development, the significance and reliability, or lack thereof, of experimental data have immediate consequences. Biopharma companies need to be able to rely on the data to determine a course of action in a research project, to shape the future of a drug discovery program, and to extrapolate doses that will be administered to humans. There is thus both a financial interest and an ethical imperative, and from the patient's perspective, an absolute requirement, to base a decision to test a compound in humans on reliable data. When the overarching goal of nonclinical pharmacology research is to bring a compound to the clinic, transitioning to an evidence-based model, using and generating evidence rated in the upper levels of the pyramid to inform decisions, would benefit discovery, and at the very least, reduce the amount of wasted experiments.

Even with high quality of evidence and better informed decision-making, it remains to be seen whether the approaches discussed here will effectively decrease the attrition rate of drug candidates and lead to more success in translating findings from nonclinical to clinical studies. There are many reasons for "failure," and only some are related to scientific rigor, reproducibility, and robustness. However, progress in understanding disease mechanisms and target tractability (https://docs. targetvalidation.org/faq/what-is-target-tractability) is linked to the ability to design experiments and clinical trials that provide reliable information. In the near future, as solutions for enhancing data access emerge and stringent reporting standards become mandatory, scientists of all sectors should be encouraged to adapt and adopt EBM principles, to better enable reliable data-driven decisions.

Acknowledgments The authors thank the members of the IMI EQIPD consortium for many thought-provoking and fruitful discussions, Catherine Deon for providing insight on liquid chromatography-coupled mass spectrometry experiments, and Stéphanie Eyquem for critically reviewing the manuscript.

References

Andrews NA, Latremoliere A, Basbaum AI et al (2016) Ensuring transparency and minimization of methodologic bias in preclinical pain research: PPRECISE considerations. Pain 157:901–909

Balshem H, Helfand M, Schünemann HJ et al (2011) GRADE guidelines: 3. Rating the quality of evidence. J Clin Epidemiol 64(4):401–406

Bannach-Brown A, Przybyła P, Thomas J et al (2019) Machine learning algorithms for systematic review: reducing workload in a preclinical review of animal studies and reducing human screening error. Syst Rev 8(1):23

Begley CG, Ellis LM (2012) Drug development: raise standards for preclinical cancer research. Nature 483(7391):531–533

Button KS, Ioannidis JP, Mokrysz C et al (2013) Power failure: why small sample size undermines the reliability of neuroscience. Nat Rev Neurosci 14:365–376

Carter A, Tilling K, Munafò MR (2017) A systematic review of sample size and power in leading neuroscience journals. https://www.biorxiv.org/content/early/2017/11/23/217596

Djulbegovic B, Guyatt GH (2017) Progress in evidence-based medicine: a quarter century on. Lancet 390(10092):415–423

Egan KJ, Vesterinen HM, Beglopoulos V, Sena ES, Macleod MR (2016) From a mouse: systematic analysis reveals limitations of experiments testing interventions in Alzheimer's disease mouse models. Evid Based Preclin Med 3(1):e00015

Goodman SN, Fanelli D, Ioannidis JP (2016) What does research reproducibility mean? Sci Transl Med 8(341):341ps12

Hair K, Macleod MR, Sena ES, IICARus Collaboration (2019) A randomised controlled trial of an Intervention to Improve Compliance with the ARRIVE guidelines (IICARus). Res Integr Peer Rev 4:12

Hooijmans CR, Rovers MM, de Vries RBM et al (2014) SYRCLE's risk of bias tool for animal studies. BMC Med Res Methodol 14:43

Hooijmans CR, de Vries RBM, Ritskes-Hoitinga M et al (2018) GRADE Working Group. Facilitating healthcare decisions by assessing the certainty in the evidence from preclinical animal studies. PLoS One 13(1):e0187271

Jin Y, Sanger N, Shams I et al (2018) Does the medical literature remain inadequately described despite having reporting guidelines for 21 years? – A systematic review of reviews: an update. J Multidiscip Healthc 11:495–510

Kilkenny C, Browne WJ, Cuthill C et al (2010) Improving bioscience research reporting: the ARRIVE guidelines for reporting animal research. PLoS Biol 8:e1000412; ARRIVE: https://www.nc3rs.org.uk/arrive-guidelines

Kimmelman J, Mogil JS, Dirnagl U (2014) Distinguishing between exploratory and confirmatory preclinical research will improve translation. PLoS Biol 12(5):e1001863

Knopp KL, Stenfors C, Baastrup C et al (2015) Experimental design and reporting standards for improving the internal validity of pre-clinical studies in the field of pain: consensus of the IMI-Europain consortium. Scand J Pain 7(1):58–70

Landis SC, Amara SG, Asadullah K et al (2012) A call for transparent reporting to optimize the predictive value of preclinical research. Nature 490:187–191; and https://www.nih.gov/research-training/rigor-reproducibility/principles-guidelines-reporting-preclinical-research

Langhof H, Chin WWL, Wieschowski S et al (2018) Preclinical efficacy in therapeutic area guidelines from the U.S. Food and Drug Administration and the European Medicines Agency: a cross-sectional study. Br J Pharmacol 175(22):4229–4238

Latta-Mahieu M, Elmer B, Bretteville A et al (2018) Systemic immune-checkpoint blockade with anti-PD1 antibodies does not alter cerebral amyloid-β burden in several amyloid transgenic mouse models. Glia 66(3):492–504

Macleod MR (2015) Prof Benchie and Dr Athena-a modern tragedy. Evid Based Preclin Med 2 (1):16–19

Macleod MR, Fisher M, O'Collins V et al (2009) Good laboratory practice: preventing introduction of bias at the bench. Stroke 40(3):e50–e52

Nosek BA, Ebersole CR, DeHaven AC et al (2018) The preregistration revolution. Proc Natl Acad Sci U S A 115(11):2600–2606

Peers IS, Ceuppens PR, Harbron C (2012) In search of preclinical robustness. Nat Rev Drug Discov 11(10):733–734

Pianta MJ, Makrai E, Verspoor KM et al (2018) Crowdsourcing critical appraisal of research evidence (CrowdCARE) was found to be a valid approach to assessing clinical research quality. J Clin Epidemiol 104:8–14

Prinz F, Schlange T, Asadullah K (2011) Believe it or not: how much can we rely on published data on potential drug targets? Nat Rev Drug Discov 10(9):712

Scott S, Kranz JE, Cole J et al (2008) Design, power, and interpretation of studies in the standard murine model of ALS. Amyotroph Lateral Scler 9(1):4–15

Sena ES, Currie GL, McCann SK et al (2014) Systematic reviews and meta-analysis of preclinical studies: why perform them and how to appraise them critically. J Cereb Blood Flow Metab 34 (5):737–742

ter Riet G, Korevaar DA, Leenaars M et al (2012) Publication bias in laboratory animal research: a survey on magnitude, drivers, consequences and potential solutions. PLoS One 7:e43404

van der Worp HB, Howells DW, Sena ES et al (2010) Can animal models of disease reliably inform human studies? PLoS Med 7(3):e1000245

Wallach JD, Boyack KW, Ioannidis JPA (2018) Reproducible research practices, transparency, and open access data in the biomedical literature, 2015–2017. PLoS Biol 11:e2006930

Wieschowski S, Chin WWL, Federico C et al (2018) Preclinical efficacy studies in investigator brochures: do they enable risk-benefit assessment? PLoS Biol 16(4):e2004879

Wilkinson MD, Dumontier M, Aalbersberg IJ et al (2016) The FAIR guiding principles for scientific data management and stewardship. Sci Data 3:160018

Wodarski R, Delaney A, Ultenius C et al (2016) Cross-centre replication of suppressed burrowing behaviour as an ethologically relevant pain outcome measure in the rat: a prospective multicentre study. Pain 157(10):2350–2365

General Principles of Preclinical Study Design

Wenlong Huang, Nathalie Percie du Sert, Jan Vollert, and Andrew S. C. Rice

Contents

Abstract

Preclinical studies using animals to study the potential of a therapeutic drug or strategy are important steps before translation to clinical trials. However, evidence has shown that poor quality in the design and conduct of these studies has not only impeded clinical translation but also led to significant waste of valuable research resources. It is clear that experimental biases are related to the poor quality seen with preclinical studies. In this chapter, we will focus on hypothesis testing type of preclinical studies and explain general concepts and principles in

W. Huang (✉)
Institute of Medical Sciences, School of Medicine, Medical Sciences and Nutrition, University of Aberdeen, Aberdeen, UK
e-mail: w.huang@abdn.ac.uk

N. Percie du Sert
NC3Rs, London, UK

J. Vollert · A. S. C. Rice
Pain Research Group, Faculty of Medicine, Department of Surgery and Cancer, Imperial College London, London, UK

© The Author(s) 2019
A. Bespalov et al. (eds.), *Good Research Practice in Non-Clinical Pharmacology and Biomedicine*, Handbook of Experimental Pharmacology 257,
https://doi.org/10.1007/164_2019_277

relation to the design of in vivo experiments, provide definitions of experimental biases and how to avoid them, and discuss major sources contributing to experimental biases and how to mitigate these sources. We will also explore the differences between confirmatory and exploratory studies, and discuss available guidelines on preclinical studies and how to use them. This chapter, together with relevant information in other chapters in the handbook, provides a powerful tool to enhance scientific rigour for preclinical studies without restricting creativity.

Keywords

Experimental bias · Hypothesis generating · Hypothesis testing · In vivo studies · Preclinical research

This chapter will give an overview of some generic concepts pertinent to the design of preclinical research. The emphasis is on the requirements of in vivo experiments which use experimental animals to discover and validate new clinical therapeutic approaches. However, these general principles are, by and large, generically relevant to all areas of preclinical research. The overarching requirement should be that preclinical research should only be conducted to answer an important question for which a robust scrutiny of the available evidence demonstrates that the answer is not already known. Furthermore, such experiments must be designed, conducted, analysed and reported to the highest levels of rigour and transparency. Assessments of research outputs should focus more on these factors and less on any apparent "novelty".

1   An Overview

Broadly, preclinical research can be classified into two distinct categories depending on the aim and purpose of the experiment, namely, "hypothesis generating" (exploratory) and "hypothesis testing" (confirmatory) research (Fig. 1). Hypothesis generating studies are often scientifically-informed, curiosity and intuition-driven explorations which may generate testable theories regarding the pathophysiology of disease and potential drug targets. The freedom of researchers to explore such innovative ideas is the lifeblood of preclinical science and should not be stifled by excessive constraints in terms of experimental design and conduct. Nevertheless, in order to subsequently assess the veracity of hypotheses generated in this way, and certainly to justify clinical development of a therapeutic target, hypothesis testing studies which seek to show reproducible intervention effects in relevant animal models must be designed, conducted, analysed and reported to the highest possible levels of rigour and transparency. This will also contribute to reducing research "waste" (Ioannidis et al. 2014; Macleod et al. 2014). Chapter "Resolving the Tension Between Exploration and Confirmation in Preclinical Biomedical Research" of the handbook will deal with exploratory and confirmatory studies in details. This chapter will only focus on general design principles for hypothesis testing studies. We will

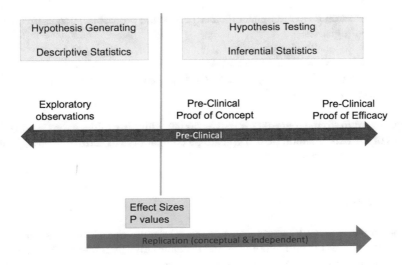

Fig. 1 Comparison of exploratory (hypothesis generating) and confirmatory (hypothesis testing) preclinical studies. Descriptive statistics describes data and provides descriptions of the population, using numerical calculations, graphs, and tables. In contrast, inferential statistics predicts and infers about a population using a sample of data from the population, therefore one can take data from samples and make generalisation about a population

address the issue of design principles for hypothesis-generating studies at the end of this chapter. We advise that when researchers design and conduct hypothesis testing in vivo studies, they should conform to the general principles for the major domains that are outlined in Sect. 4 of the chapter and incorporate these principles into a protocol that can be registered and published. The purpose of using these principles is to enhance scientific rigour without restricting creativity. It is advisable that sometimes there can be exploratory elements within the same hypothesis testing studies; therefore, extra care in terms of applying these principles to reduce experimental biases would be needed before the start of the studies. This chapter will not cover reporting, which will be detailed in chapters "Minimum Information and Quality Standards for Conducting, Reporting, and Organizing In Vitro Research", "Minimum Information in In Vivo Research", and "Quality Governance in Biomedical Research" of the handbook.

We would recommend that researchers who conduct hypothesis testing in vivo studies should prepare clear protocols, which include a statistical analysis plan, detailing how they are going to set up measures to address the major domains of experimental biases before the experiments start. Ideally, these protocols should be preregistered and/or published, so that the methods which will be used to reduce the impact of bias are documented in an a priori fashion. The process of peer review of a protocol prior to initiating experiments of course is a valuable opportunity for refinement and improvement. Registering protocols encourages rigour and transparency, even if the protocol is not peer-reviewed. Some journals are open to submissions of these types of protocols, such as BMJ Open Science, and many journals offer the Registered Reports format. In addition, there are online resources that allow researchers to preregister their experimental protocols, such as preclinical. eu and osf.io/registries.

2   General Scientific Methods for Designing In Vivo Experiments

Designing an in vivo experiment involves taking a number of decisions on different aspects of the experimental plan. Typically, a comparative experiment can be broken into several component parts.

2.1   Hypotheses and Effect Size

The objective is usually to test a hypothesis. On some occasions, two hypotheses may be postulated: the null hypothesis and the alternative hypothesis. The alternative hypothesis refers to the presumption that the experimental manipulation has an effect on the response measured; the null hypothesis is the hypothesis of no change, or no effect. In a statistical test, the p-value reports the probability of observing an effect as large or larger than the one being observed if the null hypothesis was true; the smaller the p-value, the least likely it is that the null hypothesis is true. The null hypothesis cannot be accepted or proven true. This also defines the effect of interest, i.e. the outcome that will be measured to test the hypothesis. The minimum effect size is the smallest effect the researcher designs the experiment to be able to detect and should be declared in the protocol; it is set up as the minimum difference which would be of biological relevance. The effect size is then used in the sample size calculation to ensure that the experiment is powered to detect only meaningful effects and does not generate statistically significant results that are not biologically relevant. In many cases, it will be hard to determine the minimum difference of biological relevance as for early stage experiments it might be completely unknown, or translatability between clinical relevance and experimental detection thresholds will be complex. There is no simple and easy answer to this question, but in general, a minimum effect size should be set so one can assume to have a beneficial effect for individuals rather than large cohorts, the difference must be experimentally testable and reasonable to achieve, and should have a rationale for translation into patients in the long run.

2.2   Groups, Experimental Unit and Sample Size

In comparative experiments, animals are split into groups, and each group is subjected to different interventions, such as a drug or vehicle injection, or a surgical procedure. The sample size is the number of experimental units per group; identifying the experimental unit underpins the reliability of the experiment, but it is often incorrectly identified (Lazic et al. 2018). The experimental unit is the entity subjected to an intervention independently of all other units; it must be possible to assign any two experimental units to different comparison groups. For example, if the treatment is applied to individual mice by injection, the experimental unit may be the animal, in which case the number of experimental units per group and the number of animals per group is the same. However, if there is any contamination between mice within a cage, the treatment given to one mouse might influence other mice in that cage, and it would be more appropriate to subject all mice in one cage to the same treatment and treat the cage as the experimental unit. In another example, if the treatment is added to the water in a fish tank, two fish in the same tank cannot receive different treatments; thus the experimental unit is the tank, and the sample size is the number of tanks per group. Once identified, experimental units are allocated to the different comparison groups of the desired sample size; this is done using an appropriate method of randomisation to prevent selection bias (see Sect. 3). Each comparison group will be subjected to different interventions, at least one of which will be a control. The purpose of the control group is to allow the researcher to investigate the effect of a treatment and distinguish it from other confounding experimental effects. It is therefore crucial that any control group is treated exactly in the same way as the other comparison groups. Types of control group to consider include negative control, vehicle control, positive control, sham control, comparative control and naïve control (Bate and Clark 2014).

2.3 Measurements and Outcome Measures

Measurements are taken to assess the results; these are recorded as outcome measures (also known as dependent variable). A number of outcome measures can be recorded in a single experiment, for example, if burrowing behaviour is measured, the outcome measure might be the weight of gravel displaced, or if neuronal density is measured from histological brain slides, the outcome measure might be the neuron count. The primary outcome measure should be identified in the planning stage of the experiment and stated in the protocol; it is the outcome of greatest importance, which will answer the main experimental question. The number of animals in the experiment is determined by the power needed to detect a difference in the primary outcome measure. A hypothesis testing experiment may also include additional outcome measures, i.e. secondary outcome measures, which can be used to generate hypotheses for follow-up experiments. Secondary outcome measures cannot be used to draw conclusions about the experiment if the experiment was not powered to detect a minimum difference for these outcome measures.

For the purpose of the statistical analysis, outcome measures fall into two broad categories: continuous or categorical. Continuous measures are sometimes referred to as quantitative data and are measured on a numerical scale. Continuous measures include truly continuous data but also discrete data. Examples of true continuous data include bodyweight, body temperature, blood/CSF concentration or time to event, while examples of discrete data include litter size, number of correct response or clinical score. Categorical responses are measured on a nonnumerical scale; they can be ordinal (e.g. severity score, mild/moderate/severe), nominal (e.g. behavioural response, left/middle/right arm maze) or binary (e.g. disease state, present/absent). Continuous responses may take longer to measure, but they contain more information. If possible, it is preferable to measure a continuous rather than categorical response because continuous data can be analysed using the parametric analyses, which have higher power; this reduces the sample size needed (Bate and Clark 2014).

2.4   Independent Variables and Analysis

There are many ways to analyse data from in vivo experiments; the first step in devising the analysis plan is to identify the independent variables. There can be two broad types: independent variables of interest which the researcher specifically manipulates to test the hypothesis, for example, a drug with different doses, and nuisance variables, which are other sources of variability that may impact on the outcome measure, but are not of direct interest to the researcher. Examples of nuisance variables could be the day of the experiment, if animals used on different days, or baseline body weight or locomotor activity. Every experiment has nuisance variables. Identifying them at the protocol stage and accounting for them in the design and the analysis, for example, as blocking factors, or co-variables, increase the sensitivity of the experiment to detect changes induced by the independent variable(s) of interest. The analysis plan should be established before the experiment starts and any data is collected; it should also be included in the protocol. Additional analyses can be performed on the data, but if an analysis was not planned before the data was collected, it should be clearly reported as a post hoc or exploratory analysis. Exploratory analyses are at greater risk of yielding false positive results.

3   Experimental Biases: Definitions and Methods to Reduce Them

For any researcher who intends to carry out preclinical in vivo studies, it is important to understand what experimental biases are. First, we need to know the definition of bias. It is the inadequacies in the design, conduct, analysis or reporting of an experiment that cause systematic distortion of the estimated intervention effect away from the "truth" (Altman et al. 2001; van der Worp et al. 2010), and it will significantly confound in vivo studies and reduce their internal validity. Sources of bias are multiple and in many cases context dependant. In this overview chapter, it is not possible to give an exhaustive list of potential sources of bias, and it behoves the researcher to systematically identify all potential significant sources of bias for the particular experiment being in planned and to design appropriate mitigation tactics into the protocol. Major known types of biases include selection bias, performance bias, detection bias, and attrition bias. Table 1 gives the definition of each type of bias and describe the methods to reduce them.

Table 1 Bias definition and bias-reducing methods (Lazic et al. 2018)

Name of bias	Definition of bias	Methods to reduce bias
Selection bias	Refers to the biased allocation of animals to different treatment groups, which could happen at the beginning of an animal study or at a stage where reassigning animals to different treatment groups is needed following an initial surgical procedure or treatment. Selection bias results in systematic differences in baseline characteristics between treatment groups (Higgins et al. 2011)	To avoid systematic differences between animals allocated to different treatment groups, one shall use a valid randomisation method, e.g. a randomisation software or even a simple method such as picking a number from a hat (Baastrup et al. 2010; Huang et al. 2013; Saghaei 2004). Detail for randomisation is covered in chapter "Blinding and Randomization". Note that it is also necessary to conceal the allocation sequence from experimenters who will assign animals to treatment groups until the time of assignment
Performance bias	Related to the systematic differences in the care that is provided between different treatment groups or being exposed to factors other than the treatment that could influence the performance of the animals (Higgins et al. 2011; O'Connor and Sargeant 2014; van der Worp et al. 2010). Performance bias is a result of animals being managed differently due to, e.g. housing conditions, diet, group sizes per cage, location in the animal house, and experimenters who provide the care to animals are not blinded to treatment groups	One can avoid performance bias by improving the study design, e.g. applying the same housing, diet, location conditions to all the animals and by ensuring proper blinding of the experimenters to treatment groups, which keeps the experimenters who perform the experiment, collect data and access outcomes unaware of treatment allocation. Detail for blinding is covered in chapter "Blinding and Randomization"
Detection bias	Defined as the systematic distortion of the results of a study that occurs when the experimenter assessing behavioural outcome measures has the knowledge of treatment assignment to groups (van der Worp et al. 2010). In this circumstance, experimenters measuring the outcomes may introduce differential measurement of the outcomes rather than the treatment itself due to inadvertent expectation	The only way to avoid detection bias is a complete blinding of the experimenters, including those who analyse the data, so that they are not aware which animal(s) belong to which treatment group(s). The protocol should define at what stage the blinding codes will be broken (preferably only after data analysis has been completed). Detail for blinding is covered in chapter "Blinding and Randomization"
Attrition bias	Is the unequal occurrence and handling of deviations from protocol and loss to follow-up between treatment groups (van der Worp et al. 2010). This bias can occur when animals die or are removed from the study due to adverse effects of the treatment or pre-set criteria for	Experimenters should report attrition information for each experimental group and also include outcomes that will not be affected by attrition. It is also advisable to consult a statistician to minimise the impact of attrition bias using some statistical approaches such as intention-to-treat analysis by

(continued)

Table 1 (continued)

Name of bias	Definition of bias	Methods to reduce bias
	removal before observing the outcomes; therefore, the outcomes are not observed for all animals, causing inadvertent bias (O'Connor and Sargeant 2014)	imputing the missing data. Excluding "outliers" from analysis should be only undertaken as an extremely measure and should only be done to pre-stated criteria. Detail for statistics is covered in chapter "Blinding and Randomization"

Researchers who conduct hypothesis testing in vivo animal work should understand the importance of limiting the impact of experimental biases in the design, conduct, analysis and reporting of in vivo experiments. Experimental biases can cause significant weakness in the design, conduct and analysis of in vivo animal studies, which can produce misleading results and waste valuable resources. In biomedical research, many effects of interventions are fairly small, and small effects therefore are difficult to distinguish from experimental biases (Ioannidis et al. 2014). Evidence (1960–2012 from PubMed) shows that adequate steps to reduce biases, e.g. blinded assessment of outcome and randomisation, have not been taken in more than 20% and 50% of biomedical studies, respectively, leading to inflated estimates of effectiveness, e.g. in the fields of preclinical stroke, multiple sclerosis, Parkinson's disease, bone cancer pain and myocardial infarction research (Currie et al. 2013; Macleod et al. 2008; Rooke et al. 2011; Sena et al. 2007; van Hout et al. 2016; Vesterinen et al. 2010) and consequently significant research waste (Ioannidis et al. 2014; Macleod et al. 2014, 2015). Therefore, it is imperative that biomedical researchers should spend efforts on improvements in the quality of their studies using the methods described in this chapter to reduce experimental biases which will lead to increased effect-to-bias ratio.

However, it is worth pointing out that the notion that experimental biases could significantly impact on in vivo animal studies is often assumed because they are believed to be important in clinical research. Therefore, such an assumption may be flawed, as the body of evidence showing the importance of bias-reducing methods such as randomisation, blinding, etc. for animal studies is still limited and most of the evidence is indirect. Furthermore, there may also be sources of bias which impact on preclinical studies which are currently unknown. Thus, systematic review and meta-analysis of in vivo studies have shown that papers that do not report bias-reducing methods report larger effect sizes (Vesterinen et al. 2010). However, these studies are based on reported data alone, and therefore there might be a difference between what researchers do and what they report in their publications (Reichlin et al. 2016). Reporting of the precise details of bias reduction methods is often scanty, and therefore accurate assessment of the precise method and rigour of such procedures is challenging. Moreover, those papers that do not report one bias-reducing method, e.g. randomisation, also tend to not report other bias-reducing methods, e.g. blinding and sample size calculation, suggesting that there could be interactions between these methods.

4 Experimental Biases: Major Domains and General Principles

In this section, we will describe the major domains, in other words, sources that could contribute to experimental bias if not carefully considered and if mitigating tactics are not included in the design of hypothesis testing experiments before data collection starts. These include sample size estimation, randomisation, allocation concealment, blinding, primary and secondary outcome measures and inclusion/exclusion criteria. General descriptions for these domains (Macleod et al. 2009; Rice et al. 2008; Rice 2010; van der Worp et al. 2010) are shown in the following Table 2. It is important to note that these domains are key things to be included in a protocol as mentioned in Sect. 1.

Table 2 General descriptions for the major domains that contribute to experimental biases

Major domains	General descriptions
Sample size estimation	The sample size refers to the number of experimental units (e.g. a single animal, a cage of animals) per group. In hypothesis testing experiments, it should be determined with a power calculation. Studies that are not appropriately powered are unethical, and both underpowered and overpowered studies lead to a waste of animals. The former because they produce unreliable results and the latter because they use more animals than necessary
Randomisation	Refers to the steps to reduce systematic differences between comparison groups. Failure to conduct randomisation leads to selection bias
Allocation concealment	Refers to the practice of concealment of the group or treatment assignment (i.e. the allocation) and its sequence of each experimental unit from the experimenter until the time of assignment. Failure to conceal allocation will lead to selection bias. This should not be confused with randomisation
Blinding	Refers to the practice of preventing the experimenter who administer treatments, take care of the animals, assess the responses and analyse data from knowing the test condition. Failure of appropriate blinding leads to selection, performance and detection biases
Primary and secondary outcome measures	Primary outcome measure refers to the outcome measure of most interest, and it is related to the efficacy of an intervention that has the greatest importance for a given study. Secondary outcome measure refers to the outcome measure that is related to intervention efficacy but with less importance than the primary outcome measure and is used to evaluate additional intervention effects. It is important to declare what intervention effects are in the study protocol
Inclusion/exclusion criteria	Refers to criteria by which animals will be included or excluded in a given study, e.g. due to abnormal baselines or not reaching the required change in thresholds after designed experimental insult

Table 3 General principles to prevent experimental biases in hypothesis testing in vivo studies

Major domains	General principles
Sample size estimation	A power calculation (desired power of at least 0.8, and alpha = 0.05) to estimate the experimental group size should be carried out before any hypothesis testing study using pilot data or those relevant data from the literature. This could be done by using a statistical software. Detail on this can be found in chapter "A Reckless Guide to P-Values: Local Evidence, Global Errors"
Randomisation	There are different methods available to randomly allocate animals to experimental groups such as computer-generated randomisation. One should always consider to use the most robust, appropriate and available method for randomisation. Detail on this can be found in chapter "Blinding and Randomization"
Allocation concealment	Methods should be used to conceal the implementation of the random allocation sequence (e.g. numbered cages) until interventions are assigned, so that the sequence will not be known or predictable in advance by the experimenters involved in allocating animals to the treatment groups
Blinding	Blinding procedures should be carried out, so that the treatment identity should not be disclosed until after the outcome assessments have been finished for all animals and the primary analysis have been completed. In case that one experimenter conducts the whole study, any additional steps should be taken to preserve the blinding. Detail on this can be found in chapter "Blinding and Randomization"
Primary and secondary outcome measures	Experimenters should decide the outcome of great importance regarding the treatment efficacy before any study starts as the primary outcome measure. This is also usually used in the sample size estimation. Primary outcome measure cannot be changed once the study starts and when the results are known. Experimenters should also include secondary outcome measures relating to additional effects of treatments; these may be used for new hypothesis generating
Inclusion/exclusion criteria	Experimenters should set up the exact criteria which will include and exclude animals from their studies. Every animal should be accounted for, except under these criteria. They should be determined appropriately according to the study nature before the studies commence. Once determined, they cannot be changed during the course of investigation

General principles to reduce experimental bias in each of the above-mentioned domains (Andrews et al. 2016; Knopp et al. 2015) are outlined in the following Table 3.

5   Existing Guidelines and How to Use Them

There are resources to assist investigators in designing rigorous protocols and identify sources of bias. Cross-referencing to experimental reporting guidelines and checklists (e.g. ARRIVE (NC3Rs 2018a), the NIH guidelines (NIH 2018a) and the Nature reporting of animal studies checklist (Nature 2013)) can be informative and helpful when planning an experimental protocol. However, it is important to bear in mind that these are primarily designed for reporting purposes and are not specifically designed for use in assisting with experimental design. There are more comprehensive planning guidelines specifically aiming at early experimental design stage. Henderson et al. identified 26 guidelines for in vivo experiments in animals in 2012 (Henderson et al. 2013) (and a few more have been published since, like PREPARE (Smith et al. 2018), developed by the NORECEPA (Norway's National Consensus Platform for the advancement of the 3Rs), and PPRECISE for the field of pain research (Andrews et al. 2016)). Most of them have been developed for a specific research field but carry ideas and principles that can be transferred to all forms of in vivo experiments. Notable are, for example, the very detailed Lambeth Conventions (Curtis et al. 2013) (developed for cardiac arrhythmia research), from Alzheimer's research recommendations by Shineman et al. (2011) and generally applicable call by Landis et al. (2012).

The authors of many of these guidelines state that their list might need adaption to the specific experiment. This is pointing out the general shortcoming that a fixed-item list can hardly foresee and account for any possible experimental situation and a blind ticking of boxes ticking of boxes is unlikely to improve experimental design. Such guidelines rather serve an educational purpose of making researchers aware of possible pitfalls and biases before the experimental conduct.

Two examples for a more adaptive and reactive way to serve a similar purpose should be stated: the NIH pages on rigour and reproducibility (NIH 2018b) provide in-depth information and collect important publications and workshop updates on these topics and have a funding scheme specifically for rigour and reproducibility. Second, using the Experimental Design Assistant (EDA) (NC3Rs 2018b; Percie du Sert et al. 2017) developed by the UK's National Centre for the 3Rs (NC3Rs), a free to use online platform guiding researchers through experimental planning will give researchers the opportunity to adopt guideline and rigour principles precisely to their needs. The researcher creates a flow diagram of their experimental set-up grouped in three domains: the experiment (general questions on hypotheses and aims, animals used, animal strains, etc.), the practical steps (experimental conduct, assessment, etc.) and the analysis stage (e.g. outcome measures, statistical methods, data processing). Unlike a fixed checklist, the EDA checks the specific design as presented by the experimenter within the tool using logic algorithms. The user is then faced with the flaws the EDA identified and can adjust their design accordingly. This process can go through multiple rounds, by that forming a dynamic feedback loop educating the researcher and providing more nuanced assistance than a static checklist can.

While this process, however valid, might take time, the following steps of the EDA actively guide researchers through crucial and complex questions of the experiment, by suggesting fitting methods of statistical analyses of the experiment and subsequently carrying out sample size calculations. The EDA can then also generate a randomization sequence or compile a report of the planned experiment that can, e.g. be part of a preregistration of the experimental protocol.

6 Exploratory and Confirmatory Research

It is necessary to understand that there are in general two types of preclinical research, namely, exploratory and confirmatory research, respectively. Figure 1 shows that exploratory studies mainly aim to produce theories regarding the pathophysiology of disease (hypothesis generating), while confirmatory studies seek to reproduce exploratory findings as clearly defined intervention effects in relevant animal models (hypothesis testing). The next chapter will deal with exploratory and confirmatory studies in details. Similar standards of rigour are advisable for both forms of studies; this may be achieved by conforming to the general principles for the major domains that are outlined in Table 2 and incorporating these principles into a protocol that can be registered and published. It is important to note that both exploratory and confirmatory research can be closely linked: sometimes there can be exploratory and confirmatory components within the same studies. For example, a newly generated knockout mouse model is used to examine the effect of knockout on one specific phenotype (hypothesis testing – confirmatory) but may also describe a variety of other phenotypic characteristics as well (hypothesis generating – exploratory). Therefore, extra care in terms of applying these principles to reduce experimental bias would be needed before the commence of the studies. It also worth noting that sometimes it might not be compulsory or necessary to use some of the principles during exploratory studies such as sample size estimation and blinding which are albeit of highest importance in confirmatory research.

However, it is necessary to recognise how hypothesis confirming and hypothesis generating research relate to each other: while confirmatory research can turn into exploratory (e.g. if the findings are contrary to the hypothesis, this can lead to a new hypothesis that can be tested in a separate experiment), under no circumstances exploratory findings should be disseminated as the result of hypothesis confirming research by fitting a hypothesis to your results, i.e. to your p-values (often called HARKing = hypothesising after results are known or p-hacking = sifting through a multitude of p-values to find one below 0.05).

In conclusion, this chapter provides general concepts and principles that are important for the design and conduct of preclinical in vivo experiments, including experimental biases and how to reduce these biases in order to achieve the highest levels of rigour for hypothesis generating research using animals. The chapter should be used in conjunction with other relevant chapters in the handbook such as chapters "Blinding and Randomization", "Minimum Information and Quality Standards for Conducting, Reporting, and Organizing In Vitro Research", "Minimum Information in In Vivo Research", "A Reckless Guide to P-Values: Local Evidence, Global Errors", and "Quality Governance in Biomedical Research".

References

Altman DG, Schulz KF, Moher D, Egger M, Davidoff F, Elbourne D, Gotzsche PC, Lang T, Consort G (2001) The revised CONSORT statement for reporting randomized trials: explanation and elaboration. Ann Intern Med 134:663–694

Andrews NA, Latremoliere A, Basbaum AI, Mogil JS, Porreca F, Rice AS, Woolf CJ, Currie GL, Dworkin RH, Eisenach JC, Evans S, Gewandter JS, Gover TD, Handwerker H, Huang W, Iyengar S, Jensen MP, Kennedy JD, Lee N, Levine J, Lidster K, Machin I, McDermott MP, McMahon SB, Price TJ, Ross SE, Scherrer G, Seal RP, Sena ES, Silva E, Stone L, Svensson CI, Turk DC, Whiteside G (2016) Ensuring transparency and minimization of methodologic bias in preclinical pain research: PPRECISE considerations. Pain 157:901–909. https://doi.org/10.1097/j.pain.0000000000000458

Baastrup C, Maersk-Moller CC, Nyengaard JR, Jensen TS, Finnerup NB (2010) Spinal-, brainstem- and cerebrally mediated responses at- and below-level of a spinal cord contusion in rats: evaluation of pain-like behavior. Pain 151:670–679. https://doi.org/10.1016/j.pain.2010.08.024

Bate ST, Clark RA (2014) The design and statistical analysis of animal experiments. Cambridge University Press, Cambridge

Currie GL, Delaney A, Bennett MI, Dickenson AH, Egan KJ, Vesterinen HM, Sena ES, Macleod MR, Colvin LA, Fallon MT (2013) Animal models of bone cancer pain: systematic review and meta-analyses. Pain 154:917–926. https://doi.org/10.1016/j.pain.2013.02.033

Curtis MJ, Hancox JC, Farkas A, Wainwright CL, Stables CL, Saint DA, Clements-Jewery H, Lambiase PD, Billman GE, Janse MJ, Pugsley MK, Ng GA, Roden DM, Camm AJ, Walker MJ (2013) The Lambeth Conventions (II): guidelines for the study of animal and human ventricular and supraventricular arrhythmias. Pharmacol Ther 139:213–248. https://doi.org/10.1016/j.pharmthera.2013.04.008

Henderson VC, Kimmelman J, Fergusson D, Grimshaw JM, Hackam DG (2013) Threats to validity in the design and conduct of preclinical efficacy studies: a systematic review of guidelines for in vivo animal experiments. PLoS Med 10:e1001489. https://doi.org/10.1371/journal.pmed.1001489

Higgins J, Altman DG, Sterne J (2011) Assessing risk of bias in included studies. In: Higgins J, Green S (eds) Cochrane handbook for systematic reviews of interventions. Wiley, Hoboken

Huang W, Calvo M, Karu K, Olausen HR, Bathgate G, Okuse K, Bennett DL, Rice AS (2013) A clinically relevant rodent model of the HIV antiretroviral drug stavudine induced painful peripheral neuropathy. Pain 154:560–575. https://doi.org/10.1016/j.pain.2012.12.023

Ioannidis JP, Greenland S, Hlatky MA, Khoury MJ, Macleod MR, Moher D, Schulz KF, Tibshirani R (2014) Increasing value and reducing waste in research design, conduct, and analysis. Lancet 383:166–175. https://doi.org/10.1016/S0140-6736(13)62227-8

Knopp KL, Stenfors C, Baastrup C, Bannon AW, Calvo M, Caspani O, Currie G, Finnerup NB, Huang W, Kennedy JD, Lefevre I, Machin I, Macleod M, Rees H, Rice ASC, Rutten K, Segerdahl M, Serra J, Wodarski R, Berge OG, Treedef RD (2015) Experimental design and reporting standards for improving the internal validity of pre-clinical studies in the field of pain: consensus of the IMI-Europain Consortium. Scand J Pain 7:58–70. https://doi.org/10.1016/j.sjpain.2015.01.006

Landis SC, Amara SG, Asadullah K, Austin CP, Blumenstein R, Bradley EW, Crystal RG, Darnell RB, Ferrante RJ, Fillit H, Finkelstein R, Fisher M, Gendelman HE, Golub RM, Goudreau JL, Gross RA, Gubitz AK, Hesterlee SE, Howells DW, Huguenard J, Kelner K, Koroshetz W, Krainc D, Lazic SE, Levine MS, Macleod MR, McCall JM, Moxley RT 3rd, Narasimhan K, Noble LJ, Perrin S, Porter JD, Steward O, Unger E, Utz U, Silberg SD (2012) A call for transparent reporting to optimize the predictive value of preclinical research. Nature 490:187–191. https://doi.org/10.1038/nature11556

Lazic SE, Clarke-Williams CJ, Munafo MR (2018) What exactly is 'N' in cell culture and animal experiments? PLoS Biol 16:e2005282. https://doi.org/10.1371/journal.pbio.2005282

Macleod MR, van der Worp HB, Sena ES, Howells DW, Dirnagl U, Donnan GA (2008) Evidence for the efficacy of NXY-059 in experimental focal cerebral ischaemia is confounded by study quality. Stroke 39:2824–2829. https://doi.org/10.1161/STROKEAHA.108.515957

Macleod MR, Fisher M, O'Collins V, Sena ES, Dirnagl U, Bath PM, Buchan A, van der Worp HB, Traystman R, Minematsu K, Donnan GA, Howells DW (2009) Good laboratory practice: preventing introduction of bias at the bench. Stroke 40:e50–e52. https://doi.org/10.1161/STROKEAHA.108.525386

Macleod MR, Michie S, Roberts I, Dirnagl U, Chalmers I, Ioannidis JP, Al-Shahi Salman R, Chan AW, Glasziou P (2014) Biomedical research: increasing value, reducing waste. Lancet 383:101–104. https://doi.org/10.1016/S0140-6736(13)62329-6

Macleod MR, Lawson McLean A, Kyriakopoulou A, Serghiou S, de Wilde A, Sherratt N, Hirst T, Hemblade R, Bahor Z, Nunes-Fonseca C, Potluru A, Thomson A, Baginskaite J, Egan K, Vesterinen H, Currie GL, Churilov L, Howells DW, Sena ES (2015) Risk of bias in reports of in vivo research: a focus for improvement. PLoS Biol 13:e1002273. https://doi.org/10.1371/journal.pbio.1002273

Nature (2013) Enhancing reproducibility. Nat Methods 10(5):367. https://www.nature.com/articles/nmeth.2471

NC3Rs (2018a) ARRIVE guidelines. https://www.nc3rs.org.uk/arrive-guidelines. Accessed 20 Dec 2018

NC3Rs (2018b) The experimental design assistant. https://eda.nc3rs.org.uk. Accessed 20 Dec 2018

NIH (2018a) Principles and guidelines for reporting preclinical research. https://www.nih.gov/research-training/rigor-reproducibility/principles-guidelines-reporting-preclinical-research. Accessed 20 Dec 2018

NIH (2018b) Rigor and reproducibility. https://www.nih.gov/research-training/rigor-reproducibility. Accessed 20 Dec 2018

O'Connor AM, Sargeant JM (2014) Critical appraisal of studies using laboratory animal models. ILAR J 55:405–417. https://doi.org/10.1093/ilar/ilu038

Percie du Sert N, Bamsey I, Bate ST, Berdoy M, Clark RA, Cuthill I, Fry D, Karp NA, Macleod M, Moon L, Stanford SC, Lings B (2017) The experimental design assistant. PLoS Biol 15:e2003779. https://doi.org/10.1371/journal.pbio.2003779

Reichlin TS, Vogt L, Wurbel H (2016) The researchers' view of scientific rigor-survey on the conduct and reporting of in vivo research. PLoS One 11:e0165999. https://doi.org/10.1371/journal.pone.0165999

Rice ASC (2010) Predicting analgesic efficacy from animal models of peripheral neuropathy and nerve injury: a critical view from the clinic. In: Mogil JS (ed) Pain 2010 – an updated review: refresher course syllabus. IASP Press, Seattle, pp 1–12

Rice AS, Cimino-Brown D, Eisenach JC, Kontinen VK, Lacroix-Fralish ML, Machin I, Preclinical Pain C, Mogil JS, Stohr T (2008) Animal models and the prediction of efficacy in clinical trials of analgesic drugs: a critical appraisal and call for uniform reporting standards. Pain 139:243–247. https://doi.org/10.1016/j.pain.2008.08.017

Rooke ED, Vesterinen HM, Sena ES, Egan KJ, Macleod MR (2011) Dopamine agonists in animal models of Parkinson's disease: a systematic review and meta-analysis. Parkinsonism Relat Disord 17:313–320. https://doi.org/10.1016/j.parkreldis.2011.02.010

Saghaei M (2004) Random allocation software for parallel group randomized trials. BMC Med Res Methodol 4:26. https://doi.org/10.1186/1471-2288-4-26

Sena E, van der Worp HB, Howells D, Macleod M (2007) How can we improve the pre-clinical development of drugs for stroke? Trends Neurosci 30:433–439. https://doi.org/10.1016/j.tins.2007.06.009

Shineman DW, Basi GS, Bizon JL, Colton CA, Greenberg BD, Hollister BA, Lincecum J, Leblanc GG, Lee LB, Luo F, Morgan D, Morse I, Refolo LM, Riddell DR, Scearce-Levie K, Sweeney P, Yrjanheikki J, Fillit HM (2011) Accelerating drug discovery for Alzheimer's disease: best practices for preclinical animal studies. Alzheimers Res Ther 3:28. https://doi.org/10.1186/alzrt90

Smith AJ, Clutton RE, Lilley E, Hansen KEA, Brattelid T (2018) PREPARE: guidelines for planning animal research and testing. Lab Anim 52:135–141. https://doi.org/10.1177/0023677217724823

van der Worp HB, Howells DW, Sena ES, Porritt MJ, Rewell S, O'Collins V, Macleod MR (2010) Can animal models of disease reliably inform human studies? PLoS Med 7:e1000245. https://doi.org/10.1371/journal.pmed.1000245

van Hout GP, Jansen of Lorkeers SJ, Wever KE, Sena ES, Kouwenberg LH, van Solinge WW, Macleod MR, Doevendans PA, Pasterkamp G, Chamuleau SA, Hoefer IE (2016) Translational failure of anti-inflammatory compounds for myocardial infarction: a meta-analysis of large animal models. Cardiovasc Res 109:240–248. https://doi.org/10.1093/cvr/cvv239

Vesterinen HM, Sena ES, ffrench-Constant C, Williams A, Chandran S, Macleod MR (2010) Improving the translational hit of experimental treatments in multiple sclerosis. Mult Scler 16:1044–1055. https://doi.org/10.1177/1352458510379612

Resolving the Tension Between Exploration and Confirmation in Preclinical Biomedical Research

Ulrich Dirnagl

Contents

Abstract

Confirmation through competent replication is a founding principle of modern science. However, biomedical researchers are rewarded for innovation, and not for confirmation, and confirmatory research is often stigmatized as unoriginal and as a consequence faces barriers to publication. As a result, the current biomedical literature is dominated by exploration, which to complicate matters further is often disguised as confirmation. Only recently scientists and the public have begun to realize that high-profile research results in biomedicine can often not be replicated. Consequently, confirmation has become central stage in the quest to safeguard the robustness of research findings. Research which is pushing the

U. Dirnagl (✉)
Department of Experimental Neurology and Center for Stroke Research Berlin, Charité – Universitätsmedizin Berlin, Berlin, Germany

QUEST Center for Transforming Biomedical Research, Berlin Institute of Health (BIH), Berlin, Germany
e-mail: ulrich.dirnagl@charite.de

© The Author(s) 2019
A. Bespalov et al. (eds.), *Good Research Practice in Non-Clinical Pharmacology and Biomedicine*, Handbook of Experimental Pharmacology 257,
https://doi.org/10.1007/164_2019_278

boundaries of or challenges what is currently known must necessarily result in a plethora of false positive results. Thus, since discovery, the driving force of scientific progress, is unavoidably linked to high false positive rates and cannot support confirmatory inference, dedicated confirmatory investigation is needed for pivotal results. In this chapter I will argue that the tension between the two modes of research, exploration and confirmation, can be resolved if we conceptually and practically separate them. I will discuss the idiosyncrasies of exploratory and confirmatory studies, with a focus on the specific features of their design, analysis, and interpretation.

Keywords
False negative · False positive · Preclinical randomized controlled trial · Replication · Reproducibility · Statistics

Science, the endless frontier (V. Bush)

To boldly go where no man has gone before (G. Roddenberry)

Non-reproducible single occurrences are of no significance to science (K. Popper)

1   Introduction

Scientists' and the public's view on science is based toward innovation, novelty, and discovery. Discoveries, however, which represent single occurrences that cannot be reproduced are of no significance to science (Popper 1935). Confirmation through competent replication has therefore been a founding principle of modern science since its origins in the renaissance. Most scientists are aware of the need to confirm their own or other's results and conclusions. Nevertheless, they are rewarded for innovation, and not for confirmation. Confirmatory research is often stigmatized as unoriginal and faces barriers to publication. As a result, the current biomedical literature is dominated by exploration, which is often disguised as confirmation. In fact, many published articles claim that they have discovered a phenomenon and confirmed it with the same experiment, which is a logical impossibility: the same data cannot be used to generate and test a hypothesis. In addition, exploratory results are often garnished with far-reaching claims regarding the relevance of the observed phenomenon for the future treatment or even cure of a disease. As the results of exploration can only be tentative, such claims need to be founded on confirmed results.

2   Discrimination Between Exploration and Confirmation

Recently, scientists and the public have become concerned that research results often cannot be replicated (Begley and Ellis 2012; Prinz et al. 2011; Baker 2016) and that the translation of biomedical discovery to improved therapies for patients has a very high attrition rate (Dirnagl 2016). Undoubtedly the so-called replication crisis and the translational roadblock have complex roots, including importantly the sheer complexity of the pathophysiology of most diseases and the fact that many of the "low-hanging fruits" (e.g., antibiotics, antidiabetics, antihypertensives, treatments for multiple sclerosis, Parkinson's disease, and epilepsy, to name but a few) have already been picked. Nevertheless, results which are confounded by biases and lack statistical power must often be false (Ioannidis 2005) and hence resist confirmation. I will argue that the two modes of research need to be separated and clinical translation of preclinical evidence be based on confirmation (Kimmelman et al. 2014). Thus, exploration and confirmation are equally important for the progress of biomedical research: In exploratory investigation, researchers should aim at generating robust pathophysiological theories of disease. In confirmatory investigation, researchers should aim at demonstrating strong and reproducible treatment effects in relevant animal models. In what follows I will explore what immediate and relevant consequences this has for the design of experiments, their analysis, interpretation, and publication.

3   Exploration Must Lead to a High Rate of False Positives

Research which is pushing the boundaries of or challenges what is currently known must result in a plethora of false positive results. In fact, the more original initial findings are, the less likely it is that they can subsequently be confirmed. This is the simple and straightforward result of the fact that cutting edge or even paradigm shifting (Kuhn 1962) research must operate with low base rates, that is, low prior probabilities that the tested hypotheses are actually true. Or in other words, the more mainstream and less novel research is, the likelier it is that it will find its hypotheses to be true. This can easily be framed statistically, either frequentist or Bayesian. For example, if research operates with a probability that 10% of its hypotheses are true, which is a conservative estimate, and accepts type I errors at 5% (alpha-level) and type II errors at 20% (i.e., 80% power), almost 40% of the times, it rejects the NULL hypothesis (i.e., finds a statistically significant result), while the NULL hypothesis is actually true (false positive) (Colquhoun 2014). In other words, under those conditions, which are actually unrealistic in preclinical medicine in which power is often 50% and less (Button et al. 2013; Dirnagl 2006), the positive predictive value (PPV) of results is much worse than the type I error level. This has grave consequences, in particular as many researchers confuse PPV and significance level, nurturing their delusion that they are operating at a satisfactory level of wrongly accepting their hypothesis in only 5% of the cases. As a corollary, under those conditions true effect sizes will be overestimated by almost 50%. For a detailed discussion and R-code to simulate different scenarios, see Colquhoun (2014).

4 The Garden of Forking Paths

The "Garden of Forking Paths" (1944) is a metaphor which the statisticians Gelman and Loken (2013) have borrowed from Jorge Luis Borges' (1899–1986) novel for their criticism of experimental design, analysis, and interpretation of psychological and biomedical research: In exploratory investigation, researchers trail on branched-out paths through a garden of knowledge. And as poetic as this wandering might seem, it withholds certain dangers. On his way through the labyrinth, the researcher proceeds in an inductively deterministic manner. He (or she) does not at all notice the many levels of freedom available to him. These will arise, for example, through an alternative analysis (or interpretations) of the experiment, fluke false positive or false negative results, the choice for an alternative antibody or mouse strain, or from the choice of another article as the basis for further experiments and interpretations. The labyrinth is of infinite size, and there is not only one way through it, but many, and there are many exits. And since our researcher is proceeding exploratively, he has set up no advance rules according to which he should carry out his analyses or plan further experiments. So the many other possible results escape his notice, since he is following a trail that he himself laid. In consequence, he overestimates the strength of evidence that he generates. In particular, he overestimates what a significant p-value means regarding his explorative wanderings. He should compare his results with all the alternative analyses and interpretations that he could have carried out, which is obviously an absurd suggestion. Indeed, in the Garden of Forking Paths, the classic definition of statistical significance (e.g., $p < 0.05$) does not apply, for it states that in the absence of an effect, the probability of bumping coincidentally into a similarly extreme or even more extreme result is lower than 5%. You would have to factor in all data and analyses that would be possible in the garden. Each of these other paths could also have led to statistically significant results. Such a comparison however is impossible in explorative research. If you nonetheless do generate p-values, you will, according to Gelman and Loken, get a "machine for the production and publication of random patterns." A consequence of all this is that our knowledge derived from exploration is less robust than the chain of statistically significant results might have us believe and that the use of test statistics in exploration is of little help or may even be superfluous if not even misleading.

At this point we must summarize that exploratory research, even if of the highest quality and without selective use of data, p-hacking (collecting, selecting, or analyzing data until statistically nonsignificant results become significant), or HARKING ("hypothesizing after the results are known"), leads to at best tentative results which provide the basis for further inquiry and confirmation. In this context it is sobering that Ronald Fisher, a founding father of modern frequentist statistics, considered results which are significant at the 5% level only "worth a second look" (Nuzzo 2014). Today clinical trials may be based on such preclinical findings.

5 Confirmation Must Weed Out the False Positives of Exploration

Since discovery is unavoidably linked to high false positive rates and cannot support confirmatory inference, dedicated confirmatory investigation is needed for pivotal results. Results are pivotal and must be confirmed if they are the basis for further investigation and thus drain resources, if they directly or indirectly might impact on human health (e.g., by informing the design of future clinical development, including trials in humans (Yarborough et al. 2018)), or if they are challenging accepted evidence in a field. Exploration must be very sensitive, since it must be able to faithfully capture rare but critical results (e.g., a cure for Alzheimer disease or cancer); confirmation on the other hand must be highly specific, since further research on and development of a false positive or non-robust finding is wasteful and unethical (Al-Shahi Salman et al. 2014).

6 Exact Replication Does Not Equal Confirmation

Many experimentalists routinely replicate their own findings and only continue a line of inquiry when the initial finding was replicated. This is laudable, but a few caveats apply. For one, a replication study which exactly mimics the original study in terms of its design may not be as informative as most researchers suspect. Counterintuitively, in an exact replication, with new samples but same sample size and treatment groups, and assuming that the effect found in the original experiment with $p < 0.05$ (but $p \approx 0.05$) equals the true population effect, the probability of replication (being the probability of getting again a significant result of the same or larger effect) is only 50% (Goodman 1992). In other words, the predictive value of an exact replication of a true finding that was significant close to the 5% level is that of a coin toss! Incidentally, this is the main reason why phase III clinical trials (which aim at confirmation) have much larger sample sizes than phase II trials (which aim at exploration). For further details on sample size calculation of replication studies, see Simonsohn (2015).

A second problem of exact replications, in particular if performed by the same group which has made the initial observation, is the problem of systematic errors. One of the most embarrassing and illustrative examples for a botched replication in the recent history of science relates to the discovery of neutrinos that travel faster than the speed of light. The results of a large international experiment conducted by physicists of high repute convulsed not only the field of physics; it shook the whole world. Neutrinos had been produced by the particle accelerator at CERN in Geneva and sent on a 730 km long trip. Their arrival was registered by a detector blasted through thousands of meters of rock in the Dolomites. Unexpectedly, the neutrinos arrived faster than would the photons travelling the same route. The experiment was replicated several times, and the results remained significant with a p-value of less than 3×10^{-7}. In the weeks following the publication (the OPERA Collaboration 2011) and media excitement, the physicists found that the GPS used to measure distances was not correctly synchronized and a cable was loose.

7   Design, Analysis, and Interpretation of Exploratory vs Confirmatory Studies

Thus, exploratory and confirmatory investigation necessitates different study designs. Confirmation is not the simple replication of an exploratory experiment. Exploratory and confirmatory investigation differs in many aspects. While exploration may start without any hypothesis ("unbiased"), a proper hypothesis is the obligatory starting point of any confirmation. Exploration investigates physiological or pathophysiological mechanisms or aims at drug discovery. The tentative findings of exploration, if relevant at all, need to be confirmed. Confirmation of the hypothesis is the default primary endpoint of the confirmatory investigation, while secondary endpoints may be explored. Both modes need to be of high internal validity, which means that they need to effectively control biases (selection, detection, attrition, etc.) through randomization, blinding, and prespecification of inclusion and exclusion criteria. Of note, control of bias is as important in exploration as in confirmation. To establish an experimental design and analysis plan before the onset of the study may be useful in exploration but is a must in confirmation. Generalizability is of greater importance in confirmation than in exploration, which therefore needs to be of high external validity. Depending on the disease under study, this may include the use of aged or comorbid animals. Statistical power is important in any type of experimental study, as low power results not only in a high false negative rate but also increases the number of false positive findings and leads to an overestimation of effect sizes (Button et al. 2013). However, as exploration aims at finding what might work or is "true," type II error should be minimized and therefore statistical power high. Conversely, as confirmation aims at weeding out the false positive, type I error becomes a major concern. To make sure that statistical power is sufficient to detect the targeted effect sizes, a priori sample size calculation is recommended in exploratory mode but obligatory in confirmation where achievable effect sizes and variance can be estimated from previous exploratory evidence. Due to manifold constraints, which include the fact that exploration (1) often collects many endpoints and hence multiple group comparisons are made, (2) is usually underpowered, and (3) comes with an almost unlimited degree of freedom of the researcher with respect to selection of animal strains, biologicals, and selection of outcome parameters and their analysis ("Garden of Forking Paths," see above and (Gelman and Loken 2013)), statistical significance tests (like t-test, ANOVA, etc.) are of little use; the focus in exploration should rather be on proper descriptive statistics, including measures of variance and confidence intervals. Conversely, in confirmatory mode, the prespecified analysis plan needs to describe the planned statistical significance test. To prevent outcome switching and publication bias, in confirmatory studies the hypothesis, experimental, and analysis plan should be preregistered (Nosek et al. 2018). Preregistration can be embargoed until the results of the study are published and are therefore not detrimental to intellectual property claims. Table 1 gives a tentative overview of some of the idiosyncrasies of exploratory and confirmatory investigation.

Table 1 Suggested differences between exploratory and confirmatory preclinical study designs

	Exploratory	Confirmatory
Establish pathophysiology, discover drugs, etc.	+++	(+)
Hypothesis	(+)	+++
Blinding	+++	+++
Randomization	+++	+++
External validity (aging, comorbidities, etc.)	−	++
Experimental and analysis plan established before study onset	+	+++
Primary endpoint	−	++
Inclusion/exclusion criteria (prespecified)	++	+++
Preregistration	(−)	+++
Sample size calculation	(+)	+++
Test statistics	+	+++
Sensitivity (type II error): find what might work	++	+
Specificity (type I error): weed out false positives	+	+++

Modified with permission (Dirnagl 2016), for details see text

8 No Publication Without Confirmation?

Vis a vis the current masquerading of exploratory preclinical investigations as confirmation of new mechanisms of disease or potential therapeutic breakthroughs, Mogil and Macleod went as far as proposing a new form of publication for animal studies of disease therapies or preventions, the "preclinical trial." In it, researchers besides presenting a novel mechanism of disease or therapeutic approach incorporate an independent, statistically rigorous confirmation of the central hypothesis. Preclinical trials would be more formal and rigorous than the typical preclinical testing conducted in academic labs and would adopt many practices of a clinical trial (Mogil and Macleod 2017).

It is uncertain whether scientists or journals will pick up this sensible proposal in the near future. Meanwhile, another novel type of publication, at least in the preclinical realm, is gaining traction: preregistration. When preregistering a study, the researcher commits in advance to the hypothesis that will be tested, the study design, as well as the analysis plan. This provides full transparency and prevents HARKING, p-hacking, and many other potential barriers to the interpretability and credibility of research findings. Preregistration is not limited to but ideally suited for confirmatory studies of high quality. In fact, it may be argued that journals should mandate preregistration when processing and publishing confirmatory studies.

9 Team Science and Preclinical Multicenter Trials

Confirmation lacks the allure of discovery and is usually more resource intense. It requires higher sample sizes and benefits from multilab approaches which come with considerable organizational overhead. Permission from regulatory authorities may be hard to obtain, as the repetition of animal experiments combined with upscaling of sample sizes may antagonize the goal to reduce animal experimentation. Publication of confirmatory results faces bigger hurdles than those of novel results. Failure to confirm a result may lead to tensions between the researchers who published the initial finding and those who performed the unsuccessful confirmation. Potential hidden moderators and context sensitivity of the finding dominate the resulting discussion, as does blaming those who failed confirmation of incompetence. In short, confirmatory research at present is not very attractive. This dire situation can only be overcome if a dedicated funding stream supports team science and confirmatory studies, and researchers are rewarded (and not stigmatized) for this fundamentally important scientific activity. It is equally important to educate the scientific community that results from exploratory research even of the highest quality are inherently tentative and that failure of competent replication of such results does not disqualify their work but is rather an element of the normal progression of science.

It is promising that several international consortia have teamed up in efforts to develop guidelines for international collaborative research in preclinical biomedicine (MULTIPART 2016) or to demonstrate that confirmatory preclinical trials can be conducted and published (Llovera et al. 2015) with a reasonable budget.

10 Resolving the Tension Between Exploration and Confirmation

Science advances by exploration and confirmation (or refutation). The role of exploration is currently overemphasized, which may be one reason of the current "replication crisis" and the translational roadblock. In addition, generation of postdictions is often mistaken with the testing of predictions (Nosek et al. 2018); in other words exploration is confounded with exploration. We need to leverage the complementary strengths of both modes of investigation. This will help to improve the refinement of pathophysiological theories, as well as the generation of reliable evidence in disease models for the efficacy of treatments in humans. Adopting a two-pronged approach of exploration-confirmation requires that we shift the balance which is currently biased toward exploration back to confirmation. Researchers need to be trained in how to competently engage in high-quality exploration and confirmation. Funders and institutions need to establish mechanisms to fund and reward confirmatory investigation.

References

Al-Shahi Salman R et al (2014) Increasing value and reducing waste in biomedical research regulation and management. Lancet 383:176–185
Baker M (2016) 1,500 scientists lift the lid on reproducibility. Nature 533:452–454
Begley CG, Ellis LM (2012) Drug development: raise standards for preclinical cancer research. Nature 483:531–533

Button KS et al (2013) Power failure: why small sample size undermines the reliability of neuroscience. Nat Rev Neurosci 14:365–376

Colquhoun D (2014) An investigation of the false discovery rate and the misinterpretation of P values. R Soc Open Sci 1:1–15. https://doi.org/10.1098/rsos.140216

Dirnagl U (2006) Bench to bedside: the quest for quality in experimental stroke research. J Cereb Blood Flow Metab 26:1465–1478

Dirnagl U (2016) Thomas Willis lecture: is translational stroke research broken, and if so, how can we fix it? Stroke 47:2148–2153

Gelman A, Loken E (2013) The garden of forking paths: why multiple comparisons can be a problem, even when there is no "fishing expedition" or "p-hacking" and the research hypothesis was posited ahead of time. http://www.stat.columbia.edu/~gelman/research/unpublished/p_hacking.pdf. Accessed 23 Aug 2018

Goodman SN (1992) A comment on replication, p-values and evidence. Stat Med 11:875–879

Ioannidis JPA (2005) Why most published research findings are false. PLoS Med 2:e124

Kimmelman J, Mogil JS, Dirnagl U (2014) Distinguishing between exploratory and confirmatory preclinical research will improve translation. PLoS Biol 12:e1001863

Kuhn TS (1962) The structure of scientific revolutions. The University of Chicago Press, Chicago

Llovera G et al (2015) Results of a preclinical randomized controlled multicenter trial (pRCT): anti-CD49d treatment for acute brain ischemia. Sci Transl Med 7:299ra121

Mogil JS, Macleod MR (2017) No publication without confirmation. Nature 542:409–411

MULTIPART. Multicentre preclinical animal research team. http://www.dcn.ed.ac.uk/multipart/. Accessed 23 May 2016

Nosek BA, Ebersole CR, DeHaven AC, Mellor DT (2018) The preregistration revolution. Proc Natl Acad Sci U S A 115:2600–2606

Nuzzo R (2014) Statistical errors: P values, the 'gold standard' of statistical validity, are not as reliable as many scientists assume. Nature 506:150–152

Popper K (1935) Logik der Forschung. Springer, Berlin

Prinz F, Schlange T, Asadullah K (2011) Believe it or not: how much can we rely on published data on potential drug targets? Nat Rev Drug Discov 10:712

Simonsohn U (2015) Small telescopes. Psychol Sci 26:559–569

The OPERA Collaboration et al (2011) Measurement of the neutrino velocity with the OPERA detector in the CNGS beam. https://doi.org/10.1007/JHEP10(2012)093

Yarborough M et al (2018) The bench is closer to the bedside than we think: uncovering the ethical ties between preclinical researchers in translational neuroscience and patients in clinical trials. PLoS Biol 16:e2006343

Blinding and Randomization

Anton Bespalov, Karsten Wicke, and Vincent Castagné

Contents

Abstract

Most, if not all, guidelines, recommendations, and other texts on Good Research Practice emphasize the importance of blinding and randomization. There is, however, very limited specific guidance on when and how to apply blinding and randomization. This chapter aims to disambiguate these two terms by discussing what they mean, why they are applied, and how to conduct the acts of randomization and blinding. We discuss the use of blinding and randomization as the means against existing and potential risks of bias rather than a mandatory practice that is to be followed under all circumstances and at any cost. We argue

A. Bespalov (✉)
Partnership for Assessment and Accreditation of Scientific Practice, Heidelberg, Germany

Pavlov Medical University, St. Petersburg, Russia
e-mail: anton.bespalov@paasp.net

K. Wicke
AbbVie, Ludwigshafen, Germany

V. Castagné
Porsolt, Le Genest-Saint-Isle, France

© The Author(s) 2019
A. Bespalov et al. (eds.), *Good Research Practice in Non-Clinical Pharmacology and Biomedicine*, Handbook of Experimental Pharmacology 257,
https://doi.org/10.1007/164_2019_279

81

that, in general, experiments should be blinded and randomized if (a) this is a confirmatory research that has a major impact on decision-making and that cannot be readily repeated (for ethical or resource-related reasons) and/or (b) no other measures can be applied to protect against existing and potential risks of bias.

Keywords
Good Research Practice · Research rigor · Risks of bias

'When I use a word,' Humpty Dumpty said in rather a scornful tone, 'it means just what I choose it to mean – neither more nor less.'

Lewis Carroll (1871)

Through the Looking-Glass, and What Alice Found There

1   Randomization and Blinding: Need for Disambiguation

In various fields of science, outcome of the experiments can be intentionally or unintentionally distorted if potential sources of bias are not properly controlled. There is a number of recognized risks of bias such as selection bias, performance bias, detection bias, attrition bias, etc. (Hooijmans et al. 2014). Some sources of bias can be efficiently controlled through research rigor measures such as randomization and blinding.

Existing guidelines and recommendations assign a significant value to adequate control over various factors that can bias the outcome of scientific experiments (chapter "Guidelines and Initiatives for Good Research Practice"). Among internal validity criteria, randomization and blinding are two commonly recognized bias-reducing instruments that need to be considered when planning a study and are to be reported when the study results are disclosed in a scientific publication.

For example, editorial policy of the Nature journals requires authors in the life sciences field to submit a checklist along with the manuscripts to be reviewed. This checklist has a list of items including questions on randomization and blinding. More specifically, for randomization, the checklist is asking for the following information: "If a method of randomization was used to determine how samples/animals were allocated to experimental groups and processed, describe it." Recent analysis by the NPQIP Collaborative group indicated that only 11.2% of analyzed publications disclosed which method of randomization was used to determine how samples or animals were allocated to experimental groups (Macleod, The NPQIP Collaborative Group 2017). Meanwhile, the proportion of studies mentioning randomization was much higher – 64.2%. Do these numbers suggest that authors strongly motivated to have their work published in a highly prestigious scientific journal ignore the instructions? It is more likely that, for many scientists (authors, editors, reviewers), a statement such as "subjects were randomly assigned to one of the N treatment conditions" is considered to be sufficient to describe the randomization procedure.

Table 1 Example of an allocation schedule that is a pseudo-randomization

Group A	Group B	Group C	Group D
Mouse 1	Mouse 2	Mouse 3	Mouse 4
Mouse 5	Mouse 6	Mouse 7	Mouse 8
Mouse 9	Mouse 10	Mouse 11	Mouse 12
Mouse 13	Mouse 14	Mouse 15	Mouse 16

For the field of life sciences, and drug discovery in particular, the discussion of sources of bias, their impact, and protective measures, to a large extent, follows the examples from the clinical research (chapter "Learning from Principles of Evidence-Based Medicine to Optimize Nonclinical Research Practices"). However, clinical research is typically conducted by research teams that are larger than those involved in basic and applied preclinical work. In the clinical research teams, there are professionals (including statisticians) trained to design the experiments and apply bias-reducing measures such as randomization and blinding. In contrast, preclinical experiments are often designed, conducted, analyzed, and reported by scientists lacking training or access to information and specialized resources necessary for proper administration of bias-reducing measures.

As a result, researchers may design and apply procedures that reflect *their* understanding of what randomization and blinding are. These may or may not be the correct procedures. For example, driven by a good intention to randomize 4 different treatment conditions (A, B, C, and D) applied a group of 16 mice, a scientist may design the experiment in the following way (Table 1).

The above example is a fairly common practice to conduct "randomization" in a simple and convenient way. Another example of common practice is, upon animals' arrival, to pick them haphazardly up from the supplier's transport box and place into two (or more) cages which then constitute the control and experimental group(s). However, both methods of assigning subjects to experimental treatment conditions violate the randomness principle (see below) and, therefore, should not be reported as randomization.

Similarly, the use of blinding in experimental work typically cannot be described solely by stating that "experimenters were blinded to the treatment conditions." For both randomization and blinding, it is essential to provide details on what exactly was applied and how.

The purpose of this chapter is to disambiguate these two terms by discussing what they mean, why they are applied, and how to conduct the acts of randomization and blinding. We discuss the use of blinding and randomization as the means against existing and potential risks of bias rather than a mandatory practice that is to be followed under all circumstances and at any cost.

2 Randomization

Randomization can serve several purposes that need to be recognized individually as one or more of them may become critical when considering study designs and conditions exempt from the randomization recommendation.

First, randomization permits the use of probability theory to express the likelihood of chance as a source for the difference between outcomes. In other words, randomization enables the application of statistical tests that are common in biology and pharmacology research. For example, the central limit theorem states that the sampling distribution of the mean of any independent, random variable will be normal or close to normal, if the sample size is large enough. The central limit theorem assumes that the data are sampled randomly and that the sample values are independent of each other (i.e., occurrence of one event has no influence on the next event). Usually, if we know that subjects or items were selected randomly, we can assume that the independence assumption is met. If the study results are to be subjected to conventional statistical analyses dependent on such assumptions, adequate randomization method becomes a must.

Second, randomization helps to prevent a potential impact of the selection bias due to differing baseline or confounding characteristics of the subjects. In other words, randomization is expected to transform any systematic effects of an uncontrolled factor into a random, experimental noise. A random sample is one selected without bias: therefore, the characteristics of the sample should not differ in any systematic or consistent way from the population from which the sample was drawn. But random sampling does not guarantee that a particular sample will be exactly representative of a population. Some random samples will be more representative of the population than others. Random sampling does ensure, however, that, with a sufficiently large number of subjects, the sample becomes more representative of the population.

There are characteristics of the subjects that can be readily assessed and controlled (e.g., by using stratified randomization, see below). But there are certainly characteristics that are not known and for which randomization is the only way to control their potentially confounding influence. It should be noted, however, that the impact of randomization can be limited when the sample size is low.[1] This needs to be kept in mind given that most nonclinical studies are conducted using small sample sizes. Thus, when designing nonclinical studies, one should invest extra efforts into analysis of possible confounding factors or characteristics in order to judge whether or not experimental and control groups are similar before the start of the experiment.

Third, randomization interacts with other means to reduce risks of bias. Most importantly, randomization is used together with blinding to conceal the allocation sequence. Without an adequate randomization procedure, efforts to introduce and maintain blinding may not always be fully successful.

2.1   Varieties of Randomization

There are several randomization methods that can be applied to study designs of differing complexities. The tools used to apply these methods range from random

---

[1]https://stats.stackexchange.com/questions/74350/is-randomization-reliable-with-small-samples.

number tables to specialized software. Irrespective of the tools used, reporting on the randomization schedule applied should also answer the following two questions:

- Is the randomization schedule based on an algorithm or a principle that can be written down and, based on the description, be reapplied by anyone at a later time point resulting in the same group composition? If yes, we are most likely dealing with a "pseudo-randomization" (e.g., see below comments about the so-called Latin square design).
- Does the randomization schedule exclude any subjects and groups that belong to the experiment? If yes, one should be aware of the risks associated with excluding some groups or subjects such as a positive control group (see chapter "Out of Control? Managing Baseline Variability in Experimental Studies with Control Groups").

An answer "yes" to either of the above questions does not automatically mean that something incorrect or inappropriate is being done. In fact, a scientist may take a decision well justified by their experience with and need of particular experimental situation. However, in any case, the answer "yes" to either or both of the questions above mandates the complete and transparent description of the study design with the subject allocation schedule.

2.1.1 Simple Randomization

One of the common randomization strategies used for between-subject study designs is called simple (or unrestricted) randomization. Simple random sampling is defined as the process of selecting subjects from a population such that just the following two criteria are satisfied:

- The probability of assignment to any of the experimental groups is equal for each subject.
- The assignment of one subject to a group does not affect the assignment of any other subject to that same group.

With simple randomization, a single sequence of random values is used to guide assignment of subjects to groups. Simple randomization is easy to perform and can be done by anyone without a need to involve professional statistical help. However, simple randomization can be problematic for studies with small sample sizes. In the example below, 16 subjects had to be allocated to 4 treatment conditions. Using Microsoft Excel's function RANDBETWEEN (0.5;4.5), there were 16 random integer numbers from 1 to 4 generated. Obviously, this method has resulted in an unequal number of subjects among groups (e.g., there is only one subject assigned to group 2). This problem may occur irrespective of whether one uses machine-generated random numbers or simply tosses a coin.

Subject ID	1	2	3	4	5	6	7	8	9	10	11	12	13	14	15	16
Group ID	4	1	1	3	3	1	4	4	3	4	3	3	4	2	3	1

An alternative approach would be to generate a list of all treatments to be administered (top row in the table below) and generate a list of random numbers (as many as the total number of subjects in a study) using a Microsoft Excel's function RAND() that returns random real numbers greater than or equal to 0 and less than 1 (this function requires no argument):

Treatment	1	1	1	1	2	2	2	2	3	3	3	3	4	4	4	4
Random number	0.76	0.59	0.51	0.90	0.64	0.10	0.50	0.48	0.22	0.37	0.05	0.09	0.73	0.83	0.50	0.43

The next step would be to sort the treatment row based on the values in the random number row (in an ascending or descending manner) and add a Subject ID row:

Subject ID	1	2	3	4	5	6	7	8	9	10	11	12	13	14	15	16
Treatment	3	3	2	3	3	4	2	2	4	1	1	2	4	1	4	1
Random number	0.05	0.09	0.10	0.22	0.37	0.43	0.48	0.50	0.50	0.51	0.59	0.64	0.73	0.76	0.83	0.90

There is an equal number of subjects (four) assigned to each of the four treatment conditions, and the assignment is random. This method can also be used when group sizes are not equal (e.g., when a study is conducted with different numbers of genetically modified animals and animals of wild type).

However, such randomization schedule may still be problematic for some types of experiments. For example, if the subjects are tested one by one over the course of 1 day, the first few subjects could be tested in the morning hours while the last subjects – in the afternoon. In the example above, none of the first eight subjects is assigned to group 1, while the second half does not include any subject from group 3. To avoid such problems, block randomization may be applied.

2.1.2 Block Randomization

Blocking is used to supplement randomization in situations such as the one described above – when one or more external factors change or may change during the period when the experiment is run. Blocks are balanced with predetermined group assignments, which keeps the numbers of subjects in each group similar at all times. All blocks of one experiment have equal size, and each block represents all independent variables that are being studied in the experiment.

The first step in block randomization is to define the block size. The minimum block size is the number obtained by multiplying numbers of levels of all independent variables. For example, an experiment may compare the effects of a vehicle and three doses of a drug in male and female rats. The minimum block size in such case would be eight rats per block (i.e., 4 drug dose levels × 2 sexes). All subjects can be divided into N blocks of size X∗Y, where X is a number of groups or treatment conditions (i.e., 8 for the example given) and Y – number of subjects per treatment condition per block. In other words, there may be one or more subjects per treatment condition per block so that the actual block size is multiple of a minimum block size (i.e., 8, 16, 24, and so for the example given above).

The second step is, after block size has been determined, to identify all possible combinations of assignment within the block. For instance, if the study is evaluating effects of a drug (group A) or its vehicle (group B), the minimum block size is equal to 2. Thus, there are just two possible treatment allocations within a block: (1) AB and (2) BA. If the block size is equal to 4, there is a greater number of possible treatment allocations: (1) AABB, (2) BBAA, (3) ABAB, (4) BABA, (5) ABBA, and (6) BAAB.

The third step is to randomize these blocks with varying treatment allocations:

Block number	4	3	1	6	5	2
Random number	0.015	0.379	0.392	0.444	0.720	0.901

And, finally, the randomized blocks can be used to determine the subjects' assignment to the groups. In the example above, there are 6 blocks with 4 treatment conditions in each block, but this does not mean that the experiment must include 24 subjects. This random sequence of blocks can be applied to experiments with a total number of subjects smaller or greater than 24. Further, the total number of subjects does not have to be a multiple of 4 (block size) as in the example below with a total of 15 subjects:

Block number	4				3				1				6			
Random number	0.015				0.379				0.392				0.444			
Subject ID	1	2	3	4	5	6	7	8	9	10	11	12	13	14	15	–
Treatment	B	A	B	A	A	B	A	B	A	A	B	B	B	A	A	–

It is generally recommended to blind the block size to avoid any potential selection bias. Given the low sample sizes typical for preclinical research, this recommendation becomes a mandatory requirement at least for confirmatory experiments (see chapter "Resolving the Tension Between Exploration and Confirmation in Preclinical Biomedical Research").

2.1.3 Stratified Randomization

Simple and block randomization are well suited when the main objective is to balance the subjects' assignment to the treatment groups defined by the independent variables whose impact is to be studied in an experiment. With sample sizes that are large enough, simple and block randomization may also balance the treatment groups in terms of the unknown characteristics of the subjects. However, in many experiments, there are baseline characteristics of the subjects that do get measured and that may have an impact on the dependent (measured) variables (e.g., subjects' body weight). Potential impact of such characteristics may be addressed by specifying inclusion/exclusion criteria, by including them as covariates into a statistical analysis, and (or) may be minimized by applying stratified randomization schedules.

It is always up to a researcher to decide where there are such potentially impactful covariates that need to be controlled and what is the best way of dealing with them. In case of doubt, the rule of thumb is to avoid any risk, apply stratified randomization, and declare an intention to conduct a statistical analysis that will isolate a potential contribution of the covariate(s).

It is important to acknowledge that, in many cases, information about such covariates may not be available when a study is conceived and designed. Thus, a decision to take covariates into account often affects the timing of getting the randomization conducted. One common example of such a covariate is body weight. A study is planned, and sample size is estimated before the animals are ordered or bred, but the body weights will not be known until the animals are ready. Another example is the size of the tumors that are inoculated and grow at different rates for a pre-specified period of time before the subjects start to receive experimental treatments.

For most situations in preclinical research, an efficient way to conduct stratified randomization is to run simple (or block) randomization several times (e.g., 100 times) and, for each iteration, calculate means for the covariate per each group (e.g., body weights for groups A and B in the example in previous section). The randomization schedule that yields the lowest between-group difference for the covariate would then be chosen for the experiment. Running a large number of iterations does not mean saving excessively large volumes of data. In fact, several tools used to support randomization allow to save the seed for the random number generator and re-create the randomization schedule later using this seed value.

Although stratified randomization is a relatively simple technique that can be of great help, there are some limitations that need to be acknowledged. First, stratified randomization can be extended to two or more stratifying variables. However, given the typically small sample sizes of preclinical studies, it may become complicated to implement if many covariates must be controlled. Second, stratified randomization works only when all subjects have been identified before group assignment. While this is often not a problem in preclinical research, there may be situations when a large study sample is divided into smaller batches that are taken sequentially into the study. In such cases, more sophisticated procedures such as the covariate adaptive randomization may need to be applied similar to what is done in clinical research (Kalish and Begg 1985). With this method, subjects are assigned to treatment groups by taking into account the specific covariates and assignments of subjects that have already been allocated to treatment groups. We intentionally do not provide any further examples or guidance on such advanced randomization methods as they should preferably be developed and applied in consultation with or by biostatisticians.

Table 2 A Latin square design as a common example of a pseudo-randomization

Subject	Consecutive tests (or study periods)			
	1	2	3	4
#1	A	B	C	D
#2	B	C	D	A
#3	C	D	A	B
#4	D	A	B	C

2.1.4 The Case of Within-Subject Study Designs

The above discussion on the randomization schedules referred to study designs known as between-subject. A different approach would be required if a study is designed as within-subject. In such study designs also known as the crossover, subjects may be given sequences of treatments with the intent of studying the differences between the effects produced by individual treatments. One should keep in mind that such sequence of testing always bears the danger that the first test might affect the following ones. If there are reasons to expect such interference, within-subjects designs should be avoided.

In the simplest case of a crossover design, there are only two treatments and only two possible sequences to administer these treatments (e.g., A-B and B-A). In nonclinical research and, particularly, in pharmacological studies, there is a strong trend to include at least three doses of a test drug and its vehicle. A Latin square design is commonly used to allocate subjects to treatment conditions. Latin square is a very simple technique, but it is often applied in a way that does not result in a proper randomization (Table 2).

In this example, each subject receives each of the four treatments over four consecutive study periods, and, for any given study period, each treatment is equally represented. If there are more than four subjects participating in a study, then the above schedule is copied as many times as need to cover all study subjects.

Despite its apparent convenience (such schedules can be generated without any tools), resulting allocation schedules are predictable and, what is even worse, are not balanced with respect to first-order carry-over effects (e.g., except for the first test period, D comes always after C). Therefore, such Latin square designs are not an example of properly conducted randomization.

One solution would be to create a complete set of orthogonal Latin Squares. For example, when the number of treatments equals three, there are six (i.e., 3!) possible sequences – ABC, ACB, BAC, BCA, CAB, and CBA. If the sample size is a multiple of six, then all six sequences would be applied. As the preclinical studies typically involve small sample sizes, this approach becomes problematic for larger numbers of treatments such as 4, where there are already 24 (i.e., 4!) possible sequences.

The Williams design is a special case of a Latin square where every treatment follows every other treatment the same number of times (Table 3).

The Williams design maintains all the advantages of the Latin square but is balanced (see Jones and Kenward 2003 for a detailed discussion on the Williams squares including the generation algorithms). There are six Williams squares

Table 3 An example of a Williams design

Subject	Consecutive tests (or study periods)			
	1	2	3	4
#1	A	B	C	D
#2	B	D	A	C
#3	C	A	D	B
#4	D	C	B	A

possible in case of four treatments. Thus, if there are more than four subjects, more than one Williams square would be applied (e.g., two squares for eight subjects).

Constructing the Williams squares is not a randomization yet. In studies based on within-subject designs, subjects are not randomized to treatment in the same sense as they are in the between-subject design. For a within-subject design, the treatment sequences are randomized. In other words, after the Williams squares are constructed and selected, individual sequences are randomly assigned to the subjects.

2.2 Tools to Conduct Randomization

The most common and basic method of simple randomization is flipping a coin. For example, with two treatment groups (control versus treatment), the side of the coin (i.e., heads, control; tails, treatment) determines the assignment of each subject. Other similar methods include using a shuffled deck of cards (e.g., even, control; odd, treatment), throwing a dice (e.g., below and equal to 3, control; over 3, treatment), or writing numbers of pieces of paper, folding them, mixing, and then drawing one by one. A random number table found in a statistics book, online random number generators (random.org or randomizer.org), or computer-generated random numbers (e.g., using Microsoft Excel) can also be used for simple randomization of subjects. As explained above, simple randomization may result in an unbalanced design, and, therefore, one should pay attention to the number of subjects assigned to each treatment group. But more advanced randomization techniques may require dedicated tools and, whenever possible, should be supported by professional biostatisticians.

Randomization tools are typically included in study design software, and, for in vivo research, the most noteworthy example is the NC3Rs' Experimental Design Assistant (www.eda.nc3rs.org.uk). This freely available online resource allows to generate and share a spreadsheet with the randomized allocation report after the study has been designed (i.e., variables defined, sample size estimated, etc.). Similar functionality may be provided by Electronic Laboratory Notebooks that integrate study design support (see chapter "Electronic Lab Notebooks and Experimental Design Assistants").

Randomization is certainly supported by many data analysis software packages commonly used in research. In some cases, there is even a free tool that allows to conduct certain types of randomization online (e.g., QuickCalcs at www.graphpad.com/quickcalcs/randMenu/).

Someone interested to have a nearly unlimited freedom in designing and executing different types of randomization will benefit from the resources generated by the R community (see https://paasp.net/resource-center/r-scripts/). Besides being free and supported by a large community of experts, R allows to save the scripts used to obtain randomization schedules (along with the seed numbers) that makes the overall process not only reproducible and verifiable but also maximally transparent.

2.3   Randomization: Exceptions and Special Cases

Randomization is not and should never be seen as a goal per se. The goal is to minimize the risks of bias that may affect the design, conduct, and analysis of a study and to enable application of other research methods (e.g., certain statistical tests). Randomization is merely a tool to achieve this goal.

If not dictated by the needs of data analysis or the intention to implement blinding, in some cases, pseudo-randomizations such as the schedules described in Tables 1 and 2 may be sufficient. For example, animals delivered by a qualified animal supplier come from large batches where the breeding schemes themselves help to minimize the risk of systematic differences in baseline characteristics. This is in contrast to clinical research where human populations are generally much more heterogeneous than populations of animals typically used in research.

Randomization becomes mandatory in case animals are not received from major suppliers, are bred in-house, are not standard animals (i.e., transgenic), or when they are exposed to an intervention before the initiation of a treatment. Examples of intervention may be surgery, administration of a reagent substance inducing long-term effects, grafts, or infections. In these cases, animals should certainly be randomized after the intervention.

When planning a study, one should also consider the risk of between-subject cross-contamination that may affect the study outcome if animals receiving different treatment(s) are housed within the same cage. In such cases, the most optimal approach is to reduce the number of subjects per cage to a minimum that is acceptable from the animal care and use perspective and adjust the randomization schedule accordingly (i.e., so that all animals in the cage receive the same treatment).

There are situations when randomization becomes impractical or generates other significant risks that outweigh its benefits. In such cases, it is essential to recognize the reasons why randomization is applied (e.g., ability to apply certain statistical tests, prevention of selection bias, and support of blinding). For example, for an in vitro study with multi-well plates, randomization is usually technically possible, but one would need to recognize the risk of errors introduced during manual pipetting into a 96- or 384-well plate. With proper controls and machine-read experimental readout, the risk of bias in such case may not be seen as strong enough to accept the risk of a human error.

Another common example is provided by studies where incremental drug doses or concentrations are applied during the course of a single experiment involving just one subject. During cardiovascular safety studies, animals receive first an infusion of a vehicle (e.g., over a period of 30 min), followed by the two or three concentrations of the test drug, and the hemodynamics is being assessed along with the blood samples taken. As the goal of such studies is to establish concentration-effect relationships, one has no choice but to accept the lack of randomization. The only alternatives would be to give up on the within-subject design or conduct the study over many days to allow enough time to wash the drug out between the test days. Needless to say, neither of these options is perfect for a study where the baseline characteristics are a critical factor in keeping the sample size low. In this example, the desire to conduct a properly randomized study comes into a conflict with ethical considerations.

A similar design is often used in electrophysiological experiments (in vitro or ex vivo) where a test system needs to be equilibrated and baselined for extended periods of time (sometimes hours) to allow subsequent application of test drugs (at ascending concentrations). Because a washout cannot be easily controlled, such studies also do not follow randomized schedules of testing various drug doses.

The low-throughput studies such as in electrophysiology typically go over many days, and every day there is a small number of subjects or data points added. While one may accept the studies being not randomized in some cases, it is important to stress that there should be other measures in place that control potential sources of bias. It is a common but usually unacceptable practice to analyze the results each time a new data point has been added in order to decide whether a magic P value sank below 0.05 and the experiment can stop. For example, in one recent publication, it was stated: "For optogenetic activation experiments, cell-type-specific ablation experiments, and in vivo recordings (optrode recordings and calcium imaging), we continuously increased the number of animals until statistical significance was reached to support our conclusions." Such an approach should be avoided by clear experimental planning and definition of study endpoints.

The above examples are provided only to illustrate that there may be special cases when randomization may not be done. This is usually not an easy decision to make and even more difficult to defend later. Therefore, one should always be advised to seek a professional advice (i.e., interaction with the biostatisticians or colleagues specializing in the risk assessment and study design issues). Needless to say, this advice should be obtained before the studies are conducted.

In the ideal case, once the randomization was applied to allocate subjects to treatment conditions, the randomization should be maintained through the study conduct and analysis to control against potential performance and outcome detection bias, respectively. In other words, it would not be appropriate first to assign the subjects, for example, to groups A and B and then do all experimental manipulations first with the group A and then with the group B.

3   Blinding

In clinical research, blinding and randomization are recognized as the most important design techniques for avoiding bias (ICH Harmonised Tripartite Guideline 1998; see also chapter "Learning from Principles of Evidence-Based Medicine to Optimize

Nonclinical Research Practices"). In the preclinical domain, there is a number of instruments assessing risks of bias, and the criteria most often included are randomization and blinding (83% and 77% of a total number of 30 instruments analyzed, Krauth et al. 2013).

While randomization and blinding are often discussed together and serve highly overlapping objectives, attitude towards these two research rigor measures is strikingly different. The reason for a higher acceptance of randomization compared to blinding is obvious – randomization can be implemented essentially at no cost, while blinding requires at least some investment of resources and may therefore have a negative impact on the research unit's apparent capacity (measured by the number of completed studies, irrespective of quality).

Since the costs and resources are not an acceptable argument in discussions on ethical conduct of research, we often engage a defense mechanism, called rationalization, that helps to justify and explain why blinding should not be applied and do so in a seemingly rational or logical manner to avoid the true explanation. Arguments against the use of blinding can be divided into two groups.

One group comprises a range of factors that are essentially psychological barriers that can be effectively addressed. For example, one may believe that his/her research area or a specific research method has an innate immunity against any risk of bias. Or, alternatively, one may believe that his/her scientific excellence and the ability to supervise the activities in the lab make blinding unnecessary. There is a great example that can be used to illustrate that there is no place for beliefs and one should rather rely on empirical evidence. For decades, compared to male musicians, females have been underrepresented in major symphonic orchestras despite having equal access to high-quality education. The situation started to change in the mid-1970s when blind auditions were introduced and the proportion of female orchestrants went up (Goldin and Rouse 2000). In preclinical research, there are also examples of the impact of blinding (or a lack thereof). More specifically, there were studies that reveal substantially higher effect sizes reported in the experiments that were not randomized or blinded (Macleod et al. 2008).

Another potential barrier is related to the "trust" within the lab. Bench scientists need to be explained what the purpose of blinding is and, in the ideal case, be actively involved in development and implementation of blinding and other research rigor measures. With the proper explanation and engagement, blinding will not be seen as an unfriendly act whereby a PI or a lab head communicates a lack of trust.

The second group of arguments against the use of blinding is actually composed of legitimate questions that need to be addressed when designing an experiment. As mentioned above in the section on randomization, a decision to apply blinding should be justified by the needs of a specific experiment and correctly balanced against the existing and potential risks.

3.1 Fit-for-Purpose Blinding

It requires no explanation that, in preclinical research, there are no double-blinded studies in a sense of how it is meant in the clinic. However, similar to clinical research, blinding in preclinical experiments serves to protect against two potential sources of bias: bias related to blinding of personnel involved in study conduct including application of treatments (performance bias) and bias related to blinding of personnel involved in the outcome assessment (detection bias).

Analysis of the risks of bias in a particular research environment or for a specific experiment allows to decide which type of blinding should be applied and whether blinding is an appropriate measure against the risks.

There are three types or levels of blinding, and each one of them has its use: assumed blinding, partial blinding, and full blinding. With each type of blinding, experimenters allocate subjects to groups, replace the group names with blind codes, save the coding information in a secure place, and do not access this information until a certain pre-defined time point (e.g., until the data are collected or the study is completed and analyzed).

3.1.1 Assumed Blinding

In the assumed blinding, experimenters have access to the group or treatment codes at all times, but they do not know the correspondence between group and treatment before the end of the study. With the partial or full blinding, experimenters do not have access to the coding information until a certain pre-defined time point.

Main advantage of the assumed blinding is that an experiment can be conducted by one person who plans, performs, and analyzes the study. The risk of bias may be relatively low if the experiments are routine – e.g., lead optimization research in drug discovery or fee-for-service studies conducted using well-established standardized methods.

Efficiency of assumed blinding is enhanced if there is a sufficient time gap between application of a treatment and the outcome recording/assessment. It is also usually helpful if the access to the blinding codes is intentionally made more difficult (e.g., blinding codes are kept in the study design assistant or in a file on an office computer that is not too close to the lab where the outcomes will be recorded).

If introduced properly, assumed blinding can guard against certain unwanted practices such as remeasurement, removal, and reclassification of individual observations or data points (three evil Rs according to Shun-Shin and Francis 2013). In preclinical studies with small sample sizes, such practices have particularly deleterious consequences. In some cases, remeasurement even of a single subject may skew the results in a direction suggested by the knowledge of group allocation. One should emphasize that blinding is not necessarily an instrument against the remeasurement (it is often needed or unavoidable) but rather helps to avoid risks associated with it.

3.1.2 Partial Blinding

There are various situations where blinding (with no access to the blinding codes) is implemented not for the entire experiment but only for a certain part of it, e.g.:

- No blinding during the application of experimental treatment (e.g., injection of a test drug) but proper blinding during the data collection and analysis
- No blinding during the conduct of an experiment but proper blinding during analysis

For example, in behavioral pharmacology, there are experiments where subjects' behavior is video recorded after a test drug is applied. In such cases, blinding is applied to analysis of the video recordings but not the drug application phase. Needless to say, blinded analysis has typically to be performed by someone who was not involved in the drug application phase.

A decision to apply partial blinding is based on (a) the confidence that the risks of bias are properly controlled during the unblinded parts of the experiment and/or (b) rationale assessment of the risks associated with maintaining blinding throughout the experiment. As an illustration of such decision-making process, one may imagine a study where the experiment is conducted in a small lab (two or three people) by adequately trained personnel that is not under pressure to deliver results of a certain pattern, data collection is automatic, and data integrity is maintained at every step. Supported by various risk reduction measures, such an experiment may deliver robust and reliable data even if not fully blinded.

Importantly, while partial blinding can adequately limit the risk of some forms of bias, it may be less effective against the performance bias.

3.1.3 Full Blinding

For important decision-enabling studies (including confirmatory research, see chapter "Resolving the Tension Between Exploration and Confirmation in Preclinical Biomedical Research"), it is usually preferable to implement full blinding rather than to explain why it was not done and argue that all the risks were properly controlled.

It is particularly advisable to follow full blinding in the experiments that are for some reasons difficult to repeat. For example, these could be studies running over significant periods of time (e.g., many months) or studies using unique resources or studies that may not be repeated for ethical reasons. In such cases, it is more rational to apply full blinding rather than leave a chance that the results will be questioned on the ground of lacking research rigor.

As implied by the name, full blinding requires complete allocation concealment from the beginning until the end of the experiment. This requirement may translate into substantial costs of resources. In the ideal scenario, each study should be supported by at least three independent people responsible for:

- (De)coding, randomization
- Conduct of the experiment such as handling of the subjects and application of test drugs (outcome recording and assessment)
- (Outcome recording and assessment), final analysis

The main reason for separating conduct of the experiment and the final analysis is to protect against potential unintended unblinding (see below). If there is no risk of unblinding or it is not possible to have three independent people to support the blinding of an experiment, one may consider a single person responsible for every step from the conduct of the experiment to the final analysis. In other words, the study would be supported by two independent people responsible for:

- (De)coding, randomization
- Conduct of the experiment such as handling of the subjects and application of test drugs, outcome recording and assessment, and final analysis

3.2   Implementation of Blinding

Successful blinding is related to adequate randomization. This does not mean that they should always be performed in this sequence: first randomization and then blinding. In fact, the order may be reversed. For example, one may work with an offspring of the female rats that received experimental and control treatments while pregnant. As the litter size may differ substantially between the dams, randomization may be conducted after the pups are born, and this does not require allocation concealment to be broken.

The blinding procedure has to be carefully thought through. There are several factors that are listed below and that can turn a well-minded intention into a waste of resources.

First, blinding should as far as possible cover the entire experimental setup – i.e., all groups and subjects. There is an unacceptable practice to exclude positive controls from blinding that is often not justified by anything other than an intention to introduce a detection bias in order to reduce the risk of running an invalid experiment (i.e., an experiment where a positive control failed).

In some cases, positive controls cannot be administered by the same route or using the same pretreatment time as other groups. Typically, such a situation would require a separate negative (vehicle) control treated in the same way as the positive control group. Thus, the study is only partially blinded as the experimenter is able to identify the groups needed to "validate" the study (negative control and positive control groups) but remains blind to the exact nature of the treatment received by each of these two groups. For a better control over the risk of unblinding, one may apply a "double-dummy" approach where all animals receive the same number of administrations via the same routes and pretreatment times.

Second, experiments may be unintentionally unblinded. For example, drugs may have specific, easy to observe physicochemical characteristics, or drug treatments may change the appearance of the subjects or produce obvious adverse effects. Perhaps, even more common is the unblinding due to the differences in the appearance of the drug solution or suspension dependent on the concentration. In such cases, there is not much that can be done but it is essential to take corresponding notes and acknowledge in the study report or publication. It is interesting to note that the unblinding is often cited as an argument against the use of blinding (Fitzpatrick et al. 2018); however, this argument reveals another problem – partial blinding schemes are often applied as a normative response without any proper risk of bias assessment.

Third, blinding codes should be kept in a secure place avoiding any risk that the codes are lost. For in vivo experiments, this is an ethical requirement as the study will be wasted if it cannot be unblinded at the end.

Fourth, blinding can significantly increase the risk of mistakes. A particular situation that one should be prepared to avoid is related to lack of accessibility of blinding codes in case of emergency. There are situations when a scientist conducting a study falls ill and the treatment schedules or outcome assessment protocols are not available or a drug treatment is causing disturbing adverse effects and attending veterinarians or caregivers call for a decision in the absence of a scientist responsible for a study. It usually helps to make the right decision if it is known that an adverse effect is observed in a treatment group where it can be expected. Such situations should be foreseen and appropriate guidance made available to anyone directly or indirectly involved in an experiment. A proper study design should define a backup person with access to the blinding codes and include clear definition of endpoints.

Several practical tips can help to reduce the risk of human-made mistakes. For example, the study conduct can be greatly facilitated if each treatment group is assigned its own color. Then, this color coding would be applied to vials with the test drugs, syringes used to apply the drug, and the subjects (e.g., apply solution from a green-labeled vial using a green-labeled syringe to an animal from a green-labeled cage or with a green mark on its tail). When following such practice, one should not forget to randomly assign color codes to treatment conditions. Otherwise, for example, yellow color is always used for vehicle control, green for the lowest dose, and so forth.

To sum up, it is not always lacking resources that make full blinding not possible to apply. Further, similar to what was described above for randomization, there are clear exception cases where application of blinding is made problematic by the very nature of the experiment itself.

4 Concluding Recommendations

Most, if not all, guidelines, recommendations, and other texts on Good Research Practice emphasize the importance of blinding and randomization (chapters "Guidelines and Initiatives for Good Research Practice", and "General Principles of Preclinical Study Design"). There is, however, very limited specific guidance on when and how to apply blinding and randomization. The present chapter aims to close this gap.

Generally speaking, experiments should be blinded and randomized if:

- This is a confirmatory research (see chapter "Resolving the Tension Between Exploration and Confirmation in Preclinical Biomedical Research") that has a

major impact on decision-making and that cannot be readily repeated (for ethical or resource-related reasons).
- No other measures can be applied to protect against existing and potential risks of bias.

There are various sources of bias that affect the outcome of experimental studies and these sources are unique and specific to each research unit. There is usually no one who knows these risks better than the scientists working in the research unit, and it is always up to the scientist to decide if, when, and how blinding and randomization should be implemented. However, there are several recommendations that can help to decide and act in the most effective way:

- Conduct a risk assessment for your research environment, and, if you do not know how to do that, ask for a professional support or advice.
- Involve your team in developing and implementing the blinding/randomization protocols, and seek the team members' feedback regarding the performance of these protocols (and revise them, as needed).
- Provide training not only on how to administer blinding and randomization but also to preempt any questions related to the rationale behind these measures (i.e., experiments are blinded not because of the suspected misconduct or lack of trust).
- Describe blinding and randomization procedures in dedicated protocols with as many details as possible (including emergency plans and accident reporting, as discussed above).
- Ensure maximal transparency when reporting blinding and randomization (e.g., in a publication). When deciding to apply blinding and randomization, be maximally clear about the details (Table 4). When deciding against, be open about the reasons for such decision. Transparency is also essential when conducting multi-laboratory collaborative projects or when a study is outsourced to another laboratory. To avoid any misunderstanding, collaborators should specify expectations and reach alignment on study design prior to the experiment and communicate all important details in study reports.

Blinding and randomization should always be a part of a more general effort to introduce and maintain research rigor. Just as the randomization increases the likelihood that blinding will not be omitted (van der Worp et al. 2010), other Good Research Practices such as proper documentation are also highly instrumental in making blinding and randomization effective.

To conclude, blinding and randomization may be associated with some effort and additional costs, but, under all circumstances, a decision to apply these research rigor techniques should not be based on general statements and arguments by those who do not want to leave their comfort zone. Instead, the decision should be based on the applicable risk assessment and careful review of potential implementation burden. In many cases, this leads to a relieving discovery that the devil is not so black as he is painted.

Table 4 Minimum reporting information for blinding and randomization procedures

Procedure	Technical report/laboratory notebook record	Scientific publication
Randomization	Type of randomization Block size (if applicable) Stratification variables (if applicable) Tools used for randomization Reference to the protocol followed Deviations from the protocol (if any)	Type of randomization Tools used for randomization Stratification variables (if applicable)
Blinding	Type of blinding Records of unblinding (if applicable) Reference to the protocol followed Deviations from the protocol (if any) Colleague(s) who provided blinding	Type of blinding Statement whether blinding integrity was maintained Statement whether blinding was provided by one of the co-authors

Acknowledgments The authors would like to thank Dr. Thomas Steckler (Janssen), Dr. Kim Wever (Radboud University), and Dr. Jan Vollert (Imperial College London) for reading the earlier version of the manuscript and providing comments and suggestions.

References

Carroll L (1871) Through the looking-glass, and what Alice found there. ICU Publishing

Fitzpatrick BG, Koustova E, Wang Y (2018) Getting personal with the "reproducibility crisis": interviews in the animal research community. Lab Anim 47:175–177

Goldin C, Rouse C (2000) Orchestrating impartiality: the impact of "blind" auditions on female musicians. Am Econ Rev 90:715–741

Hooijmans CR, Rovers MM, de Vries RB, Leenaars M, Ritskes-Hoitinga M, Langendam MW (2014) SYRCLE's risk of bias tool for animal studies. BMC Med Res Methodol 14:43

ICH Harmonised Tripartite Guideline (1998) Statistical principles for clinical trials (E9). CPMP/ICH/363/96, March 1998

Jones B, Kenward MG (2003) Design and analysis of cross-over designs, 2nd edn. Chapman and Hall, London

Kalish LA, Begg GB (1985) Treatment allocation methods in clinical trials a review. Stat Med 4:129–144

Krauth D, Woodruff TJ, Bero L (2013) Instruments for assessing risk of bias and other methodological criteria of published animal studies: a systematic review. Environ Health Perspect 121:985–992

Macleod MR, The NPQIP Collaborative Group (2017) Findings of a retrospective, controlled cohort study of the impact of a change in Nature journals' editorial policy for life sciences research on the completeness of reporting study design and execution. bioRxiv:187245. https://doi.org/10.1101/187245

Macleod MR, van der Worp HB, Sena ES, Howells DW, Dirnagl U, Donnan GA (2008) Evidence for the efficacy of NXY-059 in experimental focal cerebral ischaemia is confounded by study quality. Stroke 39:2824–2829

Shun-Shin MJ, Francis DP (2013) Why even more clinical research studies may be false: effect of asymmetrical handling of clinically unexpected values. PLoS One 8(6):e65323

van der Worp HB, Howells DW, Sena ES, Porritt MJ, Rewell S, O'Collins V, Macleod MR (2010) Can animal models of disease reliably inform human studies? PLoS Med 7(3):e1000245

Out of Control? Managing Baseline Variability in Experimental Studies with Control Groups

Paul Moser

Contents

Abstract

Control groups are expected to show what happens in the absence of the intervention of interest (negative control) or the effect of an intervention expected to have an effect (positive control). Although they usually give results we can anticipate, they are an essential component of all experiments, both in vitro and in vivo, and fulfil a number of important roles in any experimental design. Perhaps most importantly they help you understand the influence of variables that you cannot fully eliminate from your experiment and thus include them in

P. Moser (✉)
Cerbascience, Toulouse, France
e-mail: pmoser@cerbascience.com

© The Author(s) 2019
A. Bespalov et al. (eds.), *Good Research Practice in Non-Clinical Pharmacology and Biomedicine*, Handbook of Experimental Pharmacology 257,
https://doi.org/10.1007/164_2019_280

your analysis of treatment effects. Because of this it is essential that they are treated as any other experimental group in terms of subjects, randomisation, blinding, etc. It also means that in almost all cases, contemporaneous control groups are required. Historical and baseline control groups serve a slightly different role and cannot fully replace control groups run as an integral part of the experiment. When used correctly, a good control group not only validates your experiment; it provides the basis for evaluating the effect of your treatments.

Keywords
Baseline values · Blinding · Historical controls · Negative control groups · Positive control groups · Sham controls · Vehicle

1   What Are Control Groups?

As Donald Rumsfeld famously said about weapons of mass destruction, there are known knowns, known unknowns and unknown unknowns. This is also true of experiments. Through good experimental design, we try to eliminate as much as possible the influence of the first two, the things we know about. The purpose of the control group is to understand the influence of the third. The term sounds comforting, as if we have managed to somehow rein in the experiment and submitted the study to our will. If anything, the opposite is true: the control group is a tacit acknowledgement, not only of all the things that we can't control but of those things that we are not even aware of, the unknown unknowns.

The choice of appropriate control groups is intimately tied to the aims of your study. To unambiguously demonstrate that your experimental treatment has (or has not) had an effect in your test system, there needs to be a value against which you can compare it. A good control group allows you to do this – a bad control group means you cannot make valid comparisons to evaluate the activity of your test condition and, even worse, means you may end up drawing invalid conclusions.

Several types of control groups have been described in the literature, including positive, negative, sham, vehicle and comparative (Johnson and Besselsen 2002). These can broadly be classed into negative and positive controls. Note that in many studies these terms are used very loosely and, as Kramer and Font (2017) rightly point out, a description of what the control group is being used for is better than a label such as positive or negative which might be misleading. What are generally referred to as negative controls include vehicle and sham groups and are expected to show what happens in the absence of the intervention of interest. These controls are necessary because all studies are open to unexpected effects.

In contrast, positive controls are expected to have an effect. They are used to show that the study can detect an effect in the desired direction and thus that the experimental protocol is sensitive to interventions expected to have an effect. They might also be used to show the magnitude of effect that is possible with an active substance. Positive controls are in a comparable position to your test treatment: they need a good negative control to be of any use. Also, just like test treatments, they need to be subject to the same randomisation and blinding procedures and must be included in the experimental design and analysis.

There should also be a distinction between what we could call primary and secondary controls. Both have a value but in different areas. A primary control is what we typically think of as a control, i.e. a group that undergoes all experimental procedures except for the variable being investigated and which are specific to the experimental question being studied. In contrast, a secondary control, such as historical control values, could be used to check the conformity of the experiment or, in the case of baseline data, could be used to verify the homogeneity of the treatment groups at the start of the experiment. In almost all cases, the presence of a primary control group is essential to be able draw valid conclusions from the study.

Although this review discusses primarily in vivo experiments, almost all the points discussed apply equally to in vitro experiments. In fact, in vitro studies are in the envious position where additional control groups can be added for little additional cost and usually without invoking the ethical questions that are important to consider for in vivo studies. In all other ways, in vitro studies require the same attention to experimental design as in vivo studies as far as blinding, randomisation and statistical analysis are concerned (Festing 2001). This should also apply to any negative and positive controls included in the study.

In this brief review, I will only discuss the use of prospective control groups, which are most appropriate for non-clinical studies. Retrospective controls, which are often used in epidemiological clinical studies, to study such things as drug use during pregnancy, substance abuse etc., and where a randomised clinical trial would be unethical (e.g. Andrade 2017; Szekér et al. 2017), raise other issues which are not particularly relevant to non-clinical studies and which will not be discussed here.

2   Basic Considerations for Control Groups

2.1   Attribution of Animals to Control Groups

Correct randomisation to avoid bias is a basic but essential part of any experimental design (see chapter "Blinding and Randomization") that applies equally to control groups. Control groups should be treated like any other experimental group within an experiment. Subjects for the control groups must come from the same population as the other groups so that the experiment is carried out on a homogenous population. This means that normally they should not be historical values nor baseline values. As we will see below, there are specific circumstances where this is either not possible or where some flexibility is permitted.

2.2   What Group Size for Control Groups?

As discussed elsewhere (see chapter "Building Robustness into Translational Research"), it is fundamentally important that your experiment be adequately powered. But does changing the relative sizes of individual groups in an experiment affect our ability to detect an effect? The majority of nonclinical studies use similar group sizes for all treatment groups, including controls, and there are articles on experimental design that present only this option (e.g. Haimez 2002; Aban and George 2015; Singh et al. 2016). Bate and Karp (2014) have looked more closely at the question of relative group sizes, and they show that the traditional, balanced approach is indeed the best experimental design when all pairwise comparisons are planned. However, when the planned comparisons are of several treatment groups with a single control group, there is a small gain in sensitivity by having relatively more animals in the control group. For example, in an experiment with 30 subjects divided into 4 treatment groups and 1 control group, the power to detect a given effect size is between 2 and 5% greater if there are 10 subjects in the control group and 5 in each treatment group as compared to 5 equal groups of 6. By contrast, if the treatment groups are increased in size relative to the control group (for the same total number of subjects per experiment), there is a marked loss of power which can be 25% lower in some worst-case scenarios. It is therefore clear that the control group should never have fewer subjects than the treatment groups.

It can sometimes be tempting to add a second experiment to increase the n values or to expand the dose-range of the treatment being tested. The practice of combining experiments is relatively widespread but not always apparent from the data presentation. Combining data sets from two studies having identical treatment groups is particularly hard to spot and unless it is a part of the original experimental design should be considered a form of p-hacking (Simmons et al. 2011; Head et al. 2015), a practice where data is manipulated until it reaches significance. Somewhat easier to spot, but more dangerous, is the combining of experiments involving different experimental groups. This issue is highlighted by Lew (2008) who shows that incorrect conclusions can be drawn if the individual experiments are not analysed separately. An example of this is shown in Fig. 1a. Another issue with combining data from different experiments is the possibility of misinterpretation due to Simpson's paradox (Ameringer et al. 2009). In this case the combination of studies leads to a different conclusion to that drawn from the individual component studies analysed separately (Fig. 1b). Simpson's paradox is caused by the unequal distribution of a confounding variable between the different experiments and is often a consequence of unequal group sizes. There are few, if any, published examples from non-clinical work, but it has been highlighted as an issue in clinical trials, as in the example for the antidiabetic drug rosiglitazone described by Rucker and Schumacher (2008) where they describe a meta-analysis of several studies which reached the opposite conclusion to each individual study. Another example concerning the effectiveness of two treatments for kidney stones is presented by Julious and Mullee (1994).

Ultimately, the biggest issue with combining experiments and so inflating the n value for the control group is the same as for using historical controls (see below): it removes one of the fundamental reasons for including a control group, namely, the control of unknown variables affecting a particular study. If the number of animals in the control group is different from the treated group (as in the example described by Lew 2008) or if the df of the ANOVA do not correspond to equal groups sizes, this should be suspected. Unfortunately, I can use an example from my work where two experiments were combined to broaden the dose range, as seen with the second graph in Figure 2 of Moser and Sanger (1999). The graph shows five doses of

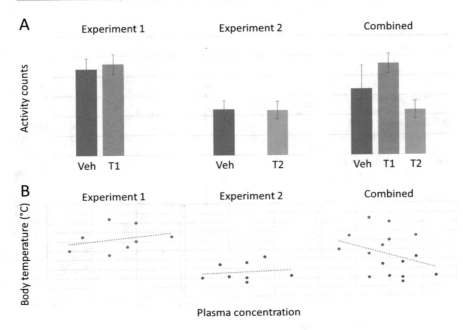

Fig. 1 Misinterpretation of experimental outcome that results from combining data from two separate experiments where the control values differ. Panel **a** shows two experiments (mean ± SD, $n = 6$) where the two treatment doses do not differ from vehicle (t-test), whereas the combined data set (now with $n = 12$ for the vehicle group) results in a significant biphasic effect (both doses $p < 0.05$ vs vehicle, Dunnett's test). Panel **b** shows two experiments correlating changes in body temperature with plasma levels of drug. In each experiment the drug shows a weak tendency to increase body temperature. However, when they are combined, the drug appears to have the opposite effect, to reduce body temperature. This is an example of Simpson's paradox. Although both sets of figures were constructed from simulated data, they highlight the importance of analysing separate experiments separately

pindolol tested in the forced swim test, but combines two experiments evaluating four doses each. In our defence, we clearly mentioned this, both in the methods and in the figure legend, and we compared the two experiments statistically for homogeneity. In retrospect, however, it would have been better to present the two experiments in separate graphs (or at least as two separate data sets on the same graph) with separate statistical analyses. Transparency in such cases is always the better option when there is nothing to hide.

2.3   Controls and Blinding

Blinding is an essential part of any experiment (see chapter "Blinding and Randomization") and must equally apply to all control groups. If you know a particular animal has received a control substance, then you will be at risk of biasing that animal's data. This was demonstrated in a study at Pfizer by Machin et al. (2009)

Fig. 2 Paw withdrawal
threshold measured using von
Frey filaments in the tibial
nerve transection model of
neuropathic pain. Vehicle
pretreatment is compared with
gabapentin (100 mg/kg PO)
under unblinded and blinded
conditions (∗: $p < 0.001$
Mann-Whitney U-test; $n = 6$).
The data shown are median
values; the bar indicates the
range. Graph drawn from data
presented in Machin et al.
(2009)

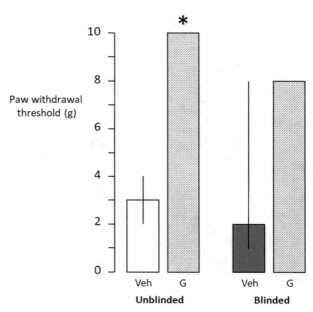

who showed that carrying out an experiment blinded greatly increased the variability
of the control group in a study of neuropathic pain where tactile analgesia was
evaluated using von Frey filaments (Fig. 2). Although the effect of the positive
control, gabapentin, appears similar in the two conditions, bear in mind that this is
largely because of a ceiling effect and does not mean that the positive control group
suffers less from such bias. Indeed, in their abstract Machin et al. (2009) state that
typical values in non-lesioned animals are around 8 g, so it is possible that carrying
out the study unblinded has exaggerated the effect of the positive control. Holman
et al. (2015) have studied the effect of blinding on effect size and shown that over
83 studies, the effect of blinding is to reduce, in some cases quite dramatically,
Hedges's g (i.e. the difference between positive and negative control means divided
by the pooled standard deviation). Their review underlines the importance of
applying blinding to all experimental groups in a study.

3   Primary Controls

3.1   Choosing Appropriate Control Treatments: Not All Negative Controls Are Equal

By definition, we expect a positive control to produce an effect in our test, but a
negative control to be essentially indistinguishable from vehicle. Indeed, vehicle
is often the negative control most people use. I would argue that a vehicle group is
not strictly a negative control group but more an evaluation of the baseline against

which to judge the effects of other control groups and treatments. It is there to evaluate the effects of things you cannot fully control or interventions you cannot avoid such as animal handling for substance administration, housing conditions, etc. One reason this neutral control cannot strictly speaking be a negative control is that other experimental variables may interact with those unavoidable interventions.

There are some documented examples of this. For example, it has been shown that elements of the surgical procedure necessary for implanting intracerebroventricular cannulae in rats were sufficient to change the behavioural response resulting from an interaction between calcium antagonists and physostigmine (Bourson and Moser 1989). These experiments showed that the potentiation of physostigmine-induced yawning by nifedipine was abolished by sham-lesioning procedures in rats, whereas the nifedipine potentiation of apomorphine-induced yawning was unaffected. The study also demonstrated that the presurgical drug treatment (i.e. desmethylimipramine and pentobarbital) or 7 days isolation was alone sufficient to reduce the yawning response to physostigmine and abolish its potentiation by nifedipine.

3.2   Vehicle Controls

These appear to be the simplest of control groups – simple administration of the vehicle used to formulate your test substance under the same conditions (pretreatment time, volume, concentration, etc.). This apparent simplicity can make it easy to overlook many of the issues surrounding the choice of vehicle. Much of the data we have for effects of vehicles has come from toxicology studies where control animals might receive the vehicle for up to 2 years in the case of carcinogenicity studies. Under these circumstances the tolerability and long-term toxicity of the vehicle are the main concerns, and there are several publications indicating the maximum tolerated doses of a wide range of potential vehicles (ten Tije et al. 2003; Gad et al. 2006). However, while these concentrations may be tolerated, that does not mean they are without behavioural effects. Castro et al. (1995) examined the effects of several commonly used vehicles on locomotion in mice and found marked effects of Tween, DMSO and ethanol-containing vehicles at levels well below those indicated by Gad et al. (2006) as being well-tolerated. Matheus et al. (1997) looked at the effects of Tween, propylene glycol and DMSO on elevated plus-maze behaviour in rats following their injection into the dorsal periaqueductal grey. Interestingly, whereas Castro et al. (1995) found DMSO to reduce locomotion, Matheus et al. (1997) found it to increase arm entries. DMSO, at concentrations above 15%, has also been found to modify sleep architecture in rats (Cavas et al. 2005).

Food reward is widely used as a motivating factor in many behavioural studies, particularly those studying operant behaviour. Modifying feeding conditions to modulate motivation has been shown to affect response rate in an operant discrimination task (Lotfizadeh et al. 2012), and it would therefore be expected that the use of high-calorie vehicles such as oils could have a similar effect. Although there do not appear to be any published examples of this, it is something I have observed in rats trained on a delayed match to position task. The use of an oil vehicle almost completely abolished responding (Moser unpublished observation) although we did not establish if this was due to the high-calorie content of the vehicle or the effects of indigestion after a large volume of oil administered directly into the stomach.

In addition to intrinsic behavioural effects, many of these vehicles also interfere with the pharmacokinetics of the drugs being tested. ten Tije et al. (2003) have reviewed clinical and non-clinical effects of vehicles on the pharmacokinetics of co-administered chemotherapy agents, and Kim et al. (2007) have reported marked effects of vehicles on the ADME properties of the insecticide deltamethrin in rats. Some of the more striking examples include a 16-fold increase in danazol bioavailability in dogs when formulated in Tween 80 compared to a commercial formulation (Erlich et al. 1999) and up to a 40% decrease in the blood/plasma ratio for paclitaxel when formulated in cremephor EL (Loos et al. 2002).

3.3   Sham Controls

The term sham control is usually employed when it is considered that there is a part of the experimental protocol that we expect to have an impact on the outcome. Although it could be argued that the handling and injection procedure associated with administration of a drug vehicle could be expected to have some effect, the term sham control is usually applied when there is a surgical intervention of some sort. If, for example, the experiment involves a chemical lesion of the brain, the sham controls will undergo a similar surgical procedure without injection of the toxin, but including anesthesia, placement in a stereotaxic frame, incision of the scalp, trepanation, lowering of a dummy cannula into the brain (this is often omitted), suturing of the wound and appropriate postsurgical care. If the intervention is the surgical removal of tissue, then the sham control animals will be anesthetized and opened but without internal intervention and then sutured and given the same postsurgical care as the lesion group. Such groups are essential, as the anesthesia and postsurgical pain can be very stressful to laboratory animals (Hüske et al. 2016). In addition, anesthesia has also been shown to induce long-lasting effects on memory in rats (Culley et al. 2003). As part of a study to understand the processes involved in liver regeneration, Werner et al. (2014) reported on the effects of sham surgery procedures and anesthesia on the expression pattern of microRNAs in rat liver as compared to partial hepatectomy. They found 49 microRNAs modified by hepatectomy and 45 modified by sham laparotomy, with 10 microRNAs showing similar changes after both real and sham surgery. Anesthesia alone had much less effect, with only one microRNA changing in the same direction as surgery. The impact of sham surgery has also been highlighted by Cole et al. (2011) who compared the effects of standard sham procedures used in research on traumatic brain injury (craniotomy by drill or manual trepanation) with the effects of anesthesia alone. They found that the traditional sham control induced significant pro-inflammatory, morphological and behavioural changes and that these could confound interpretation in brain injury models.

3.4   Non-neutral Control Groups

Many experimental designs require more than one control group. There are many situations that require a neutral control group (sham, untreated etc.), a group undergoing an intervention (drug challenge, lesion etc.) and a positive control or comparator group (a treatment known to reverse the effects of the intervention – see below). Comparison of the neutral control with the intervention control shows that the intervention has had an effect – and the intervention group then becomes the point of comparison for treatments aimed at reversing its effects. This could be the effect of a lesion on behaviour (e.g. Hogg et al. 1998), a drug challenge such as amphetamine to increase locomotion (e.g. Moser et al. 1995) or an environmental change such as housing conditions, diet, light cycle, etc. (e.g. He et al. 2010). It is important to be able to demonstrate that this intervention has produced a reliable change compared to the neutral baseline or, in some cases, to a sham control. As discussed above, these controls are no longer neutral, and only through appropriate preliminary experiments can you determine how much these interventions (sham or non-sham) interfere with the primary purpose of your study. Such preliminary experiments ultimately help to reduce animal numbers as otherwise it might be necessary to include neutral, sham and intervention controls in every experiment instead of just sham and intervention. The number of times you expect to use a particular experimental design should help guide you to the optimal solution for limiting animal use.

3.5   Controls for Mutant, Transgenic and Knockout Animals

This is a vast topic that deserves a full review to itself (see chapter "Genetic Background and Sex: Impact of Generalizability of Research Findings in Pharmacology Studies") and will only be covered superficially here. The use of mutant animals is increasing compared to non-genetically altered animals (e.g. UK Home Office statistics, 2017: https://www.gov.uk/government/statistics/statistics-of-scientific-procedures-on-living-animals-great-britain-2017), and they raise particular issues relating to the choice of control groups. If transgenic animals are maintained as a homozygous colony, they will gradually show genetic drift compared to the original non-modified founder line. The recommendation is to breed from heterozygotes and use wild-type littermates as controls for homozygous animals (Holmdahl and Malissen 2012). The alternative, using animals from the founder strain, is fraught with difficulty due to the multiplication of strains, as highlighted by Kelmensen (https://www.jax.org/news-and-insights/jax-blog/2016/june/there-is-no-such-thing-as-a-b6-mouse). Furthermore, background strain is known to have a major impact on the phenotype of transgenic animals (e.g. Brayton et al. 2012; Fontaine and Davis 2016). Jackson Laboratories have a page on their website dedicated to helping researchers choose an appropriate control for transgenic animals: https://www.jax.org/jax-mice-and-services/customer-support/technical-support/breeding-and-husbandry-support/considerations-for-choosing-controls. The use of conditional expression systems, where the mutation can be expressed or not, can greatly improve the pertinence of control groups in transgenic animals (e.g. Justice et al. 2011).

Use of littermates is not always possible, particularly in the case of inbred mutant animals such as the spontaneously hypertensive rat (SHR). In the case of SHRs, the practice is to use the Wistar-Kyoto rat as a normotensive control as it is derived from the same ancestral Wistar line but its suitability as a control is questioned (Zhang-James et al. 2013).

4   Positive Controls

Whereas a negative control has the single job of providing a baseline from which to judge other interventions, positive controls can wear several hats at once. They can (a) demonstrate that the experimental conditions are sensitive to an active intervention, (b) provide a reference effect-size against which other interventions can be judged and (c) provide a reference effect in order to judge the conformity of that particular experiment to historical studies. If your positive control does not differ from the negative (e.g. vehicle) control group, then you need to study the data before making any interpretation of the results with novel test substances. Figure 3 illustrates some of the potential outcomes of an experiment with a negative and a positive control, suggesting how you should interpret the results.

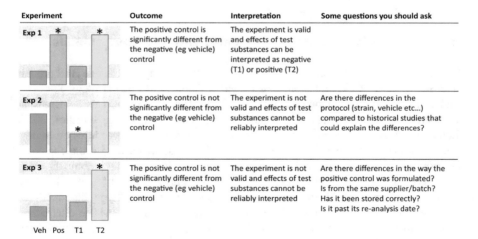

Fig. 3 Interpreting results when the positive and negative controls are not significantly different. The first case, Experiment 1, has a vehicle group (Veh) and a positive control (Pos) that are within their historical ranges (horizontal light blue and green shaded boxes, respectively). The asterisk indicates significant difference compared to the vehicle. In Experiments 2 and 3, the positive and negative controls are outside their expected range. How the effects of the test substances T1 and T2 are interpreted, as well as the questions that should be asked, are different between these scenarios. This figure is inspired by, and adapted from, an article from PAASP (https://paasp.net/heads-i-win-tails-you-lose/)

Proving that the experimental conditions are sensitive to an active intervention (typically a clinically active substance) is the main reason for using a positive control. Just as we do not necessarily know all the variables that can affect the response in negative control groups, similarly we cannot know all the variables that affect the response to an intervention we expect to work. Sometimes experiments just don't seem to work as we expect. A classic example of this is the effect of noise on the development of tolerance to benzodiazepines. Normally, chronic treatment with benzodiazepines results in tolerance developing to their anxiolytic and sedative effects. However, a series of studies carried out during a period of laboratory renovation failed to show this effect, and subsequent, more controlled studies demonstrated that under conditions of noise stress, tolerance developed to the sedative effects of benzodiazepines but not to their anxiolytic effects (File and Fernandes 1994).

The second and third uses of positive controls have an element of quality control about them. During drug development it is often necessary to compare a novel substance with one that has already demonstrated an effect in the clinic or with the lead substance of a competitor. This is particularly true when substances with similar mechanisms of action are being studied. However, some caution should be used with novel mechanisms of action. Experimental conditions are often 'optimised' to detect known compounds, and this may reduce their ability to detect compounds with novel mechanisms of activity. Models of anxiety provide some examples of this: all anxiolytic tests detect benzodiazepines, for many years the only available treatment for anxiety, but they are typically less responsive to putative anxiolytics with other mechanisms of action such as 5-HT_{1A} partial agonists (e.g. buspirone), which have shown positive effects in the clinic but are inconsistent at best in animal models (e.g. Moser 1989; Moser et al. 1990).

When the positive control does not work as expected, it is important to check historical values to make sure that it is not the negative control that has produced an aberrant response. In any such situation, it is important to be clear what we mean by 'work'. Many drugs can have side-effects which are rate-dependent (e.g. increase behaviour when activity is low but decrease it when activity is high) which could mask or enhance another effect (e.g. Sanger and Blackman 1976). Being aware of such possibilities in a particular experimental set-up can help you better understand what has really happened in your experiment.

Finally, it is important that all the groups in the experiment are treated equally. By that I mean they are an integral part of the initial experimental design, subject to the same blinding and randomisation as other groups and included in the analysis of the experiment. There can be a temptation to first analyse the positive and negative controls using a simple two-group test and, if significant, to declare that the experiment has worked. A subsequent multigroup comparison would then analyse the effects of the test substance against the negative control. Such an analysis is not comparing the effects of the positive control and the test substance under the same conditions. This introduces a bias towards declaring experiments to have 'worked' (i.e. the positive control is significantly different compared to the negative control) when a multigroup comparison including all experimental groups might be non-significant. In some cases where the positive effect might be a unilateral lesion, the bias is aggravated by the use of a within-subject test to validate the experiment but a between subject test to evaluate the test substance. Any such analysis that does not include all experimental groups together should be viewed with caution.

5 Secondary Controls

5.1 Can Baseline Values Be Used as Control?

Baseline values can be very useful points of comparison, but they are not the same as controls for a subsequent treatment or intervention. If we consider that controls are designed to measure the impact of extraneous variables in an experiment, then the baseline, by definition, cannot control for subsequent changes over time. Many physiological parameters change over time, such as locomotor activity and body temperature which show marked circadian variation (e.g. Moser and Redfern 1985). There are also many other factors which can modify drug effects and which vary over time, such as activity of metabolising enzymes (Redfern and Moser 1988).

These variations can modify the effects of drugs. This can be demonstrated by a simple example of measuring body temperature. Rodent body temperature varies over 24 h, with a trough during the light period and a peak during the dark period. Depending on when we administered a drug during the day, we might conclude that it either increased or decreased body temperature when compared against baseline when, in reality, it has no effect (e.g. Redfern and Moser 1988). The opposite is also possible: a real effect of the test substance could be masked or enhanced by the circadian variation when compared against baseline.

Measurement of baseline values is often a first step in making experimental groups homogeneous. It is essential that the groups be made as identical as possible using suitable randomisation and the spread of baseline values should not be too great. Stratified randomisation (e.g. Altman 1999) could be used if there is a wide range of baseline values but it is very rare that experimental effects will be independent of the baseline. Many drug effects are baseline- or rate-dependent such as drug effects on locomotor activity (e.g. Glick and Milloy 1973), and there is always the risk of effects being due to regression towards the mean (e.g. Morton and Torgerson 2005).

One common practice that uses baseline as a type of control value is the use of 'change from baseline' as the main variable. This can be useful when the underlying variable has low variance (such as body temperature in untreated animals at a given time of day) but dangerously confounding when the measure shows large variance and regression towards the mean (such as locomotor activity in untreated animals at a given time of day). Analysing original data or the change from baseline measure can result in different statistical outcomes (e.g. Le Cudennec and Castagné 2014). Change from baseline might be useful as an exploratory analysis but still needs very careful interpretation: a repeat measures analysis would probably be an altogether more satisfactory approach in most circumstances.

Thus, under very specific conditions, the baseline can be used as a control value. However, those conditions require that the baseline values be homogenous and with low variance and that the experiment be carried out over a very short period of time. This may be very restrictive for many types of study, but, when appropriate, the use of baseline as control can be a good way to reduce the number of animals used in a study and increase statistical power (in part due to the necessity for low variation but also because of the use of within-subject statistics), both ethically desirable.

5.2 Historical Control Values

Like any group in an experiment, the control groups can give a spurious result simply as a result of randomly selecting data from a population. Differences relative to this control group could then be interpreted as an effect of treatment in the other groups when in fact it is the control group that has created the difference. The frequency of this occurrence can be limited by powering your study correctly, but it may still happen from time to time. This can be controlled for in experiments that are carried out frequently, such as screening studies, by comparing your control group against historical controls.

As the objective of control groups is to provide a contemporaneous point of comparison for your experimental treatment, it is clear that historical values cannot fulfill all the functions of a control group. In particular, historical controls cannot control for the possible impact of those unknown unknowns specific to a given experiment. Indeed, it has been suggested that historical controls may be responsible for the unreliable results obtained with some test substances in the SOD-1 model of ALS (Scott et al. 2008) and Papageorgiou et al. (2017) have also demonstrated how using historical controls instead of concurrent controls in clinical studies can introduce bias.

At the same time, historical controls can provide an additional point of comparison to provide some validity for the experimental conduct. If your control values deviate from historical reference values, it allows you to determine if the data may be unreliable and potentially identify additional, previously unknown, variables that help you to improve an experimental protocol. Contemporaneous controls are not designed to provide a 'true' value that is somehow independent of the current study but are there to allow you to evaluate your current study correctly.

However, under certain conditions *or* if you are prepared to relax some of the requirements of a true control group *and* if you have a procedure that is regularly used, there are situations where historical control values can be better than concurrent controls and integrated into your experiment.

One of these situations, which has a relatively long history, is in the context of toxicology studies. For example, there are rare tumours that can occur spontaneously in rats at a rate below that which is reliably detectable in any reasonably sized control group. If such tumours are observed in a treated group, it is possibly only by chance, but comparison with the control group, in which no such tumours occurred, will not help determine if this is, or is not, the case. This is a serious issue in toxicology testing, and the Society for Toxicologic Pathology set up a Historical Control Data Working Group to examine the issue in some depth. Their full discussion of the issues and recommendations for best practice in the case of proliferative lesions in rodents has been presented by Keenan et al. (2009). It is a similar story for in vitro toxicity studies, such as the in vitro micronucleus test where historical controls are regarded as important for evaluating data quality and interpreting potential positive results, an approach enshrined in the OECD guidelines on genetic toxicology testing (Lovell et al. 2018).

It has been suggested that for often-repeated studies historical controls could potentially replace contemporaneous controls (e.g. Festing and Altman 2002). Kramer and Font (2017) make a strong case for considering historical controls a replacement for contemporaneous controls and present a number of simulations of typical conditions showing how they could be effectively used to reduce animal usage.

6　When Are Control Groups Not Necessary?

The short answer is almost never. However, there may be circumstances where rare or anecdotal events can be reported as preliminary evidence for an effect. Two well-known and somewhat tongue-in-cheek examples include the protective effects of umbrellas against lion attacks (Anderson 1991) and parachute use to prevent injury when falling (Smith and Pell 2003). Despite their humorous approach, both papers make serious points, and the latter has been extensively cited as presenting a situation analogous to some medical practices, in which the benefits of a treatment are so obvious that they do not need testing. However, a recent analysis has found that this analogy is frequently false: when actually evaluated in randomised controlled clinical trials, only a modest proportion (27%) of such 'obvious' findings showed a significant benefit (Hayes et al. 2018). Thus, even if you think the outcome is so obvious that a proper control group is not necessary, be aware that you are more likely than not to be wrong. Furthermore, if this is frequently true of clinical hypotheses, I suspect that for non-clinical studies, it is likely to be true in almost all cases.

7　Conclusion

Good control groups are the ground upon which your study stands. Without them, who knows where you will fall.

References

Aban IB, George B (2015) Statistical considerations for preclinical studies. Exp Neurol 270:82–87
Altman DG (1999) How to randomize. BMJ 319:703–704

Ameringer S, Serlin RC, Ward S (2009) Simpson's paradox and experimental research. Nurs Res 58:123–127

Anderson DR (1991) Umbrellas and lions. J Clin Epidemiol 44:335–337

Andrade C (2017) Offspring outcomes in studies of antidepressant-treated pregnancies depend on the choice of control group. J Clin Psychiatry 78:e294–e297

Bate S, Karp NA (2014) Common control group – optimising the experiment design to maximise sensitivity. PLoS One 9(12):e114872. https://doi.org/10.1371/journal.pone.0114872

Bourson A, Moser PC (1989) The effect of pre- and post-operative procedures on physostigmine- and apomorphine-induced yawning in rats. Pharmacol Biochem Behav 34:915–917

Brayton CF, Treuting PM, Ward JM (2012) Pathobiology of aging mice and GEM: background strains and experimental design. Vet Pathol 49:85–105

Castro CA, Hogan JB, Benson KA, Shehata CW, Landauer MR (1995) Behavioral effects of vehicles: DMSO, ethanol, Tween-20, Tween-80, and emulphor-620. Pharmacol Biochem Behav 50:521–526

Cavas M, Beltran D, Navarro JF (2005) Behavioural effects of dimethyl sulfoxide (DMSO): changes in sleep architecture in rats. Toxicol Lett 157:221–232

Cole JT, Yarnell A, Kean WS, Gold E, Lewis B, Ren M, McMullen DC, Jacobowitz DM, Pollard HB, O'Neill JT, Grunberg NE, Dalgard CL, Frank JA, Watson WD (2011) Craniotomy: true sham for traumatic brain injury, or a sham of a sham? J Neurotrauma 28:359–369

Culley DJ, Baxter M, Yukhananov R, Crosby G (2003) The memory effects of general anesthesia persist for weeks in young and aged rats. Anesth Analg 96:1004–1009

Erlich L, Yu D, Pallister DA, Levinson RS, Gole DG, Wilkinson PA, Erlich RE, Reeve LE, Viegas TX (1999) Relative bioavailability of danazol in dogs from liquid-filled hard gelatin capsules. Int J Pharm 179:49–53

Festing MF (2001) Guidelines for the design and statistical analysis of experiments in papers submitted to ATLA. Altern Lab Anim 29:427–446

Festing MF, Altman DG (2002) Guidelines for the design and statistical analysis of experiments using laboratory animals. ILAR J 43:244–258

File SE, Fernandes C (1994) Noise stress and the development of benzodiazepine dependence in the rat. Anxiety 1:8–12

Fontaine DA, Davis DB (2016) Attention to background strain is essential for metabolic research: C57BL/6 and the International Knockout Mouse Consortium. Diabetes 65:25–33

Gad SC, Cassidy CD, Aubert N, Spainhour B, Robbe H (2006) Nonclinical vehicle use in studies by multiple routes in multiple species. Int J Toxicol 25:499–521

Glick SD, Milloy S (1973) Rate-dependent effects of d-amphetamine on locomotor activity in mice: possible relationship to paradoxical amphetamine sedation in minimal brain dysfunction. Eur J Pharmacol 24:266–268

Haimez C (2002) How much for a star? Elements for a rational choice of sample size in preclinical trials. Trends Pharmacol Sci 23:221–225

Hayes MJ, Kaestner V, Mailankody S, Prasad V (2018) Most medical practices are not parachutes: a citation analysis of practices felt by biomedical authors to be analogous to parachutes. CMAJ Open 6:E31–E38

He M, Su H, Gao W, Johansson SM, Liu Q, Wu X, Liao J, Young AA, Bartfai T, Wang M-W (2010) Reversal of obesity and insulin resistance by a non-peptidic glucagon-like peptide-1 receptor agonist in diet-induced obese mice. PLoS One 5:e14205

Head ML, Holman L, Lanfear R, Kahn AT, Jennions MD (2015) The extent and consequences of p-hacking in science. PLoS Biol 13:e1002106

Hogg S, Sanger DJ, Moser PC (1998) Mild traumatic lesion of the right parietal cortex in the rat: characterisation of a conditioned freezing deficit and its reversal by dizocilpine. Behav Brain Res 93:157–165

Holman L, Head ML, Lanfear R, Jennions MD (2015) Evidence of experimental bias in the life sciences: why we need blind data recording. PLoS Biol 13:e1002190

Holmdahl R, Malissen B (2012) The need for littermate controls. Eur J Immunol 42:45–47

Hüske C, Sander SE, Hamann M, Kershaw O, Richter F, Richter A (2016) Towards optimized anesthesia protocols for stereotactic surgery in rats: analgesic, stress and general health effects of injectable anesthetics. A comparison of a recommended complete reversal anesthesia with traditional chloral hydrate monoanesthesia. Brain Res 1642:364–375

Johnson PD, Besselsen DG (2002) Practical aspects of experimental design in animal research. ILAR J 43:202–206

Julious SA, Mullee MA (1994) Confounding and Simpson's paradox. BMJ 309:1480

Justice MJ, Siracusa LD, Stewart AF (2011) Technical approaches for mouse models of human disease. Dis Model Mech 4:305–310

Keenan C, Elmore S, Francke-Carroll S, Kemp R, Kerlin R, Peddada S, Pletcher J, Rinke M, Schmidt SP, Taylor I, Wolf DC (2009) Best practices for use of historical control data of proliferative rodent lesions. Toxicol Pathol 37:679–693

Kim KB, Anand SS, Muralidhara S, Kum HJ, Bruckner JV (2007) Formulation-dependent toxicokinetics explains differences in the GI absorption, bioavailability and acute neurotoxicity of deltamethrin in rats. Toxicology 234:194–202

Kramer M, Font E (2017) Reducing sample size in experiments with animals: historical controls and related strategies. Biol Rev 92:431–445

Le Cudennec C, Castagné V (2014) Face-to-face comparison of the predictive validity of two models of neuropathic pain in the rat: analgesic activity of pregabalin, tramadol and duloxetine. Eur J Pharmacol 735:17–25

Lew MJ (2008) On contemporaneous controls, unlikely outcomes, boxes and replacing the 'Student': good statistical practice in pharmacology, problem 3. Br J Pharmacol 155:797–803

Loos WJ, Szebeni J, ten Tije AJ, Verweij J, van Zomeren DM, Chung KN, Nooter K, Stoter G, Sparreboom A (2002) Preclinical evaluation of alternative pharmaceutical delivery vehicles for paclitaxel. Anti-Cancer Drugs 13:767–775

Lotfizadeh AD, Redner R, Edwards TL, Quisenberry AJ, Baker LE, Poling A (2012) Effects of altering motivation for food in rats trained with food reinforcement to discriminate between d-amphetamine and saline injections. Pharmacol Biochem Behav 103:168–173

Lovell DP, Fellows M, Marchetti F, Christiansen J, Elhajouji A, Hashimoto K, Kasamoto S, Li Y, Masayasu O, Moore MM, Schuler M, Smith R, Stankowski LF Jr, Tanaka J, Tanir JY, Thybaud V, Van Goethem F, Whitwell J (2018) Analysis of negative historical control group data from the in vitro micronucleus assay using TK6 cells. Mutat Res 825:40–50

Machin I, Gurrel R, Corradini L (2009) Impact of study blinding on outcome of behavioural studies in pain research. Proceedings of the British Pharmacological Society. http://www.pa2online. org/abstracts/1Vol7Issue3abst002P.pdf

Matheus MG, de-Lacerda JC, Guimarães FS (1997) Behavioral effects of "vehicle" microinjected into the dorsal periaqueductal grey of rats tested in the elevated plus maze. Braz J Med Biol Res 30:61–64

Morton V, Torgerson DJ (2005) Regression to the mean: treatment effect without the intervention. J Eval Clin Pract 11:59–65

Moser PC (1989) An evaluation of the elevated plus-maze test using the novel anxiolytic buspirone. Psychopharmacology 99:48–53

Moser PC, Redfern PH (1985) Circadian variation in behavioural responses to 5-HT receptor stimulation. Psychopharmacology 86:223–227

Moser PC, Sanger DJ (1999) 5-HT1A receptor antagonists neither potentiate nor inhibit the effects of fluoxetine and befloxatone in the forced swim test in rats. Eur J Pharmacol 372:127–134

Moser PC, Moran PM, Frank RA, Kehne JH (1995) Reversal of amphetamine-induced behaviours by MDL 100,907, a selective 5-HT2A antagonist. Behav Brain Res 73:163–167

Moser PC, Tricklebank MD, Middlemiss DN, Mir AK, Hibert MF, Fozard JR (1990) Characterization of MDL 73005EF as a 5-HT1A selective ligand and its effects in animal models of anxiety: comparison with buspirone, 8-OH-DPAT and diazepam. Br J Pharmacol 99:343–349

Papageorgiou SN, Koretsi V, Jäger A (2017) Bias from historical control groups used in orthodontic research: a meta-epidemiological study. Eur J Orthod 39:98–105

Redfern PH, Moser PC (1988) Factors affecting circadian variation in responses to psychotropic drugs. Ann Rev Chronopharmacol 4:107–136

Rucker G, Schumacher M (2008) Simpson's paradox visualized: the example of the rosiglitazone meta-analysis. BMC Med Res Methodol 8:34

Sanger DJ, Blackman DE (1976) Rate-dependent effects of drugs: a review of the literature. Pharmacol Biochem Behav 4:73–83

Scott S, Kranz JE, Cole J, Lincecum JM, Thompson K, Kelly N, Bostrom A, Theodoss J, Al-Nakhala BM, Vieira FG, Ramasubbu J, Heywood JA (2008) Design, power, and inter-pretation of studies in the standard murine model of ALS. Amyotroph Lateral Scler 9:4–15

Simmons JP et al (2011) False-positive psychology: undisclosed flexibility in data collection and analysis allows presenting anything as significant. Psychol Sci 22:1359–1366

Singh VP, Pratap K, Sinha J, Desiraju K, Bahal D, Kukreti R (2016) Critical evaluation of chal-lenges and future use of animals in experimentation for biomedical research. Int J Immunopathol Pharmacol 29:551–561

Smith GCS, Pell JP (2003) Parachute use to prevent death and major trauma related to gravitational challenge: systematic review of randomised controlled trials. BMJ 327:1459

Szekér S, Fogarassy G, Vathy-Fogarassy A (2017) Comparison of control group generating methods. Stud Health Technol Inform 236:311–318

ten Tije AJ, Verweij J, Loos WJ, Sparreboom A (2003) Pharmacological effects of formulation vehicles. Clin Pharmacokinet 42:665–685

Werner W, Sallmon H, Leder A, Lippert S, Reutzel-Selke A, Morgül MH, Jonas S, Dame C, Neuhaus P, Iacomini J, Tullius SG, Sauer IM, Raschzok N (2014) Independent effects of sham laparotomy and anesthesia on hepatic microRNA expression in rats. BMC Res Notes 7:702

Zhang-James Y, Middleton FA, Faraone SV (2013) Genetic architecture of Wistar-Kyoto rat and spontaneously hypertensive rat substrains from different sources. Physiol Genomics 45:528–538

Quality of Research Tools

Dario Doller and Paul Wes

Contents

D. Doller (✉)
Alcyoneus/ScienceWorks, Sparta, NJ, USA

P. Wes
Centers for Therapeutic Innovation, Pfizer Inc., New York, NY, USA
e-mail: paul.wes@pfizer.com

© The Author(s) 2019
A. Bespalov et al. (eds.), *Good Research Practice in Non-Clinical Pharmacology and Biomedicine*, Handbook of Experimental Pharmacology 257,
https://doi.org/10.1007/164_2019_281

Abstract

Drug discovery research is a complex undertaking conducted by teams of scientists representing the different areas involved. In addition to a strong familiarity with existing knowledge, key relevant concepts remain unknown as activities start. This is often an accepted risk, mitigated by gaining understanding in real time as the project develops. Chemicals play a role in all biology studies conducted in the context of drug discovery, whether endogenously or exogenously added to the system under study. Furthermore, new knowledge often flourishes at the interface of existing areas of expertise. Due to differences in their training, adding a chemist's perspective to research teams would at least avoid potentially costly mistakes and ideally make any biology research richer. Thus, it would seem natural that one such team member be a chemist. Still, as that may not always be the case, we present some suggestions to minimize the risk of irreproducibility due to chemistry-related issues during biology research supporting drug discovery and make these efforts more robust and impactful. These include discussions on identity and purity, target and species selectivity, and chemical modalities such as orthosteric or allosteric small molecules or antibodies. Given the immense diversity of potential chemical/biological system interactions, we do not intend to provide a foolproof guide to conduct biological experimentation. Investigate at your own peril!

Keywords

Antibodies · Biological hypothesis · Biologics · Chemical biology · Chemical modalities · Chemical probes · Critical path · Identity · Purity · Quality of drug discovery · Selectivity · Small molecule · Species selectivity

1   Introduction

Drug discovery research is a quintessential example of a team activity, given the complexity and breadth of the many scientific areas involved. The integrity, quality, and impact of new knowledge emerging from scientific research are based on individual and collective adherence to core values of objectivity, accountability, and stewardship (The National Academies of Sciences, Engineering, and Medicine 2017). This is of particular importance in life sciences and drug discovery, as we strive *simultaneously* to develop new knowledge to enable new discoveries that lead project teams to the invention of efficacious therapies that cause life-altering impact on patients suffering from devastating diseases. Not a small task!

Drug discovery may become a very costly undertaking, and investments in an ill-conceived strategy may be devastating for an organization, regardless of whether they are industrial or academic, for-profit or nonprofit, or well-established or starting up. Finding a causative relationship between a single drug target and a disease state remains a formidable exercise in understanding human biology and pathophysiology. As it is broadly acknowledged, significant knowledge gaps in these areas exist today.

Seeking to maximize the chances of success, every project in its conception is linked to a *biological hypothesis of disease*: The exact pronunciation of the scientific

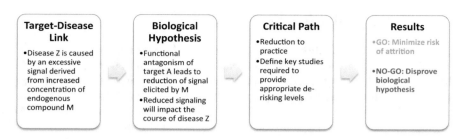

Fig. 1 Example of a biological hypothesis and its relationship to the target-disease link

basis of how a concept is thought to lead to disease treatment. During project execution, teams aim either at disproving or adding support to such hypotheses. It may be argued that every nonredundant, reasonable study designed to deliver supporting evidence or rule out the central biological hypothesis must be conducted. However, reality dictates that resources are always finite. As a consequence, teams reduce the biological hypothesis to practice and define the smallest set of studies required to provide appropriate de-risking levels (which vary with every one of us) and support reasonably well the biological hypothesis, called the *critical path*. Therefore, establishing a sharply defined biological hypothesis linking the target to the disease is a key to framing the scope of the project team (Fig. 1).

The choice of the biological target linked to the disease to be treated is a necessary – but not sufficient – condition for a successful project. Work on an ill-conceived hypothesis or a poorly executed strategy will inevitably lead to a failure due to lack of efficacy – barring serendipity. Thus, the work of most effective project teams will find a way to deliver consensus-derived, logical, and sensible project milestones that either establish evidence that argues against the biological hypothesis and recommend the termination of the project or succeed in minimizing the risks moving forward into clinical validation. The reader will realize that assuring clinical efficacy is not within the realm of possible outcomes at this time.

Chemicals play a role in all biology studies conducted in the context of drug discovery, whether endogenously or exogenously added to the system under study. During the execution of the project's strategy, multiple compounds will be acquired, by either purchase or chemical synthesis, and studied experimentally to illuminate the decisions made by project teams. These will be chemical probes, radioligands, imaging agents, and drug candidates. Thus, while it would seem natural to enlist a chemist as a team member, unfortunately that is not always the case for a number of reasons. The chemist's contributions are often incorrectly perceived as not adding other values than providing the test article. While apparently adding efficiencies, this separation of tasks limits the potential synergies across science disciplines, which are key to develop new knowledge.

Be that as it may, this chapter aims to discuss some aspects of the qualification that different compounds, as well as other research reagents and tools, would meet in order to maximize the quality of the science and minimize the changes of misinterpretation of experiments. We do not intend to provide a foolproof guide to conduct biological experimentation, since the diversity of potential chemical/biological system interactions precludes us from such a goal.

2   Drugs in the Twenty-First Century

The concept of "druggability" was introduced around the turn of the last century as a way to qualify the perceived likelihood that a ligand could be found for a given binding site in a biological target, and it was initially conceived thinking of small molecules as drugs, generally compliant with Lipinski rule of 5 (Ro5), and proteins as biological targets (Hopkins and Groom 2003; Workman 2003). Today, the number and nature of chemical modalities that have led to the appropriate modulation of pathophysiology for therapeutic use have increased remarkably. These range, on one end of the spectrum, from small molecules designed with parameters increasingly beyond those set by the Ro5 [e.g., oxysterols (Blanco et al. 2018), cyclic peptides, millamolecular chemistry (millamolecular chemistry refers to macrocycles with a molecular weight between 500 and 1,000 daltons), PROTACs (Churcher 2017)] to molecular entities grouped under the name "biologics," such as antibodies, antibody-drug conjugates, peptides, and oligonucleotides.

With so many different chemical modalities having been incorporated into the arsenal of scientists, it behooves drug discovery project teams to secure experimental support, in the way of preponderance of evidence, for the concept that the drug is acting according to the mechanism of action at the core of the biological hypothesis. In other words, secure mechanistic evidence that the drug is doing its job "as advertised." This task includes a thorough characterization and qualification of the chemical probes used during biology studies.

2.1   Chemical Tools Versus Drugs

Drug repurposing consists of the use of advanced clinical compounds or approved drugs to treat different diseases. It takes advantage of work previously done on the active pharmaceutical ingredient, leading to reduced development costs and timelines, and has recently become an area of active interest (Doan et al. 2011). Often drugs already in clinical use are used as chemical probes based on their mode of action. It is important to note that having been approved by a regulatory agency for clinical or veterinary treatment does not necessarily qualify a compound as a high-quality chemical probe. For example, a chemical probe must meet very high standards in terms of its selectivity toward a biological target. On the other hand, for a drug aimed for clinical use, lack of selectivity for a molecular target (known as polypharmacology) may not only not be an issue, but actually provide the basis for its efficacy and differentiation from similar drugs.

3   First Things First: Identity and Purity

Where do the samples of chemical probes and drug candidates used for biological testing come from? Historically, these compounds used to be synthesized by medicinal chemists at intramural laboratories. The final step before submitting newly made compounds for biological testing was to determine that their identity and purity were within expected specifications.

Compound management departments collected these samples and were responsible for managing corporate compound collections that grew in size and structural diversity over the years. They would also deliver compounds for testing to the corresponding laboratory. This process ensured the right compound was delivered in the right amount to the right destination.

However, things have changed. Most drug research organizations today take advantage to some extent of external, independent, contract research organizations (CROs) to synthesize the small molecules for their research activities or commercial suppliers of compound libraries for their hit-finding activities. Indeed, a fair number of start-up biotechnology organizations lack laboratories and conduct 100% of their experimentation at off-site laboratories. Often, compounds travel thousands of kilometers to reach their testing destinations after being synthesized. How does one assure the identity of the compound used for testing is correct? How does all the travel and manipulation impact the purity of the test article and its potential degradation?

An obvious yet often overlooked step when requesting a chemical tool for a biology study without the help of a trained chemist – who would typically communicate using unambiguous chemical structures – is to make sure the compound ordered *is* the actual chemical intended to be used in the research. In other words, unequivocal identification of the research tools, including their sources (if commercial suppliers, include catalog number and batch number) and assays used in the characterization of the compound, has a major impact on the reproducibility of biological studies by providing a well-defined starting point.

Names given to chemicals may be ambiguous. Some drugs are given an official generic and nonproprietary name, known as international nonproprietary name (INN), with the exact goal of making communication more precise by providing a unique standard name for each active ingredient, to avoid prescribing errors. However, most compounds used in research do not get to receive an INN from regulatory agencies. For example, a SciFinder search for the chemicals "cholesterol" shows a number of "other names" for it, while the drug known as Prozac™ (fluoxetine) has more than 50 different names, some of them included in Table 1.

Using a compound's Chemical Abstracts Service Registry Number (CAS RN) is likely the best way to avoid errors communicating the name of a compound (https://www.cas.org/support/documentation/chemical-substances/faqs). The CAS Registry is considered the most authoritative collection of disclosed chemical substance information, covering substances identified from the scientific literature from 1957 to the present, with additional substances going back to the early 1900s. This database is updated daily with thousands of new substances. Essentially, a CAS RN is a unique numeric identifier that may contain up to ten digits, divided by hyphens into three parts. It designates only one substance. The numerical sequence has no chemical significance by itself. For example, the CAS RN of cholesterol is 57-88-5. However, CAS RN are not fool proof either. For example, fluoxetine free base and its hydrochloride salt have different CAS RN (54910-89-3 and 56296-78-7,

Table 1 Different names for cholesterol (left) or Prozac™ (right) found in SciFinder

Cholesterol (8CI)	Benzenepropanamine, N-methyl-γ-[4-(trifluoromethyl)phenoxy]-, hydrochloride (9CI)
(3β)-Cholest-5-en-3-ol	(±)-N-Methyl-3-phenyl-3-[4-(trifluoromethyl)phenoxy] propylamine hydrochloride
(−)-Cholesterol	Affectine
3β-Hydroxycholest-5-ene	Deproxin
5:6-Cholesten-3β-ol	Fluoxac
Cholest-5-en-3β-ol	Fluoxeren
Cholesterin	Fluoxetine hydrochloride
Dythol	Fluoxil
Lidinit	LY 110140
Lidinite	Lilly 110140
NSC 8798	N-Methyl-3-(4-trifluoromethylphenoxy)-3-phenylpropylamine hydrochloride
Marine cholesterol	N-Methyl-3-[4-(trifluoromethyl)phenoxy]-3-phenylpropanamine hydrochloride
Cholesteryl alcohol	Profluzac
Provitamin D	Prozac
SyntheChol	Prozac 20
Δ5-Cholesten-3β-ol	... etcetera

respectively). Thus, it is recommended to consult with a chemist to avoid costly mistakes.

In terms of assessing the purity of a chemical sample, a growing number of organizations manage the on-demand syntheses of compounds at CROs by designated medicinal chemists. These two teams must collaborate closely to support project activities. Analogs made are not accepted for delivery unless extensive characterization with unambiguous analytical data exists consistent with the correct chemical structure and purity (often referred to as Certificate of Analysis or CoA). Typically, an elemental analysis is provided, which should be consistent with the compound molecular formula within experimental error, as well as a set of ultraviolet (UV), infrared, H-1 or C-13 nuclear magnetic resonance spectra, and a chromatographic trace with different detection methods (e.g., total ion current, UV absorption). Appearance (oil, powder, crystal, color) and expected purity (as a %) are also stated.

Catalog chemicals, acquired either as singletons or as part of compound libraries, are previously made and are not always accompanied by a current CoA or set of spectroscopic data. The reasons for this vary. So we would rather focus on ways to make sure the compound being tested is the right one.

Furthermore, compound management is also available from CROs, either at the synthesis site or not. Compounds bought are received, barcoded, stored, weighed, cherry-picked, and shipped for testing. When compound management occurs at extramural organizations requiring shipping or when the last chemical analysis was done some time in the past (e.g., a year), it is strongly recommended that *before testing* quality control be conducted to confirm that the compound's identity is correct and its purity is within acceptable levels, ideally similar to those obtained when the sample was originally synthesized.

3.1 The Case of Evans Blue

Evans blue (1) is a chemical used to assess permeability of the blood–brain barrier (BBB) to macromolecules. First reported in 1914 (Evans et al. 1914; Saunders et al. 2015), this compound is extensively used for mechanistic studies seeking to interrogate the structural and functional integrity of the BBB. The basic principle is that serum albumin cannot cross a healthy BBB, and virtually all Evans blue is bound to albumin, leaving the brain unstained. When the BBB has been compromised, albumin-bound Evans blue enters the central nervous system (CNS), generating a typical blue color (Fig. 2). However, a reappraisal of the compound properties (as available from commercial sources, Fig. 3) under the scrutiny of modern techniques reveals caveats. For example, one source lists the chemical purity as 75%, with an elemental analysis differing significantly from the theoretical one, and with a variable solid state nature (crystal, amorphous) (www.sigmaaldrich.com/life-science.html). Differences between crystalline states may impact drug solubility, and the presence of up to 25% of unknown impurities introduces random factors in experiments conducted with this material. These factors are not aligned with experimental best practice and may be easily corrected by a skillful chemist in a research team.

1

The purity of the reagents used in biology research is highly linked to their quality and reproducibility. For example, lipopolysaccharides (LPS) are large molecules found in the outer membrane of Gram-negative bacteria. Chemically, they are a mixture formed by a diversity of lipids and polysaccharides joined by a covalent bond (Moran 2001). LPS is a microbe-associated molecular pattern that potently activates innate immune cells. Peripheral administration of LPS is used as an immunological challenge in animal models of inflammation. However, commercially available samples of LPS are subject to significant variability and typically display a range of potencies, in part due to the wide range of possible methods used to purify LPS (Darveau and Hancock 1983; Johnson and Perry 1976; Apicella

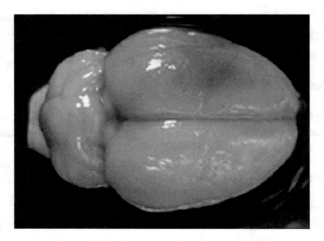

Fig. 2 A rodent brain showing the effects of Evans blue staining

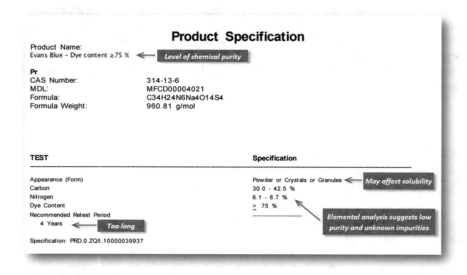

Fig. 3 Example of technical sheet for a commercial sample of Evans blue

2008). Indeed, some preparations may contain large amounts of nucleic acid contaminants, which activate different innate immunity receptors, resulting in different biological effects. This causes variability in the results from biological studies using LPS, limiting their utility (Ray et al. 1991; Lee et al. 2013). A superior form of LPS for biological research is Control Standard Endotoxin (CSE), which is an LPS preparation whose activity is standardized using coagulation of horseshoe crab blood cells as a functional assay. Indeed, analysis of mouse brain and plasma cytokines and

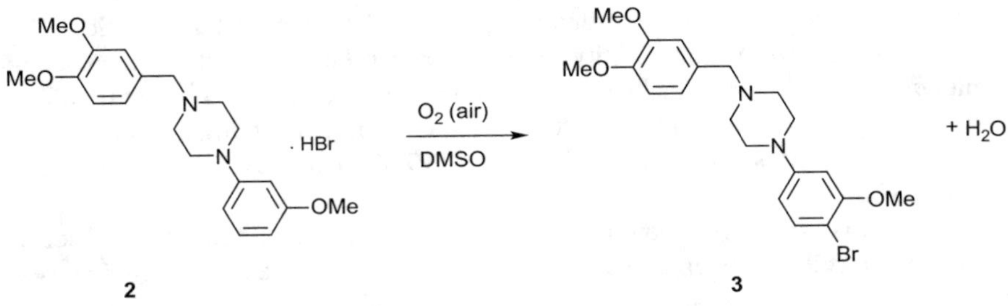

Fig. 4 Transformation of arylpiperazine **2** as an HBr salt into bromopiperazine 3 during storage (Moran 2001)

kynurenine pathway metabolites isolated upon CSE treatment suggests improvements in the quality of results over LPS in an in vivo neuroinflammation model (Lee et al. 2013).

Even when the purity and identity of compounds used in testing are assessed right after synthesis, some may deteriorate over time. This is true for solid samples, as well as those stored as solutions in dimethyl sulfoxide (DMSO). Cycles of freezing and thawing exacerbate deterioration, especially if the sample is exposed to air. For example, the arylpiperazine **2** was thought to be a hit in a high-throughput screen conducted to find melanin-concentrating hormone receptor 1 (MCHR1) antagonists. However, upon confirmation of the hit via resynthesis, it was discovered that the actual structure of the compound was **3** (Tarrant et al. 2017). The reason for this was that at the end of the preparation, compounds were isolated by precipitating them as their hydrobromide salts and then stored in DMSO at a 10 μM concentration. However, in the presence of oxygen (from the air), the thawed solutions reacted according to the chemical reaction shown in Fig. 4.

In summary, confirming the identity and the purity of all compounds used in any experiment is a simple and necessary step to get the most out of the efforts, conduct rigorous scientific experimentation, and avoid costly mistakes.

3.2 Identity and Purity of Research Reagents

Not only is it important to unequivocally confirm the identity and purity of compounds, but it is essential to do the same for all the biological tools used in studies. For in vitro work, cell lines should be authenticated for their identity and lack of mycoplasma infection. It is surprisingly common in research for scientists to work on cell lines that are not what they think they are, resulting in spurious published reports, irreproducible data, and wasted resources (Chatterjee 2007; Drexler et al. 2002; Masters 2000). In a notorious set of studies, it was found that a large proportion of cell lines being used worldwide were HeLa cells (Nelson-Rees et al. 1974, 1981), the first human cancer cell line developed. In other studies, of cell lines submitted for banking, 17–36% were found to be different from what was claimed and were even of a different species (Markovic and Markovic 1998; MacLeod et al. 1999). The identity of cell lines can be evaluated by microscopic evaluation, growth curve analysis, and karyotyping. However, the most definitive test to confirm a cell lines identity is to perform DNA fingerprinting by short tandem repeat profiling (Masters 2000; MacLeod et al. 1997). This service is provided by a number of CROs.

In order to ensure reproducibility of published work, it is important to include all details of the cell culture experiments. This includes the source of any cell line used, including suppliers and catalog numbers, where applicable. It may also be relevant to specify the range of passage numbers used in experiments, as this may affect functional outcomes (Briske-Anderson et al. 1997; Esquenet et al. 1997; Yu et al. 1997; Wenger et al. 2004). The same applies to all culture conditions, including seeding density, time of culture, media composition, and whether antibiotics are used.

Another common pitfall in cell biology research is infection of cells by mycoplasma (Drexler and Uphoff 2002). *Mycoplasma* are a genus of bacteria that lack a cell wall and are resistant to common antibiotics. Mycoplasma infection can affect cell behavior and metabolism in many ways and therefore confound any results (Drexler et al. 2002; Kagemann et al. 2005; Lincoln and Gabridge 1998). *Mycoplasma* are too small to detect by conventional microscopy and are generally screened for in laboratories by DNA staining, PCR, or mycoplasmal enzyme activity (Drexler and Uphoff 2002; Lawrence et al. 2010). This should be done routinely, and with the availability of rapid kits it is relatively painless and cost-effective.

Many people use antibiotics in cell culture. However, it is a better practice to avoid antibiotics. Antibiotics mask errors in aseptic technique and quality control and select for antibiotic-resistant bacteria, including mycoplasma (Lincoln and Gabridge 1998). Furthermore, small quantities of bacteria killed by antibiotics may still have effects on cells due to microbe-associated molecules, such as LPS, that are potent activators of innate immune cells (Witting and Moller 2011).

For in vivo work, care should be taken to ensure that the correct mouse strains and lines are being used. Full genetic nomenclature and sources should be tracked and reported. Mouse strain can differ substantially depending on which vendor it is obtained from due to spontaneous mutations and founder effects. For instance, it was discovered that the commonly used C57BL/6J inbred mouse strain from one vendor contained a spontaneous deletion of the gene for α-synuclein, while the same strain from another vendor did not (Specht and Schoepfer 2001). Mutations in α-synuclein are a genetic cause of Parkinson's disease, and many researchers in the field had been inadvertently using this strain, possibly confounding their results.

4 Drug Specificity or Drug Selectivity?

First off: drugs and chemical probes do not act specifically at any target. They can't be, as even high-affinity drugs, at some concentration will start interacting with secondary targets. So, at best, some compounds are "highly selective."

Typical ways to establish selectivity of a compound against a number of antitargets or subtypes are based on in vitro assays where individual test articles are tested for measures of binding affinity or functional activity at a target. The selectivity at a given target is usually qualified by the ratio between quantitative measures of drug effects at such targets. A fair number of these panels are available from commercial organizations, as well as government-funded agencies and academic institutes. However large as the number of such counterscreens may be, there are always targets that remain unknown or cannot be tested or modalities for which the in vitro test does not exist. For example, depending on the magnitude of their binding cooperativity, a compound may or may not show a signal competing with the orthosteric ligand in a radioligand binding assay because it binds to an allosteric site. In this case, a functional screen might appear to be a better choice (Christopoulos and Kenakin 2002). However, ruling out the possibility the compound acts as a silent allosteric modulator would require the combination of binding and functional screens. These may not always be available off-the-shelf, thus requiring additional research (Gregory et al. 2010).

Second, the selective actions of a compound depend on the concentration at which the test is conducted. This is a particularly significant issue when using cell-permeable chemical inhibitors to explore protein function, such as protein kinase inhibitors. In this area, determining compound selectivity is a difficult undertaking given the large number (>500) of such proteins encoded by the human genome and the highly conserved ATP binding site of protein kinases, with which most inhibitors interact. Due to the high importance of this field, a fairly large number of such inhibitors have become available from commercial suppliers and, most notably, offered online by academic organizations (http://www.kinase-screen.mrc.ac.uk/; http://www.chemicalprobes.org/; https://probeminer.icr.ac.uk/#/). Often, claims of "specificity" made for a compound toward a given kinase tested using broad panels have shown to be unrealistic, leading to erroneous conclusions regarding the participation of a kinase in a certain mechanism. Suggested criteria have been proposed for publication of studies using protein kinase inhibitors in intact cells (Cohen 2009). These include screening against large protein kinase panels (preferably >100), confirming functional effects using inhibitors in at least two distinct structural chemotypes, demonstrating that the effective concentrations are commensurate with those that prevent phosphorylation of an established physiological target, meaningful rank ordering of analogs, and replication at a different laboratory. For example, let's say compound A inhibits kinase X with an intrinsic affinity $K_i = 1$ nM and a 1,000-fold selectivity over undesired antitarget kinase Z ($K_i' = 1$ μM). Conducting an in vitro study in a comparable matrix at inhibitor concentration of 30 nM will selectively occupy the binding site of target X over the antitarget Z, and the effects measured may be considered as derived from kinase X. On the other hand, conducting the same study at inhibitor concentration of 10 μM will produce results derived from inhibiting kinases X (completely) and Z (highly) (Smyth and Collins 2009).

It is important to understand that therapeutic drugs used in the clinic do not need to be selective. Indeed, most are not, and the effects derived from cross-reactivity may even be beneficial to the therapeutic properties. However, during target identification and validation efforts, where a biological hypothesis is under examination, the risk of misinterpreting observations from a study due to cross-reactivity issues can easily be de-risked. In spite of efforts from the chemical biology community, a recent publication discusses the evidence of widespread continuing misuse of chemical probes and the challenges associated with the selection and use of tool compounds and suggests how biologists can and should be more discriminating in the probes they employ (Blagg and Workman 2017).

5 Species Selectivity

The term species selectivity refers to the observation that the effects of a compound may vary depending on the biological origin of the system where the test is conducted. The mechanistic origin of the observed differences may vary, including virtually every aspect of drug discovery, from lack of binding to the biological target due to differences in amino acid sequence, differences in nonspecific binding to matrix components, dissimilar drug absorption, stability to matrix components (e.g., plasma hydrolases) or drug metabolism by homologous (or even not expressed) proteins like cytochrome P450 or aldehyde oxidase, or simply differences in physiology across species (e.g., Norway rats vs. aged C57BL/6J mice) or even strains of the same animal species (e.g., Wistar vs. Sprague Dawley rats).

5.1 Animal Strain and Preclinical Efficacy Using In Vivo Models

Pharmacological responses to the action of a compound or the efficacious range of doses or exposures linked to effects observed during in vivo studies in preclinical species may vary when different strains of the same species are used. For example, C57BL/6J mice showed greater preference for saccharin and less avoidance of a cocaine-paired saccharin cue when compared with DBA/2J mice (Freet et al. 2013a). And in studies using opioid agonists in mice and rats, strain differences in the nociceptive sensitivity have been reported (Freet et al. 2013b). Wistar and Sprague-Dawley are most frequently the rat strains chosen in life sciences research, yet other outbred or inbred strains are sporadically used (Freet et al. 2013b; Festing 2014). A systematic study was recently conceived to evaluate the impact of rat strain (Lewis, Fischer F344, and Wistar Kyoto), as well as other important parameters, such as investigator, vendor, and pain assay, on the effects of morphine, a broadly studied analgesic used as a prototype in this work. Three experimental protocols were studied: hot plate, complete Freund's adjuvant (CFA)-induced inflammatory hyperalgesia, and locomotor activity. Findings revealed strain- and vendor-dependent differences in nociceptive thresholds and sensitivity to morphine – both before and after the inflammatory injury. The authors conclude that the translational value of work conducted using a specific strain or preclinical model is limited and propose ways to mitigate this risk (Hestehave et al. 2019).

5.2 Differences in Sequence of Biological Target

Modifications in the chemical nature of the building blocks of a target receptor may lead to major differences in the biological activity of chemical probes – affinity or efficacy. Best known examples come from the area of small molecules acting at protein targets. Such alterations may be due to mutations and manifest themselves as loss-of-function or gain-of-function mutations. For example, autosomal dominant inherited mutations in the gene encoding leucine-rich repeat kinase 2 (LRRK2) are the most common genetic causes of Parkinson's disease (Rideout 2017), while some have been linked to rare diseases (Platzer et al. 2017).

Species selectivity is a significant, yet not unsurmountable, challenge to drug discovery programs. For example, receptors in the purinergic family (including the adenosine (ARs), P2Y, and P2X receptors) are notorious for their proclivity to display reduced activity in rat compared with mouse or human receptors. Subtype selectivity values reported for some of the early tool compounds were revised following thorough pharmacological characterization across preclinical species. For example, tool compound **4**, broadly used to study the activation of the A_{2A} AR, has high AR subtype selectivity in rat and mouse, but it is reduced at human ARs (Jacobson and Muller 2016).

4

These observations are not rare, unfortunately. This suggests that best practice requires alignment and consistency when testing the activity for drug candidates or tool compounds using receptors corresponding to different relevant species and highlights the risk of extrapolating biological activity across species without proper experimental confirmation.

5.3 Metabolism

Lu AF09535 (**5**) is an mGluR5 negative allosteric modulator studied for the potential treatment of major depressive disorders. During early clinical development, an unanticipated low exposure of the drug was observed in humans, both by conventional bioanalytical methods and the highly sensitive microdosing of ^{14}C-labeled drug. This observation was attributed to extensive metabolism through a human-specific metabolic pathway since a corresponding extent of metabolism had not been seen in the preclinical species used (rat and dog). A combination of in vitro and in vivo models, including chimeric mice with humanized livers compared with control animals, showed that aldehyde oxidase (AO) was involved in the biotransformation of Lu AF09535 (Jensen et al. 2017). There is no equivalent protein to AO expressed in rat or dog. Cynomolgus monkey has been recommended as a suitable surrogate to study potential human AO metabolism (Hutzler et al. 2014).

5

6   What We Dose Is Not Always *Directly* Responsible for the Effects We See

Often during research using chemical tools, especially during in vivo studies, the experimental observations are interpreted as derived from the compound being dosed. However, this is not always the case, and thorough research requires establishing this direct link between the parent compound and the biological target in agreement with the hypothesis under testing.

As an example, compound **6** is an MCHR1 antagonist with potent in vitro activity. When tested in vivo in a sub-chronic diet-induced obesity model (DIO), it showed efficacy. However, a metabolite identification study indicated that significant amounts of primary alcohol **7** remained in circulation. Alcohol **7** was synthesized and it demonstrated potent in vitro inhibition of MCHR1 effects. In vivo, rat pharmacokinetics was very good and the compound crossed the BBB. As anticipated, when tested in a DIO model, it showed efficacy superior to its methoxy precursor **6**. Given its favorable physicochemical properties, compound **7** became NGD-4715 and reached Phase 1 clinical tests before being discontinued due to the observation of mechanistic effects altering sleep architecture (Moran 2001).

6

7

Furthermore, for compounds that are "well-behaved," the unbound concentrations measured at the hypothetical site of action should be commensurate with affinity or efficacy obtained using in vitro binding or functional assays. It must be reminded that, at the end of the day, these drug concentrations represent a certain receptor occupancy, which is expected to be consistent across different tests used to assess, and ideally translate, to the clinic.

Oftentimes these relationships are visualized through "exposuregrams," graphics where efficacious different in vitro, ex vivo, and in vivo tests are compared (Fig. 5).

6.1 Conditions Where In Vitro Potency Measures Do Not Align

Occasionally, in vitro measurements of compound potency derived using recombinant receptor protein will not overlap perfectly with those obtained using a cellular matrix. Due to the increased chemical and structural complexity of the cellular assay matrix compared with the recombinant milieu, often lower potency measures are determined. These tend to be attributed to poor cell membrane permeability, reduction in unbound concentrations due to increased nonspecific binding, or simply differences in concentrations of relevant binding partners between the two assays (e.g., ATP concentrations too high for kinase inhibitors).

On the other hand, occasionally a compound's potency increases in a cellular matrix compared with the recombinant assay. This rare effect may be explained by the formation of active drug metabolites or posttranslational modifications (cellular systems are metabolically able), the existence of unknown protein−protein interactions (purified recombinant systems ignore cellular localization and avoid contacts with other cell components such as proteins or nucleic acids), or intracellular localization of compound driven by transporter systems. Interpretation of these shifts is not always feasible, as target occupancy is not routinely established in cellular assays. For the case of allosteric drugs, a left shift in a cellular system suggests the presence of an endogenous component with positive cooperativity with the test article.

7 Chemical Modalities: Not All Drugs Are Created Equal

Target de-risking efforts benefit from experimentation with chemical tools of different nature. Due to advances in technology, probe compounds belonging to the group of biologics (e.g., proteins, oligonucleotides) tend to be faster to develop up to a quality good enough to conduct early testing of the biological hypothesis. On the other hand, developing a small molecule with the desired activity at the target and selectivity against antitargets usually requires a significant investment of financial and human resources. Table 2 compares some of the characteristics of these chemical modalities.

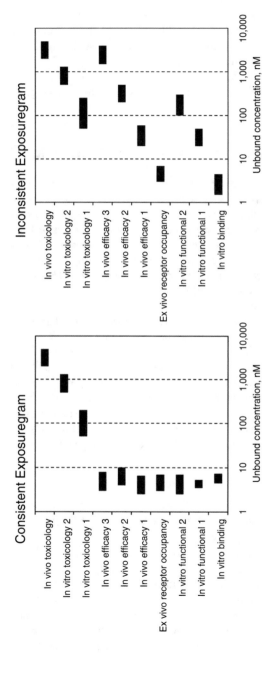

Fig. 5 Comparison of "exposuregrams" for two different compounds. The left graphic shows consistent efficacious unbound drug concentrations and a reasonable separation from unbound concentrations showing toxic effects. The graphic on the right shows inconsistent values for efficacious drug concentrations and overlapping with unbound concentrations leading to toxic effects

Table 2 Some characteristics of small-molecule drugs and biologics

	Small-molecule drugs	Biological drugs
Size	– Small (single molecule) – Low molecular weight (up to around 1,000 amu)	– Large (mixture of related molecules) – High molecular weight ($>>$1,000 amu)
Structure	Simple, well defined, independent of manufacturing process	Complex (heterogeneous), defined by the exact manufacturing process
Modification	Well defined	Many options
Manufacturing	– Produced by chemical synthesis – Predictable chemical process – Identical copy can be made – Low cost of goods	– Produced in living cell culture – Difficult to control from starting material to final API – Currently very difficult to ensure identical copy – High cost to produce
Characterization	Easy to characterize completely at chemical and physical levels	Not completely characterized in terms of chemical composition and heterogeneity
Stability	Relatively stable	Relatively unstable, sensitive to external conditions
Immunogenicity	Mostly non-immunogenic	Immunogenic
Brain penetration	Ways to establish brain penetration and receptor occupancy are well understood	Bifunctional antibodies developed for brain targets; for other types of biologics, establishing brain target engagement is still a challenge
Plasma half-life	Usually hours	From days up to months
Dosing regime	Mostly 1–2 times a day	May be longer than a month

Adapted from Doller (2017)

8   Receptor Occupancy and Target Engagement

As discussed in the introduction, the ultimate test of any drug discovery project is the exploration of the clinical hypothesis in humans. Ideally, the drug will show efficacy and safety and eventually become marketed. However, most drug discovery projects fail to meet these objectives. The second best outcome is then being able to rule out the biological hypothesis. Logically, this requires, at a minimum, being able to demonstrate sufficient clinical receptor occupancy (i.e., the drug is binding to the biological target in the tissue linked to the pathophysiology) at the site of action and ulterior target engagement (i.e., the expected functional effects derived from receptor occupancy are seen).

Oftentimes target engagement is inferred from a functional observation in a biological system upon treatment with a probe compound. This may be reasonable when studying systems with extensive prior biochemical or behavioral phenotype knowledge. However, when conducting research on novel biological systems and targets, an actual determination of the degree of receptor occupancy provides a much more robust line of support to conclude that the phenotype observed is produced by a molecular interaction between the test article and the receptor. For example, GPCR antagonists or agonists may elicit their functional effects (functional inhibition or activation, respectively) at very different degrees of receptor occupation. For the former, receptor occupancy typically parallels level of inhibition of a receptor, whereas agonists may exert functional effects occupying a very small fraction of a receptor (Michel and Seifert 2015), sometimes so low that it is hard to measure practically after consideration of the experimental error (Finnema et al. 2015).

9   Radioligands and PET Ligands as Chemical Tools

Chemical tools labeled with radioactive atoms are often used in drug discovery projects in a number of tasks, including binding affinity measurements, drug pharmacokinetics, ex vivo or in vivo receptor occupancy, or BBB permeability, among other uses.

PET agents are generally synthesized containing carbon-11 ($t_{1/2} = 20.4$ min) and/or fluorine-18 ($t_{1/2} = 109.7$ min) as radioisotope. Radioligands most often contain tritium ($t_{1/2} = 12.3$ years), carbon-14 ($t_{1/2} = 5,730$ years), phosphorous-32 ($t_{1/2} = 14$ days), sulfur-35 ($t_{1/2} = 87$ days), or iodine-125 ($t_{1/2} = 60$ days).

As explained for "cold" chemical tools (non-radiolabeled), it is important to understand that there is no hierarchy established between different types of radioligands. Each tool is designed with a fit-for-purpose mentality. In other words, a great human PET agent may or may not be acceptable for preclinical use, and the short half-life would most likely preclude its use as, for example, a radioligand binding assay.

Criteria have been developed to aid the discovery of high-quality novel PET ligands based on the physicochemical properties, brain permeability, and nonspecific binding for 62 clinically successful PET ligands and 15 unsuccessful radioligands considered as controls for undesired properties (Zhang et al. 2013). Properties chosen in this analysis are cLogP, cLogD, MW, TPSA, HBD, and pKa. It should be taken into consideration that this is just one approach to PET ligand design developed by one research group. Different approaches have been tested with successful results too. Target localization in the brain and its specific identity are important parameters to consider (Van de Bittner et al. 2014). Non-labeled ligands have also been used to measure in vivo receptor occupancy taking advantage of the increasing sensitivity of liquid chromatography coupled to mass spectrometry methods (LC-MS/MS). In a retrospective analysis, brain penetration, binding potential, and brain exposure kinetics were established for a number of non-labeled PET ligands using in vivo LC-MS/MS and compared with PET ligand performance in nonhuman primates and humans (Joshi et al. 2014).

A key parameter in the quality of a PET ligand is its selectivity at the biological target of interest. Best practice requires that this selectivity be thoroughly established using in vitro tests first and in vivo systems afterward to minimize the risk of misinterpreting experimental observations. An interesting example was recently reported for validation studies for the compound [^{11}C]-JNJ-42491293 (**8**), a PET ligand candidate to study the metabotropic glutamate 2 receptor (mGluR2), a drug target for CNS diseases (Leurquin-Sterk et al. 2017). Compound **8** has high affinity for the human mGluR2 receptor (IC_{50} around 9 nM). Preclinical studies conducted in Wistar rats demonstrated moderate brain penetration followed by distribution in brain regions consistent with those expected based on known expression of mGluR2. However, an additional unexpected observation indicated high retention in heart tissue. In order to explore this issue, the team conducted comparative PET studies using wild-type rats with their mGluR2 knockout counterparts. These studies indicated off-target binding in vivo to a yet-unidentified molecular target and highlight the importance of conducting in vitro and in vivo comparative studies to conduct a rigorous validation of PET radioligands.

8   **9**

10   Monoclonal Antibodies as Target Validation Tools

Monoclonal antibodies (mAbs) offer a compelling alternative to small molecules as tools in support of target validation. mAbs can potentially be generated with high affinity and selectivity against targets of interest with relative speed compared to small molecules. mAbs also have an advantage over small molecules, as they have the potential to disrupt thermodynamically stable large molecule interactions. While this effect has also been accomplished by small-molecule allosteric modulators, the task of compound optimization may be complex (Watson et al. 2005). In vivo, mAbs generally exhibit much longer half-lives than small molecules and therefore may be more practical in long-term target validation experiments as they can be administered infrequently.

While mAbs are usually used as antagonists to test the target of interest, it is also possible to generate agonistic mAbs. The use and potential challenges of immune agonist antibody design and their development as potential therapies for cancer treatment have been reviewed (Mayes et al. 2018).

10.1   Targets Amenable to Validation by mAbs

Since antibodies are large molecules normally secreted by B cells into the circulation, the target repertoire of mAbs is traditionally thought to be confined to extracellular or cell surface proteins. In more recent years, it has become apparent, somewhat surprisingly, that mAbs may also possess the ability to target intracellular proteins, including cytosolic proteins. The mechanism for this is still unclear, as mAbs exist in a topologically distinct compartment from the cytosol. Many cells express receptors for the Fc constant region of antibodies, allowing their uptake into the endolysosomal compartment of the cell. Certain cytosolic antigens may be targeted to the endolysosomal compartment by autophagocytosis. For instance, this may be true for the neuronal protein, tau, which aggregates and becomes targeted for autolysosomal degradation in the pathological state (Congdon et al. 2013; Sankaranarayanan et al. 2015).

In addition, cells express a high-affinity Fc receptor in the cytosol called TRIM21, which is important in mediating intracellular immunity against viral infections (Mallery et al. 2010). TRIM21 also contains a ubiquitin ligase domain, allowing antigen/antibody complexes to be targeted to the proteasome for degradation. Antibody tools may be used to target endogenous proteins for degradation by this pathway in target validation experiments (Yanamandra et al. 2013). The mechanisms by which mAbs enter the cytosol remain poorly understood.

10.2 The Four Pillars for In Vivo Studies

The concept of the four pillars of drug action (Bunnage et al. 2013) applies to antibodies as well as small molecules. For a soluble target, it is important to determine the concentration of the target in the target tissue, as well as the soluble ("free") fraction of the antibody achieved in that tissue. The free mAb concentration must significantly exceed the affinity of the mAb for its targets in order to ensure that the target is saturated by the mAb. For a cell surface target, it should be determined whether the target is released from the cell as part of the disease process. The soluble form of the target could act as a sink to sequester the mAb, and this will not necessarily be measurable by determining mAb concentration in the tissue of interest. Similarly, if the cell surface target is internalized upon mAb binding, this could rapidly deplete the mAb. However, this could be determined in an appropriate pharmacokinetic time-course dose-response study.

Another potential challenge is the development of immunogenicity to the mAb. That is, the host may raise antibodies against the test mAb, thereby neutralizing its activity and efficacy. Therefore, it is necessary to monitor an immune response against the biologics, particularly if no efficacy is observed. To reduce the risk of immunogenicity, the host of the test mAb should be the same species as the animal model (i.e., a mouse mAb should be used in a mouse model, but not a rat model, and vice versa).

10.3   Quality Control of Antibody Preparation

There are a variety of methods by which antibodies can be generated and purified. mAbs may be produced by hybridoma clones or by recombinant expression in an immortalized cell line. Purification of mAbs from supernatants can be performed with Protein A or G affinity columns and/or by size-exclusion chromatography (SEC). mAbs purified by Protein A/G columns alone may not be completely pure, as additional serum proteins may stick to the columns or antibodies. Furthermore, if regular serum is used to grow the cells, the Protein A/G column will also bind to the endogenous antibodies in the serum. This will make it more difficult to determine the concentration of the antibody of interest. Another potential pitfall is leaching of Protein A/G into the antibody preparation. This should be addressed directly.

SEC-purified antibodies are generally purer and more reliable. Not only does SEC separate mAbs from other contaminants but also allows for the purification of mAb monomers away from potential mAb aggregates. Aggregated antibody can potentially lead to artifacts, perhaps due to increased avidity for low-affinity antigens. This could result in off-target reactivity in binding assays (e.g., immunohistochemistry, ELISA) or even in in vitro or in vivo functional assays.

Another very important contaminant to consider is LPS or endotoxin. As discussed above, endotoxin is a component of the Gram-negative bacteria cell wall and a potent activator of mammalian immune cells. Sterilization does not remove endotoxin, so unless equipment and reagents are expressly endotoxin (or "pyrogen-")-free, antibody preparations may contain endotoxin, even if sterile (Witting and Moller 2011; Weinstein et al. 2008). Endotoxin will certainly interfere with any immune-based endpoints and may also affect other endpoints in vivo, such as cognitive or motor tests, if the animal is suffering from inflammation-induced sickness behavior (Weinstein et al. 2008; Remus and Dantzer 2016).

10.4   Isotype

The choice of antibody isotype may profoundly impact the efficacy of an antibody in a validation experiment. Different isotypes have different effector functions due to their differential effects on Fcγ receptors (Jonsson and Daeron 2012; Wes et al. 2016). Engaging Fcγ receptors trigger a variety of cellular responses, such as phagocytosis and proinflammatory cytokine release. In some cases, effector function is needed in order to see efficacy, such as when clearance of the antigen by immune cells is desirable. In other cases, effector function is not important, such as when the antibody effect is driven by occluding interaction of a ligand with its receptor. Effector function may also confound interpretation of results by causing an inflammatory response at the site of action. In these cases, using effector function null antibodies may be desirable.

It is important to ensure that any effect of the antibody is not due to effector function rather than engagement of the antigen. For instance, if the mAb target is expressed on an immune cell that also expresses Fcγ receptors, modulation of the cell may be via the Fcγ receptor rather than the target. Therefore, negative control antibodies should always be of the same isotype.

10.5   Selectivity

In order to correctly interpret a target validation experiment using a mAb, it is of course critical to use a mAb that is highly selective for the target. Selectivity is often determined by testing the antibody on a Western blot following SDS-PAGE and observing a band of the correct size. However, SDS-PAGE Western blots detect denatured proteins, and therefore there may be additional proteins that the antibody recognizes in the native state. Also, a band of the correct size is not definitive proof that the band is the target of interest. One approach to test for specificity is to compete the signal using peptides of the antigenic sequence. However, cross-reactivity may be due to a shared epitope, and an antigen peptide may compete this out as well. The best control is to test the antibody in a knockout animal or, if not available, ablate the target in cells using CRISPR, shRNA, or other knockdown technologies. Indeed, there have been examples of mAbs that recognized a band of the correct size on a Western blot, but when knockouts of the target became available, the band did not disappear. The best approach to be sure that you are correctly interrogating the target of interest is to use to multiple independent mAbs, ideally with different epitopes, against your target.

GPCRs, ligand-gated ion channels (LGICs), and transporters are key cell membrane-bound biological targets often studied for a number of potential CNS treatments. Biological research with these receptors often uses antibodies targeting them, even though their usefulness has been questioned on the basis of their selectivity. An increasing number of reports suggest lack of selectivity for a number of GPCRs and LGICs, including those from commercial sources (Michel et al. 2009; Berglund et al. 2008). Two often-applied criteria to assess antibody selectivity are the disappearance of staining upon addition of blocking peptide and observing distinct staining patters in tissues when testing antibodies against different receptor subtypes. While reasonable, these may be insufficient to assert antibody selectivity. Criteria thought to be reliable enough to demonstrate selectivity of GPCR antibodies have been proposed:

1. The staining disappears in immunohistochemical studies or immunoblots of tissues from genetic animals not expressing the receptor.
2. A major reduction of staining by a given antibody when using animals or cell lines treated with genetic tools to knockdown expression of a given receptor.
3. The target receptor yields positive staining when using transfection of multiple subtypes of a given receptor into the same host cell line, and it does not in the related subtypes.
4. Multiple antibodies against different epitopes of a GPCR (e.g., N-terminus, intracellular loop and C-terminus) show a very similar staining pattern in immunohistochemistry or immunoblotting (Bradbury and Plückthun 2015). The

issue of antibody quality and its impact on reproducibility of biological studies seem to be rather broad.

A report from 2008 states that less than half of ca. 5,000 commercial antibodies acted with the claimed selectivity at their specified targets. In addition, some manufacturers deliver consistently good antibodies, while others do not (Berglund et al. 2008). It is proposed that "if all antibodies were defined by their sequences and made recombinantly, researchers worldwide would be able to use the same binding reagents under the same conditions." To execute this strategy, two steps would be required. First is obtaining the sequences for widely used hybridoma-produced monoclonal antibodies. Indeed, it is proposed that polyclonal antibodies should be phased out of research entirely. Second, the research community should turn to methods that directly yield recombinant binding reagents that can be sequenced and expressed easily (Bradbury and Plückthun 2015). The practicality of this proposal appears a challenge in the short term.

In summary, while antibodies are broadly used as chemical tools to aid target validation, they are not free from issues that may lead to reproducibility issues when they are not optimally characterized.

11 Parting Thoughts

Over the last decade, major progress has been witnessed in the life sciences area. This could not be achieved without increased sophistication and understanding aiding the design of high-quality probe compounds, in an increasing number of chemical modalities. In turn, improved tools enabled the formulation of new questions to interrogate novel hypotheses, leading to heightened understanding of fundamental biology and pathophysiology, a key step in the discovery of new therapies for the treatment of diseases. Chemistry is part of the solution to disease treatment (no pun intended). Better chemistry understanding leads to better drugs.

Drug discovery projects require major commitments from society. Scientists dedicate decades of efforts and sacrifices. Investors risk billions of dollars. Patients are waiting and deserve the most ethical behaviors from all of us seeking to find new palliatives to ease their suffering. The right path forward is one of high-quality and rigorous scientific research.

References

Apicella MA (2008) In: DeLeo FR, Otto M (eds) Bacterial pathogenesis: methods and protocols. Humana Press, pp 3–13

Berglund L, Bjorling E, Oksvold P, Fagerberg L, Asplund A, Szigyarto CA, Persson A, Ottosson J, Wernerus H, Nilsson P, Lundberg E, Sivertsson A, Navani S, Wester K, Kampf C, Hober S, Ponten F, Uhlen M (2008) A genecentric human protein atlas for expression profiles based on antibodies. Mol Cell Proteomics 7:2019–2027. https://doi.org/10.1074/mcp.R800013-MCP200

Blagg J, Workman P (2017) Choose and use your chemical probe wisely to explore cancer biology. Cancer Cell 32:268–270. https://doi.org/10.1016/j.ccell.2017.07.010

Blanco M-J, La D, Coughlin Q, Newman CA, Griffin AM, Harrison BL, Salituro FG (2018) Breakthroughs in neuroactive steroid drug discovery. Bioorg Med Chem Lett 28:61–70

Bradbury A, Plückthun A (2015) Reproducibility: standardize antibodies used in research. Nature 518:27–29. https://doi.org/10.1038/518027a

Briske-Anderson MJ, Finley JW, Newman SM (1997) The influence of culture time and passage number on the morphological and physiological development of Caco-2 cells. Proc Soc Exp Biol Med 214:248–257

Bunnage ME, Chekler EL, Jones LH (2013) Target validation using chemical probes. Nat Chem Biol 9:195–199. https://doi.org/10.1038/nchembio.1197

Chatterjee R (2007) Cell biology. Cases of mistaken identity. Science 315:928–931. https://doi.org/10.1126/science.315.5814.928

Christopoulos A, Kenakin T (2002) G protein-coupled receptor allosterism and complexing. Pharmacol Rev 54:323–374

Churcher I (2017) Protac-induced protein degradation in drug discovery: breaking the rules or just making new ones? J Med Chem 61:444–452

Cohen P (2009) Guidelines for the effective use of chemical inhibitors of protein function to understand their roles in cell regulation. Biochem J 425:53–54. https://doi.org/10.1042/bj20091428

Congdon EE, Gu J, Sait HB, Sigurdsson EM (2013) Antibody uptake into neurons occurs primarily via clathrin-dependent Fcgamma receptor endocytosis and is a prerequisite for acute tau protein clearance. J Biol Chem 288:35452–35465. https://doi.org/10.1074/jbc.M113.491001

Darveau RP, Hancock R (1983) Procedure for isolation of bacterial lipopolysaccharides from both smooth and rough Pseudomonas aeruginosa and Salmonella typhimurium strains. J Bacteriol 155:831–838

Doan TL, Pollastri M, Walters MA, Georg GI (2011) In: Annual reports in medicinal chemistry, vol 46. Elsevier, pp 385–401

Doller D (2017) Allosterism in drug discovery, RSC drug discovery series no. 56. Royal Society of Chemistry, Cambridge

Drexler HG, Uphoff CC (2002) Mycoplasma contamination of cell cultures: incidence, sources, effects, detection, elimination, prevention. Cytotechnology 39:75–90. https://doi.org/10.1023/A:1022913015916

Drexler HG, Uphoff CC, Dirks WG, MacLeod RA (2002) Mix-ups and mycoplasma: the enemies within. Leuk Res 26:329–333

Esquenet M, Swinnen JV, Heyns W, Verhoeven G (1997) LNCaP prostatic adenocarcinoma cells derived from low and high passage numbers display divergent responses not only to androgens but also to retinoids. J Steroid Biochem Mol Biol 62:391–399

Evans HM, Bowman FB, Winternitz M (1914) An experimental study of the histogenesis of the miliary tubercle in vitally stained rabbits. J Exp Med 19:283–302

Festing MF (2014) Evidence should trump intuition by preferring inbred strains to outbred stocks in preclinical research. ILAR J 55:399–404. https://doi.org/10.1093/ilar/ilu036

Finnema SJ, Scheinin M, Shahid M, Lehto J, Borroni E, Bang-Andersen B, Sallinen J, Wong E, Farde L, Halldin C, Grimwood S (2015) Application of cross-species PET imaging to assess neurotransmitter release in brain. Psychopharmacology 232:4129–4157. https://doi.org/10.1007/s00213-015-3938-6

Freet CS, Arndt A, Grigson PS (2013a) Compared with DBA/2J mice, C57BL/6J mice demonstrate greater preference for saccharin and less avoidance of a cocaine-paired saccharin cue. Behav Neurosci 127:474–484. https://doi.org/10.1037/a0032402

Freet CS, Wheeler RA, Leuenberger E, Mosblech NA, Grigson PS (2013b) Fischer rats are more sensitive than Lewis rats to the suppressive effects of morphine and the aversive kappa-opioid agonist spiradoline. Behav Neurosci 127:763–770. https://doi.org/10.1037/a0033943

Gregory KJ, Malosh C, Turlington M, Morrison R, Vinson P, Daniels JS, Jones C, Niswender CM, Conn PJ, Lindsley CW, Stauffer SR (2010) Probe reports from the NIH Molecular Libraries Program. National Center for Biotechnology Information

Hestehave S, Abelson KSP, Bronnum Pedersen T, Munro G (2019) The analgesic efficacy of morphine varies with rat strain and experimental pain model: implications for target validation efforts in pain drug discovery. Eur J Pain 23(3):539–554. https://doi.org/10.1002/ejp.1327

Hopkins A, Groom C (2003) Small molecule—protein interactions. Springer, pp 11–17

Hutzler JM, Cerny MA, Yang YS, Asher C, Wong D, Frederick K, Gilpin K (2014) Cynomolgus monkey as a surrogate for human aldehyde oxidase metabolism of the EGFR inhibitor BIBX1382. Drug Metab Dispos 42:1751–1760. https://doi.org/10.1124/dmd.114.059030

Jacobson KA, Muller CE (2016) Medicinal chemistry of adenosine, P2Y and P2X receptors. Neuropharmacology 104:31–49. https://doi.org/10.1016/j.neuropharm.2015.12.001

Jensen KG, Jacobsen AM, Bundgaard C, Nilausen DO, Thale Z, Chandrasena G, Jorgensen M (2017) Lack of exposure in a first-in-man study due to aldehyde oxidase metabolism: investigated by use of 14C-microdose, humanized mice, monkey pharmacokinetics, and in vitro methods. Drug Metab Dispos 45:68–75. https://doi.org/10.1124/dmd.116.072793

Johnson K, Perry M (1976) Improved techniques for the preparation of bacterial lipopolysaccharides. Can J Microbiol 22:29–34

Jonsson F, Daeron M (2012) Mast cells and company. Front Immunol 3:16. https://doi.org/10.3389/fimmu.2012.00016

Joshi EM, Need A, Schaus J, Chen Z, Benesh D, Mitch C, Morton S, Raub TJ, Phebus L, Barth V (2014) Efficiency gains in tracer identification for nuclear imaging: can in vivo LC-MS/MS evaluation of small molecules screen for successful PET tracers? ACS Chem Neurosci 5:1154–1163. https://doi.org/10.1021/cn500073j

Kagemann G, Henrich B, Kuhn M, Kleinert H, Schnorr O (2005) Impact of mycoplasma hyorhinis infection on L-arginine metabolism: differential regulation of the human and murine iNOS gene. Biol Chem 386:1055–1063. https://doi.org/10.1515/bc.2005.121

Lawrence B, Bashiri H, Dehghani H (2010) Cross comparison of rapid mycoplasma detection platforms. Biologicals 38:218–223. https://doi.org/10.1016/j.biologicals.2009.11.002

Lee A, Budac D, Charych E, Bisulco S, Zhou H, Moller T, Campbell B (2013) Control standard endotoxin as a potential tool to reduce the variability observed with lipopolysaccharide. Brain Behav Immun 32:e27

Leurquin-Sterk G, Celen S, Van Laere K, Koole M, Bormans G, Langlois X, Van Hecken A, Te Riele P, Alcazar J, Verbruggen A, de Hoon J, Andres JI, Schmidt ME (2017) What we observe in vivo is not always what we see in vitro: development and validation of 11C-JNJ-42491293, a novel Radioligand for mGluR2. J Nucl Med 58:110–116. https://doi.org/10.2967/jnumed.116.176628

Lincoln CK, Gabridge MG (1998) Cell culture contamination: sources, consequences, prevention, and elimination. Methods Cell Biol 57:49–65

MacLeod RA, Dirks WG, Reid YA, Hay RJ, Drexler HG (1997) Identity of original and late passage Dami megakaryocytes with HEL erythroleukemia cells shown by combined cytogenetics and DNA fingerprinting. Leukemia 11:2032–2038

MacLeod RA, Dirks WG, Matsuo Y, Kaufmann M, Milch H, Drexler HG (1999) Widespread intraspecies cross-contamination of human tumor cell lines arising at source. Int J Cancer 83:555–563

Mallery DL, McEwan WA, Bidgood SR, Towers GJ, Johnson CM, James LC (2010) Antibodies mediate intracellular immunity through tripartite motif-containing 21 (TRIM21). Proc Natl Acad Sci U S A 107:19985–19990. https://doi.org/10.1073/pnas.1014074107

Markovic O, Markovic N (1998) Cell cross-contamination in cell cultures: the silent and neglected danger. In Vitro Cell Dev Biol Anim 34:1–8. https://doi.org/10.1007/s11626-998-0040-y

Masters JR (2000) Human cancer cell lines: fact and fantasy. Nat Rev Mol Cell Biol 1:233–236. https://doi.org/10.1038/35043102

Mayes PA, Hance KW, Hoos A (2018) The promise and challenges of immune agonist antibody development in cancer. Nat Rev Drug Discov 17:509–527. https://doi.org/10.1038/nrd.2018.75

Michel MC, Seifert R (2015) Selectivity of pharmacological tools: implications for use in cell physiology. A review in the theme: cell signaling: proteins, pathways and mechanisms. Am J Physiol Cell Physiol 308:C505–C520. https://doi.org/10.1152/ajpcell.00389.2014

Michel MC, Wieland T, Tsujimoto G (2009) Springer

Moran AP (2001) In: Mobley HLT, Mendz GL, Hazell SL (eds) Helicobacter pylori: physiology and genetics. ASM Press

Nelson-Rees WA, Flandermeyer RR, Hawthorne PK (1974) Banded marker chromosomes as indicators of intraspecies cellular contamination. Science 184:1093–1096

Nelson-Rees WA, Daniels DW, Flandermeyer RR (1981) Cross-contamination of cells in culture. Science 212:446–452

Platzer K, Yuan H, Schutz H, Winschel A, Chen W, Hu C, Kusumoto H, Heyne HO, Helbig KL, Tang S, Willing MC, Tinkle BT, Adams DJ, Depienne C, Keren B, Mignot C, Frengen E, Stromme P, Biskup S, Docker D, Strom TM, Mefford HC, Myers CT, Muir AM, LaCroix A, Sadleir L, Scheffer IE, Brilstra E, van Haelst MM, van der Smagt JJ, Bok LA, Moller RS, Jensen UB, Millichap JJ, Berg AT, Goldberg EM, De Bie I, Fox S, Major P, Jones JR, Zackai EH, Abou Jamra R, Rolfs A, Leventer RJ, Lawson JA, Roscioli T, Jansen FE, Ranza E, Korff CM, Lehesjoki AE, Courage C, Linnankivi T, Smith DR, Stanley C, Mintz M, McKnight D, Decker A, Tan WH, Tarnopolsky MA, Brady LI, Wolff M, Dondit L, Pedro HF, Parisotto SE, Jones KL, Patel AD, Franz DN, Vanzo R, Marco E, Ranells JD, Di Donato N, Dobyns WB, Laube B, Traynelis SF, Lemke JR (2017) GRIN2B encephalopathy: novel findings on phenotype, variant clustering, functional consequences and treatment aspects. J Med Genet 54:460–470. https://doi.org/10.1136/jmedgenet-2016-104509

Ray A, Redhead K, Selkirk S, Poole S (1991) Variability in LPS composition, antigenicity and reactogenicity of phase variants of Bordetella pertussis. FEMS Microbiol Lett 79:211–218

Remus JL, Dantzer R (2016) Inflammation models of depression in rodents: relevance to psychotropic drug discovery. Int J Neuropsychopharmacol 19:pyw028. https://doi.org/10.1093/ijnp/pyw028

Rideout HJ (2017) Neuronal death signaling pathways triggered by mutant LRRK2. Biochem Soc Trans 45:123–129. https://doi.org/10.1042/bst20160256

Sankaranarayanan S, Barten DM, Vana L, Devidze N, Yang L, Cadelina G, Hoque N, DeCarr L, Keenan S, Lin A, Cao Y, Snyder B, Zhang B, Nitla M, Hirschfeld G, Barrezueta N, Polson C, Wes P, Rangan VS, Cacace A, Albright CF, Meredith J Jr, Trojanowski JQ, Lee VM, Brunden KR, Ahlijanian M (2015) Passive immunization with phospho-tau antibodies reduces tau pathology and functional deficits in two distinct mouse tauopathy models. PloS One 10: e0125614. https://doi.org/10.1371/journal.pone.0125614

Saunders NR, Dziegielewska KM, Møllgård K, Habgood MD (2015) Markers for blood-brain barrier integrity: how appropriate is Evans blue in the twenty-first century and what are the alternatives? Front Neurosci 9:385

Smyth LA, Collins I (2009) Measuring and interpreting the selectivity of protein kinase inhibitors. J Chem Biol 2:131–151. https://doi.org/10.1007/s12154-009-0023-9

Specht CG, Schoepfer R (2001) Deletion of the alpha-synuclein locus in a subpopulation of C57BL/6J inbred mice. BMC Neurosci 2:11–11. https://doi.org/10.1186/1471-2202-2-11

Tarrant J, Hodgetts K, Chenard B, Krause J, Doller D (2017) The discovery of the MCH-1 receptor antagonist NGD-4715 for the potential treatment of obesity. Compr Med Chem III:488–515. https://doi.org/10.1016/B978-0-12-409547-2.13785-0

The National Academies of Sciences, Engineering, and Medicine (2017) Fostering integrity in research. National Academies Press

Van de Bittner GC, Ricq EL, Hooker JM (2014) A philosophy for CNS radiotracer design. Acc Chem Res 47:3127–3134. https://doi.org/10.1021/ar500233s

Watson C, Jenkinson S, Kazmierski W, Kenakin T (2005) The CCR5 receptor-based mechanism of action of 873140, a potent allosteric noncompetitive HIV entry inhibitor. Mol Pharmacol 67:1268–1282. https://doi.org/10.1124/mol.104.008565

Weinstein JR, Swarts S, Bishop C, Hanisch U-K, Möller T (2008) Lipopolysaccharide is a frequent and significant contaminant in microglia-activating factors. Glia 56:16–26. https://doi.org/10.1002/glia.20585

Wenger SL, Senft JR, Sargent LM, Bamezai R, Bairwa N, Grant SG (2004) Comparison of established cell lines at different passages by karyotype and comparative genomic hybridization. Biosci Rep 24:631–639. https://doi.org/10.1007/s10540-005-2797-5

Wes PD, Sayed FA, Bard F, Gan L (2016) Targeting microglia for the treatment of Alzheimer's Disease. Glia 64:1710–1732. https://doi.org/10.1002/glia.22988

Witting A, Moller T (2011) Microglia cell culture: a primer for the novice. Methods Mol Biol 758:49–66. https://doi.org/10.1007/978-1-61779-170-3_4

Workman P (2003) Overview: translating Hsp90 biology into Hsp90 drugs. Curr Cancer Drug Targets 3:297–300

Yanamandra K, Kfoury N, Jiang H, Mahan TE, Ma S, Maloney SE, Wozniak DF, Diamond MI, Holtzman DM (2013) Anti-tau antibodies that block tau aggregate seeding in vitro markedly decrease pathology and improve cognition in vivo. Neuron 80:402–414. https://doi.org/10.1016/j.neuron.2013.07.046

Yu H, Cook TJ, Sinko PJ (1997) Evidence for diminished functional expression of intestinal transporters in Caco-2 cell monolayers at high passages. Pharm Res 14:757–762

Zhang L, Villalobos A, Beck EM, Bocan T, Chappie TA, Chen L, Grimwood S, Heck SD, Helal CJ, Hou X, Humphrey JM, Lu J, Skaddan MB, McCarthy TJ, Verhoest PR, Wager TT, Zasadny K (2013) Design and selection parameters to accelerate the discovery of novel central nervous system positron emission tomography (PET) ligands and their application in the development of a novel phosphodiesterase 2A PET ligand. J Med Chem 56:4568–4579. https://doi.org/10.1021/jm400312y

Genetic Background and Sex: Impact on Generalizability of Research Findings in Pharmacology Studies

Stacey J. Sukoff Rizzo, Stephanie McTighe, and David L. McKinzie

Contents

Abstract

Animal models consisting of inbred laboratory rodent strains have been a powerful tool for decades, helping to unravel the underpinnings of biological problems and employed to evaluate potential therapeutic treatments in drug discovery. While inbred strains demonstrate relatively reliable and predictable responses, using a single inbred strain alone or as a background to a mutation is analogous to running a clinical trial in a single individual and their identical twins. Indeed, complex etiologies drive the most common human diseases, and a single inbred strain that is a surrogate of a single genome, or data generated from a single sex, is not representative of the genetically diverse patient populations. Further, pharmacological and toxicology data generated in otherwise healthy animals may not

S. J. Sukoff Rizzo (✉)
University of Pittsburgh School of Medicine, Pittsburgh, PA, USA
e-mail: rizzos@pitt.edu

S. McTighe
Sage Therapeutics, Cambridge, MA, USA

D. L. McKinzie
Indiana University School of Medicine, Indianapolis, IN, USA

© The Author(s) 2019   147
A. Bespalov et al. (eds.), *Good Research Practice in Non-Clinical Pharmacology
and Biomedicine*, Handbook of Experimental Pharmacology 257,
https://doi.org/10.1007/164_2019_282

translate to disease states where physiology, metabolism, and general health are compromised. The purpose of this chapter is to provide guidance for improving generalizability of preclinical studies by providing insight into necessary considerations for introducing systematic variation within the study design, such as genetic diversity, the use of both sexes, and selection of appropriate age and disease model. The outcome of implementing these considerations should be that reproducibility and generalizability of significant results are significantly enhanced leading to improved clinical translation.

Keywords
Animal models · Genetic diversity · Pharmacodynamics · Pharmacokinetics · Sex

1   Introduction

There are many perspectives on what defines an "animal model," but at the most fundamental level, it reflects an animal with a disease or condition with either face or construct validity to that observed in humans. Spontaneous animal models represent the truest form of the definition and are best exemplified by cross-species diseases such as cancer and diabetes where a particular species naturally develops the condition as observed in humans. However even in these models with high face, and seemingly construct validity, care must be taken when extrapolating from the animal phenotype to the human disease as the underlying mechanisms driving the disease may not be identical across species.

Animal models serve two primary purposes. The first use of animal models is to elucidate biological mechanisms and processes. A key assumption in this approach is that the animal species being examined has comparable enough physiology to reasonably allow for extrapolation to human biology and disease states. An extension of the first purpose is to use animal models for estimating efficacy and safety of new therapeutic treatments for alleviating human disorders. In both of these uses, the fidelity of the animal model is critically dependent upon the homology of the physiology between the animal model and human. The best model for human is human, and the greater divergence from human across the phylogenetic scale (e.g., nonhuman primates > rodents > zebrafish > drosophila) introduces increasingly larger gaps in genetic and physiological homology. For complex human-specific disorders such as schizophrenia or Alzheimer's disease, our confidence in findings from animal models must be guarded as there is not a spontaneous animal model of these human conditions. For instance, besides humans, there is no animal that spontaneously exhibits Aβ plaques and neurofibrillary tangles that define the pathology of Alzheimer's disease. Moreover, the complex spectrum of cognitive dysfunction and neuropsychiatric comorbidities that these diseases produce cannot be fully recapitulated or assessed (e.g., language impairment) in lower animal species. In such cases, animal models are relegated to attempts in simulating specific symptoms of the disorder (e.g., increasing striatal dopamine in rodents to model the striatal hyperdopaminergia observed in schizophrenia patients and thought to underlie the emergence of positive symptoms) or to model specific pathological processes observed in the human disease (e.g., generation of amyloid precursor protein overexpressing mice to model the Aβ deposition seen in Alzheimer's disease patients). In this latter example, it is important to note that the translation of transgenic mice Aβ deposition and mechanisms that reduce its accumulation have translated well into human AD patients; however, because this is an incomplete representation of the disease, agents that reduce Aβ deposition in both animals and human AD patients have yet to prove successful in delaying disease progression.

Reproducibility and generalizability are two aspects of preclinical research that have come under much scrutiny over the last several years. Examples of failures to reproduce research findings, even in high-impact journals, are numerous and well described in the literature (Jarvis and Williams 2016). Perhaps the most obvious factor impacting across-lab reproducibility are deficiencies to note important methodological variables of the study. As we discuss later in this chapter, it is surprising how often key experimental variables such as specific strain or sex of animal used are omitted in the methods section. In a direct attempt to improve scientific reporting practices, initiatives such as use of the ARRIVE guidelines (Kilkenny et al. 2010) have been instituted across the majority of scientific journals. Such factors can also affect intra-lab reproducibility; for instance, when a particular student that ran the initial study has left the lab or the lab itself has relocated to another institution and the primary investigator reports challenges in reestablishing the model.

Challenges in generalizability of research findings are best exemplified by noted failures in the realm of drug development in which a novel compound exhibits robust efficacy in putative animal models of a human condition but fails to demonstrate therapeutic benefit in subsequent clinical trials. Indeed, medication development for Alzheimer's disease has a remarkable failure rate of over 99% with a large percentage of drug development terminations attributed to lack of efficacy in Phase II or Phase III clinical trials (Cummings et al. 2014).

It is interesting to speculate that improvements in reproducibility of preclinical research may not necessarily translate into improved generalizability to the human condition (see Würbel 2002). For instance, close adherence in using the same age, substrain of mouse, husbandry conditions, and experimental details should improve the likelihood of reproducing another lab's findings. However, it also follows that if a reported research finding is highly dependent upon a specific experimental configuration and the finding is lost with subtle procedural variations, then how likely is the finding to translate to the human condition? Humans are highly heterogeneous in their genetic composition and environmental determinants, often resulting in subpopulations of patients that are responsive to certain treatments and others that are described as treatment resistant. In preclinical research the best balance of improving both reproducibility and generalizability is to institute the inclusion of both sexes and incorporation of another strain or species. This approach will most certainly reduce the number of positive findings across these additional variables, but those findings that are consistent and robust will likely result in increased reproducibility across labs and to translate into clinical benefit. In the sections that follow, we highlight the importance of genetic background and sex in conducting preclinical research.

2   Genetic Background: The Importance of Strain and Substrain

Dating back to the early 1900s, researchers have recognized the value of genetic uniformity and stability of inbred strains, which have provided such benefits as reducing study variability and needed samples sizes and improving reliability of data. To date more than 20 Nobel Prizes have resulted from work in inbred strains, and this knowledge has provided significant medical and health benefits (Festing 2014). Certainly, it continues to be an acceptable strategy to conduct research on a single inbred strain of mice, provided that the context of the results is reported to not suggest that the data are generalizable to other strains and species (e.g., humans). A single inbred strain is not representative of the genetically diverse patient populations and is instead representative of a single genome. Moreover, even different substrains of a common strain of mice (e.g., C57BL/6J, C57BL/6N, C57BL/6NTac) exhibit unique genetic dispositions resulting in surprisingly divergent phenotypes (reviewed in Casellas 2011). Therefore, a major constraint in translational research has been the common practice of limiting preclinical pharmacology studies to that of a single strain of mice.

Within the context of rodent studies, one example where lack of generalizability of strain, substrain, and sex has been well documented is the rodent experimental autoimmune encephalomyelitis (EAE) model of multiple sclerosis (MS). MS is an autoimmune disease caused by demyelination in the CNS, which results in a spectrum of clinical presentations accompanied by progressive neuromuscular disorders and paralysis (reviewed in Summers deLuca et al. 2010). In mice, immunization with myelin/oligodendrocyte glycoprotein peptide induces EAE; however the variability of disease presentation across mouse models has been a major hindrance for facilitating drug development. In line with the genetic contributions to MS in human patients, mouse strains and substrains are genetically and phenotypically divergent which introduces heterogeneous loading of risk alleles and variations in phenotypes that contribute to the variability in disease onset and severity (Guapp et al. 2003). The MS field is not unique to the challenges of experimental variability resulting from the choice of genetic background in their rodent model and has been documented in most fields of study (Grarup et al. 2014; Jackson et al. 2015; Loscher et al. 2017; Nilson et al. 2000). While known for decades that mouse substrains are genetically and phenotypically diverse from each other, many in the research community are still not aware of this important caveat and the implication on experimental findings.

Case in point, the C57BL/6 mouse strain is one of the most common and widely used inbred strains with many substrains derived from the original lineage and now maintained as separate substrain colonies. The C57BL/6J line originated at the Jackson Laboratory by C.C. Little in the 1920s, and in the 1950s a cohort of mice were shipped to the National Institutes of Health where a colony was established and aptly named C57BL/6N (the suffix "N" refers to the NIH colony, while the "J" suffix refers to the Jackson Laboratory colony) (reviewed in Kiselycznyk and Holmes 2011). At some point spontaneous mutations (i.e., genetic drift) occurred in each of these colonies resulting in these two substrains becoming genetically distinct from each other with recent reports citing >10,000 putative and 279 confirmed variant differences as well as several phenotypic differences between C57BL/6 substrains (Keane et al. 2011; Simon et al. 2013). These genetic and phenotypic differences between substrains are not unique to C57BL/6 as 129 substrains, among others, and also have similar genetic diversity issues that must be considered when reporting and extrapolating research (Kiselycznyk and Holmes 2011). Important to note is that substrain nomenclature alone is not the sole information that identifies genetic and phenotypic diversity. Individual or private colonies established for >20 generations either at a commercial vendor or an academic institution are considered a substrain and hence must adhere to the guidelines for nomenclature of mouse and rat strains as established by the International Committee on Standardized Genetic Nomenclature for Mice (reviewed in Sundberg and Schofield 2010). Laboratory code which follows substrain notation annotates for strain/substrain source including commercial vendor (e.g., C57BL/6NHsd and C57BL/6NTac, respectively, for Harlan and Taconic) and is a critical piece of information to researchers that a substrain may have further genetic variation, as in the case for C57BL/6N, than the original NIH colony. The implication on research findings where failure to understand the role of substrain differences, as well as failures to prevent inadvertent backcrossing of substrains, has been highlighted recently (Mahajan et al. 2016; Bourdi et al. 2011; McCracken et al. 2017). In one example, Bourdi and colleagues reported that JNK2−/− knockout mice were more susceptible than their WT controls to acetaminophen-induced liver injury which was in contrast to findings from other laboratories demonstrating that JNK2−/− and inhibitors of JNK were protective from acetaminophen-induced liver injury (Bourdi et al. 2011). Through careful retrospective analysis, the researchers were able to determine that backcrossing on two different background substrains conferred either toxicity or protective effects (Bourdi et al. 2011).

This issue of genetic drift is not unique to mice. For instance, in the study of hypertension and attention-deficit/hyperactivity disorder (ADHD), one of the most studied rat models are the spontaneously hypertensive (SHR) and Wistar Kyoto (WKY) ratlines. In terms of ADHD, the SHR rats display symptoms of inattention, hyperactivity, and impulsiveness in various behavioral paradigms (Sagvolden et al. 2009). However like the C57BL/6 substrains, numerous SHR and WKY substrains have been generated over the years. The SHR ratline was derived originally from a WKY male with marked hypertension and a female with moderate blood pressure elevations. Brother-sister matings continued with selection pressure for spontaneous hypertension. The SHR line arrived at the National Institutes of Health (NIH) in 1966 from the Kyoto School of Medicine. From the NIH colony (SHR/N), SHR lines were derived by Charles River, Germany (SHR/NCrl), and the Møllegaard Breeding Centre, Denmark (SHR/NMol), as well as other institutions over the years. The SHR rat strains exhibit an ADHD-like phenotype, whereas the WKY line serves as a normative control. A problem exists, in that, while the WKY strain was established from the same parental Wistar stock as the SHR line, there is considerable genetic variability among WKY strains because the WKY breeding stock was not fully inbred prior to being distributed to different institutions for breeding which resulted in accelerated genetic drift. A further issue for using the WKY strain as a genetic and behavioral control for the SHR strain is that the inbreeding for the WKY strain was initiated over 10 years later than that of the SHR strain which calls into question the validity of the WKY rats as a proper control for findings in SHR rats (Louis and Howes 1990). As one might expect from such genetic diversity in SHR and WKY lines, findings from both cardiovascular blood pressure and ADHD phenotypes have at times been contradictory, and much commentary has been made about the appropriate selection of controls when studying phenotypes associated with these strains of rats (St. Lezin et al. 1992).

3   Importance of Including Sex as a Variable

The X and Y chromosomes are not the only difference that separates a female from a male. In preclinical studies there has been a pervasively, flawed assumption that male and female rodents have similar phenotypes. Publications that include such general statements as "data were combined for sex since no sex effect was observed" without the inclusion of the analysis, or simply reporting "data not shown" for the evaluation of effects of sex, are unacceptable. From basic physiological phenotypes (e.g., body weight, lean and fat mass) to any number of neuroendocrine, immune, and behavioral phenotypes beyond reproductive behaviors, males and females differ (reviewed in Hughes 2007; Karp et al. 2017). Furthermore, many human diseases affect males and females differently, whereas the influence of sex can affect disease susceptibility, symptom presentation and progression, and treatment outcomes. Well-documented sex differences exist for cardiovascular disease, autoimmune diseases, chronic pain, and neuropsychiatric disorders with females generally having greater incidences than males (reviewed in Regitz-Zagrosek 2012; IOM 2011). Therefore, ignorance of sex-specific effects in study design, phenotypes, pharmacokinetics, pharmacodynamic measures, or interpretation of data without sex as a covariate are failing to provide accuracy in reporting of the data. To this end, in 2014 the NIH issued a directive to ensure that both male and female subjects are represented in preclinical studies, an extension of the 1993 initiative to include women as participants in clinical trials receiving NIH funding (Clayton and Collins 2014).

With respect to animal models used in pharmacology experiments, sex differences in disease presentation and progression have also been reported. For example, while women have a higher prevalence of chronic pain and related disorders, preclinical studies have largely focused on male subjects. Problematically, after hundreds of studies historically employed male mice to study nociceptive responses mediated by the toll-like 4 receptor (TLR4), and subsequent pharmacology studies targeting TLR4 for analgesia, it was later discovered that the involvement of TLR4 in pain behaviors in male mice was dependent on testosterone (Sorge et al. 2011). Therefore, these results and any potential therapeutics for the treatment of pain with a mechanism of action targeting TLR4 could not be generalized to both sexes (Sorge et al. 2011). In another example, the NOD mouse model of Type 1 diabetes has a higher incidence and an earlier onset of diabetes symptoms in females than males (Leiter 1997). Consequently, female NOD mice are much more widely used than males although the incidence in the clinic is nearly 1:1 for males/ females which may present a conundrum when potential novel treatments are only studied in a single sex as in the TLR4 experiments highlighted above. Furthermore, in neuropsychiatric disorders whereas major depressive disorder, for example, has a higher incidence in females than males, preclinical studies have largely used only males for testing – even though sex differences in rodent emotional behavior exist (Dalla et al. 2010; Kreiner et al. 2013; Kokras et al. 2015; Laman-Maharg et al. 2018). One of the more common arguments made for not including female subjects in preclinical studies is that they have larger variability, likely contributed to by the estrus cycle. However, a meta-analysis of 293 publications revealed that variability in endpoints using female mice was not greater than those in males, inclusive of variations in the estrus cycle as a source of variability in the females (Becker et al. 2016; Prendergast et al. 2014; Mogil and Chanda 2005). There are, however, baseline differences for males versus females across behavioral phenotypes that further highlight the need to study both sexes and with data analyzed within sex when drug treatment is evaluated in both sexes.

4 Pharmacokinetic and Pharmacodynamic Differences Attributable to Sex

In addition to the observation of sex differences across disease and behavioral phenotypes, sex differences are also commonly observed in pharmacokinetic (PK) and drug efficacy studies; yet for many years test subjects in both clinical and preclinical studies have most commonly been male. A survey of the pain and neuroscience field in the early 1990s revealed that only 12% of published papers had used both male and female subjects and 45% failed to reveal the sex of the subjects included in the studies (Berkley 1992). A later study building on this revealed that between 1996 and 2005 although researchers now reliably reported the sex of their preclinical subjects (97%), most studies (79%) were still performed on male animals (Mogil and Chanda 2005). Although the translatability of preclinical sex differences to human may not always be clear-cut, assessments of these parameters in both sexes can provide additional information during phenotyping and genetic studies, as well as the drug discovery and development process.

In the drug discovery field, there are multiple examples in the clinical literature of sex differences in both measured exposure and pharmacological effect in response to novel drugs. A meta-analysis of 300 new drug applications (NDAs) reviewed by the FDA between 1995 and 2000 showed that 163 of these included a PK analysis by sex. Of these 163, 11 studies showed greater than 40% difference in PK parameters between males and females (Anderson 2005). There are important implications for sex differences in exposure levels. For example, zolpidem (Ambien[®]) results in exposure levels 40–50% higher in females when administered sublingually (Greenblatt et al. 2014). These sex differences in exposure levels for zolpidem were also observed in rats, with maximal concentration (Cmax) and area under the curve (AUC) both significantly higher in females relative to males (Peer et al. 2016). While Ambien was approved in 1992, in 2013 the FDA recommended decreasing the dose by half for females due to reports of greater adverse events including daytime drowsiness observed in female patients (United States Food and Drug Agency 2018).

While any aspect of a drug's pharmacokinetic properties could potentially lead to sex differences in measured drug exposure, sexually divergent differences in metabolism appear to be the most concerning (Waxman and Holloway 2009). In multiple species, enzymes responsible for drug metabolism show sexually dimorphic expression patterns that affect the rate of metabolism of different drugs. In humans, females show higher cytochrome p450 (CYP) 3A4 levels in the liver as measured by both mRNA and protein (Wolbold et al. 2003). Studies have also observed higher activity of this enzyme in females (Hunt et al. 1992). In rodents, both the mouse (Clodfelter et al. 2006, 2007; Yang et al. 2006) and rat (Wautheir and Waxman 2008) liver show a large degree of sexually dimorphic gene expression. For instance, rats exhibit a male-specific CYP2C11 expression pattern, whereas CYP2C12 shows a female-specific one (Shapiro et al. 1995). While rodent sex differences may not necessarily translate into similar patterns in humans, the complexity of metabolic pathways underscore the importance of understanding drug exposure in each sex, at the relevant time point, and in the relevant tissue when making pharmacodynamic measurements.

With respect to pharmacodynamics, sex differences exist in functional outcome measures, both with respect to baseline activity in the absence of drug, and in response to treatments. As critically highlighted in the field of preclinical pain research, a meta-analysis reported sex differences in sensitivity to painful stimuli in acute thermal pain and in chemically induced inflammatory pain (Mogil and Chanda 2005). For example, in a study by Kest and colleagues, baseline nociceptive responses and sensitivity to thermal stimuli were examined across males and females of 11 inbred mouse strains (Kest et al. 1999). Results of this study not only revealed divergent phenotypic responses across genotypes for pain sensitivity but also sex by genotype interactions. Moreover, when morphine was administered directly into the CNS, the analgesic effects varied across both strain and sex, further highlighting the importance of including both sexes in pharmacodynamic studies, as well as considering subject populations beyond a single inbred strain. These sex differences are not specific to morphine as they have also been demonstrated in rats and mice for sensitivity to the effects of other mu opioid receptor agonists (Dahan et al. 2008). Importantly, both sexually dimorphic circuitry and differences in receptor expression levels mediating pain perception and pharmacological responses, likely driven by genetics, are suggested to contribute to these differences (Mogil and Bailey 2010).

In clinical pain research, sex differences in pharmacodynamic responses have been highlighted by reports from clinical trials with MorphiDex, a potential medication for the treatment of pain that combined an NMDA antagonist with morphine (Galer et al. 2005). While many preclinical studies demonstrated robust and reliable efficacy, these reports were almost exclusively conducted in male subjects. During clinical trials where both men and women were included, the drug failed to produce any clinical benefit over standard pain medications (Galer et al. 2005). Intriguingly, it was later determined that while the drug was efficacious in men, it was ineffective in women with retrospective experiments in female mice corroborating these data (discussed in IOM 2011; Grisel et al. 2005). Overall, while we may not fully understand the biological underpinnings of sex differences in responses to pharmacology, profiling both sexes in preclinical pharmacology studies should provide insight into the differences and potentially enable better clinical trial design.

5   Improving Reproducibility Through Heterogeneity

While the major attention on the "reproducibility crisis" in biomedical research has generally been focused on the lack of translation related to issues with experimental design and publication bias, recent literature has provided insight to the concept that researchers might be practicing "overstandardization" as good research practices. For example, the considerations for controlling as much as possible within an experiment (i.e., sex, strain, vendor, housing conditions, etc.), and across experiments within a given laboratory in order to enable replication (i.e., same day of week, same technician, same procedure room), have not necessarily been previously considered an issue with respect to contributing to lack of reproducibility. However, as recently highlighted by several publications, this "standardization fallacy" suggests that the more control and homogeneity given to an experiment within a laboratory may lead to the inability for others to reproduce the findings given the inherent differences in environment that cannot be standardized across laboratories (Würbel 2000; Voelkl et al. 2018; Kafkafi et al. 2018). In this respect, there is indeed value in applying various levels of systematic variation to address a research question, both through intra- and interlaboratory experiments. One approach to improve heterogeneity beyond including both sexes within an experiment and extending experimental findings to multiple laboratories (interlaboratory reproducibility) is to also introduce genetic diversity. While it may be cost prohibitive to engineer genetic mutations across multiple lines of mouse strains in a given study, one could alternatively employ strategically developed recombinant mouse populations such as the Collaborative Cross (CC) (Churchill et al. 2004). The CC are recombinant inbred mouse strains that were created by cross breeding eight different common inbred strains resulting in increased genetic and phenotypic diversity. CC lines include contributions from the common inbred C57BL/6J strain as well as two inbred strains with high susceptibility for Type I and Type II diabetes, two inbred strains with high susceptibility for developing cancers (129S1/SvlmJ and A/J), and three wild-derived strains (Srivastava et al. 2017). A recent study from Nachshon et al. (2016) highlighted the value of using a CC population for studying the impact of genetic variation on drug metabolism, while Mosedale et al. (2017) have demonstrated the utility of CC lines for studying potential toxicological effects of drugs on genetic variation in kidney disease (Nachshon et al. 2016; Mosedale et al. 2017).

6   Good Research Practices in Pharmacology Include Considerations for Sex, Strain, and Age: Advantages and Limitations

Improving translation from mouse to man requires selection of the appropriate animal model, age, and disease-relevant state. Behavioral pharmacology studies with functional outcome measures that planned for enablement of translational efficacy studies should include pharmacokinetics and PK/PD modeling in the animal model at the pathologically relevant age. It should not be expected that PK data in young, healthy subjects would generalize to PK data in aged, diseased subjects or across both sexes. Similarly, pharmacodynamic measures including behavior, neuroendocrine, immune, metabolic, cardiovascular, and physiology may not generalize across age, sex, or disease state. Figure 1 depicts a sample flow diagram of experimental design parameters required for deliberation where species, strain, substrain, age, sex, and disease state are crucial considerations.

7   Conclusions and Recommendations

Drug discovery in both preclinical studies and in the clinic has only begun to harness the power of genetic diversity. Large-scale clinical trials have focused on recruitment of patients (i.e., enrollment metrics) based on "all comers" symptom presentation for enrollment. It is tempting to believe that at least some of the high clinical attrition of new therapeutic agents can be attributable to a failure to consider patient heterogeneity. It is a common adage that a rule of thirds exist in patient treatment response to a medication: a third of patients show robust efficacy, a third exhibit partial benefit to the agent, and a third are termed "treatment resistant." One reason that much of the pharmaceutical industry has moved away from developing antidepressant medications is that established antidepressant medications, such as SSRIs, when used as a positive control, do not separate from placebo in 30–50% of the trials, resulting in a "busted" clinical trial (reviewed in Mora et al. 2011). Importantly, the preclinical studies that have enabled these trials have largely used male subjects and frequently in otherwise healthy mice of a single inbred strain such as C57BL/6J mice (reviewed in Caldarone et al. 2015; reviewed in Belzung 2014). It is possible that preclinical studies focused on treatment response in both sexes and in genetically divergent populations with face and construct validity would be in a better position to translate to a heterogeneous treatment resistant clinical population.

Within the last decade, however, as the genetic contributions of diseases become known, precision medicine approaches that recruit patients with specific genetic factors (e.g., ApoE4 carriers at risk for Alzheimer's disease) to test specific mechanisms of action will continue to evolve over recruitment for "all comers" patients with a diagnosis of Alzheimer's disease (Watson et al. 2014). In this respect, in animal studies, analogous genetic factors (e.g., mouse model homozygous for the Apoe4 allele), and at an analogous mouse to human age comparison, to test a similar hypothesis are critical.

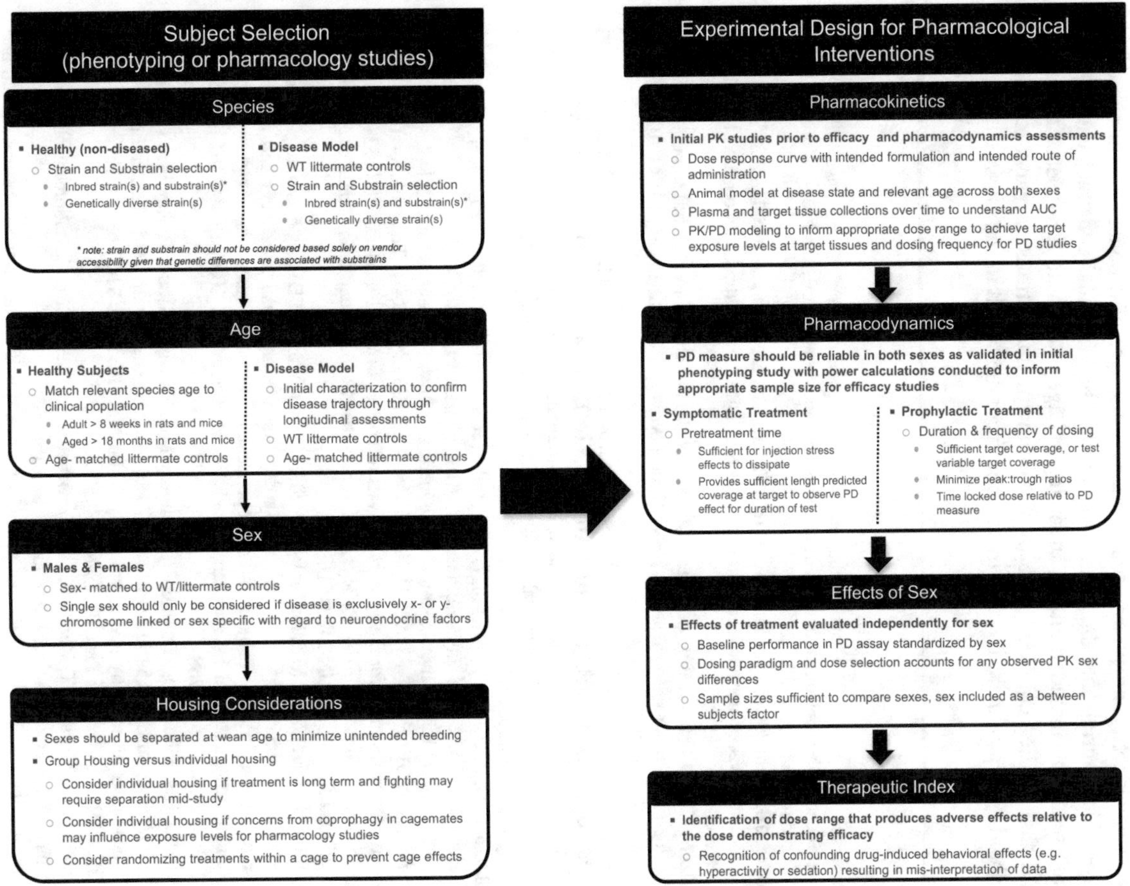

Fig. 1 Example flow diagram for preclinical study design

As previously stated above, the best model for human is human. In drug discovery prior to the FDA enabling clinical trials in humans, it is critical that the best approach to translation is the design and rigorous execution of preclinical pharmacology studies that best mirror the intended patient population. In this respect, for pharmacokinetic and pharmacodynamics studies, careful consideration should be taken for ensuring that the animal model used has face and construct validity, that both sexes are included and at an analogous age relevant to the disease trajectory, and that studies consider gene by environment interactions as ways to improve reliability, reproducibility, and translation from the bench to the clinic.

References

Anderson GD (2005) Sex and racial differences in pharmacological response: where is the evidence? Pharmacogenetics, pharmacokinetics, and pharmacodynamics. J Womens Health (Larchmt) 14(1):19–29. Review

Becker JB, Prendergast BJ, Liang JW (2016) Female rats are not more variable than male rats: a meta-analysis of neuroscience studies. Biol Sex Differ 7:34. https://doi.org/10.1186/s13293-016-0087-5

Belzung C (2014) Innovative drugs to treat depression: did animal models fail to be predictive or did clinical trials fail to detect effects? Neuropsychopharmacology 39(5):1041–1051. https://doi.org/10.1038/npp.2013.342

Berkley KJ (1992) Vive la différence! Trends Neurosci 15(9):331–332. Review. PMID: 1382330

Bourdi M, Davies JS, Pohl LR (2011) Mispairing C57BL/6 substrains of genetically engineered mice and wild-type controls can lead to confounding results as it did in studies of JNK2 in Acetaminophen and Concanavalin a liver injury. Chem Res Toxicol 24(6):794–796. https://doi.org/10.1021/tx200143x

Caldarone BJ, Zachariou V, King SL (2015) Rodent models of treatment-resistant depression. Eur J Pharmacol 753:51–65. https://doi.org/10.1016/j.ejphar.2014.10.063

Casellas J (2011) Inbred mouse strains and genetic stability: a review. Animal 5:1–7

Churchill GA, Airey DC, Allayee H, Angel JM, Attie AD et al (2004) The collaborative cross, a community resource for the genetic analysis of complex traits. Nat Genet 36:1133–1137

Clayton JA, Collins FS (2014) Policy: NIH to balance sex in cell and animal studies. Nature 509 (7500):282–283. PMID: 24834516

Clodfelter KH, Holloway MG, Hodor P, Park SH, Ray WJ, Waxman DJ (2006) Sex-dependent liver gene expression is extensive and largely dependent upon signal transducer and activator of transcription 5b (STAT5b): STAT5b-dependent activation of male genes and repression of female genes revealed by microarray analysis. MolEndocrinol 20(6):1333–1351. Epub 2006 Feb 9. PMID: 16469768

Clodfelter KH, Miles GD, Wauthier V, Holloway MG, Zhang X, Hodor P, Ray WJ, Waxman DJ (2007) Role of STAT5a in regulation of sex-specific gene expression in female but not male mouse liver revealed by microarray analysis. Physiol Genomics 31(1):63–74. Epub 2007 May 29. PMID: 17536022; PMCID: PMC2586676

Cummings JL, Morstorf T, Zhong K (2014) Alzheimer's disease drug-development pipeline: few candidates, frequent failures. Alzheimers Res Ther 6:37

Dahan A, Kest B, Waxman AR, Sarton E (2008) Sex-specific responses to opiates: animal and human studies. Anesth Analg 107(1):83–95. https://doi.org/10.1213/ane.0b013e31816a66a4. Review. PMID: 18635471

Dalla C, Pitychoutis PM, Kokras N, Papadopoulou-Daifoti Z (2010) Sex differences in animal models of depression and antidepressant response. Basic Clin Pharmacol Toxicol 106 (3):226–233. https://doi.org/10.1111/j.1742-7843.2009.00516.x. PMID: 20050844

Festing MF (2014) Evidence should trump intuition by preferring inbred strains to outbred stocks in preclinical research. ILAR J 55(3):399–404. https://doi.org/10.1093/ilar/ilu036

Galer BS, Lee D, Ma T, Nagle B, Schlagheck TG (2005) MorphiDex (morphine sulfate/dextromethorphan hydrobromide combination) in the treatment of chronic pain: three multicenter, randomized, double-blind, controlled clinical trials fail to demonstrate enhanced opioid analgesia or reduction in tolerance. Pain 115(3):284–295. Epub 2005 Apr 20

Guapp S, Pitt D, Kuziel WA, Cannella B, Raine CS (2003) Experimental autoimmune encephalomyelitis (EAE) in CCR2($-/-$) mice: susceptibility in multiple strains. Am J Pathol 162:139–150

Grarup N, Sandholt CH, Hansen T, Pedersen O (2014) Genetic susceptibility to type 2 diabetes and obesity: from genome-wide association studies to rare variants and beyond. Diabetologia 57:1528–1541

Greenblatt DJ, Harmatz JS, Singh NN, Steinberg F, Roth T, Moline ML, Harris SC, Kapil RP (2014) Gender differences in pharmacokinetics and pharmacodynamics of zolpidem following sublingual administration. J Clin Pharmacol 54(3):282–290. https://doi.org/10.1002/jcph.220. Epub 2013 Nov 27. PMID: 24203450

Grisel JE, Allen S, Nemmani KV, Fee JR, Carliss R (2005) The influence of dextromethorphan on morphine analgesia in Swiss Webster mice is sex-specific. Pharmacol Biochem Behav 81 (1):131–138

Hughes RN (2007) Sex does matter: comments on the prevalence of male-only investigations of drug effects on rodent behaviour. Behav Pharmacol 18:583–589

Hunt CM, Westerkam WR, Stave GM (1992) Effect of age and gender on the activity of human hepatic CYP3A. Biochem Pharmacol 44(2):275–283. PMID: 1642641

IOM (Institute of Medicine US) (2011) Forum on neuroscience and nervous system disorders. Sex differences and implications for translational neuroscience research: workshop summary. National Academies Press, Washington DC

Jackson HM, Onos KD, Pepper KW, Graham LC, Akeson EC, Byers C, Reinholdt LG, Frankel WN, Howell GR (2015) DBA/2J genetic background exacerbates spontaneous lethal seizures but lessens amyloid deposition in a mouse model of Alzheimer's disease. PLoS One 10: e0125897

Jarvis MF, Williams M (2016) Irreproducibility in preclinical biomedical research: perceptions, uncertainties, and knowledge gaps. Trends Pharmacol Sci 37:290–302

Kafkafi N, Agassi J, Chesler EJ, Crabbe JC, Crusio WE, Eilam D, Gerlai R, Golani I, Gomez-Marin A, Heller R, Iraqi F, Jalijuli I, Karp NA, Morgan H, Nicholson G, Pfaff DW, Richter H, Stark PB, Stiedl O, Stodden V, Tarantino LM, Tucci V, Valdar W, Williams RW, Wurbel H, Benjamini Y (2018) Reproducibility and replicability of rodent phenotyping in preclinical studies. Neurosci Biobehav Rev 87:218–232. https://doi.org/10.1016/j.neubiorev.2018.01.003

Karp NA, Mason J, Beaudet AL et al (2017) Prevalence of sexual dimorphism in mammalian phenotypic traits. Nat Commun 8:15475. https://doi.org/10.1038/ncomms15475

Keane TM, Goodstadt L, Danecek P, White MA, Wong K, Yalcin B, Heger A, Agam A, Slater G, Goodson M, Furlotte NA, Eskin E, Nellåker C, Whitley H, Cleak J, Janowitz D, Hernandez-Pliego P, Edwards A, Belgard TG, Oliver PL, McIntyre RE, Bhomra A, Nicod J, Gan X, Yuan W, van der Weyden L, Steward CA, Bala S, Stalker J, Mott R, Durbin R, Jackson IJ, Czechanski A, Guerra-Assunção JA, Donahue LR, Reinholdt LG, Payseur BA, Ponting CP, Birney E, Flint J, Adams DJ (2011) Mouse genomic variation and its effect on phenotypes and gene regulation. Nature 477(7364):289–294. https://doi.org/10.1038/nature10413

Kest B, Wilson SG, Mogil JS (1999) Sex differences in supraspinal morphine analgesia are dependent on genotype. J Pharmacol Exp Ther 289(3):1370–1375. PMID: 10336528

Kilkenny C, Browne WJ, Cuthill IC, Emerson M, Altman DG (2010) Improving bioscience research reporting: the ARRIVE guidelines for reporting animal research. PLoS Biol 8: e1000412

Kiselycznyk C, Holmes A (2011) All (C57BL/6) mice are not created equal. Front Neurosci 5:10. https://doi.org/10.3389/fnins.2011.00010

Kokras N, Antoniou K, Mikail HG, Kafetzopoulos V, Papadopoulou-Daifoti Z, Dalla C (2015) Forced swim test: what about females? Neuropharmacology 99:408–421. https://doi.org/10.1016/j.neuropharm.2015.03.016. Epub 2015 Apr 1. Review. PMID: 25839894

Kreiner G, Chmielarz P, Roman A, Nalepa I (2013) Gender differences in genetic mouse models evaluated for depressive-like and antidepressant behavior. Pharmacol Rep 65(6):1580–1590. Review. PMID: 24553006

Laman-Maharg A, Williams AV, Zufelt MD, Minie VA, Ramos-Maciel S, Hao R, Ordoñes Sanchez E, Copeland T, Silverman JL, Leigh A, Snyder R, Carroll FI, Fennell TR, Trainor BC (2018) Sex differences in the effects of a kappa opioid receptor antagonist in the forced swim test. Front Pharmacol 9:93. https://doi.org/10.3389/fphar.2018.00093. PMID: 29491835

Leiter EH (1997) The NOD mouse: a model for insulin-dependent diabetes mellitus. Curr Protoc Immunol 24(Suppl):15.9.1–15.9.23

Loscher W, Ferland RJ, Ferraro TN (2017) The relevance of inter- and intrastrain differences in mice and rats and their implications of seizures and epilepsy. Epilepsy Behav 73:214–235

Louis WJ, Howes LG (1990) Genealogy of the spontaneously hypertensive rat and Wistar-Kyoto rat strains: implications for studies of inherited hypertension. J Cardiovasc Pharmacol 16 (Suppl 7):S1–S5

Mahajan VS, Demissie E, Mattoo H, Viswanadham V, Varki A, Morris R, Pillai S (2016) Striking immune phenotypes in gene-targeted mice are driven by a copy-number variant originating from a commercially available C57BL/6 strain. Cell Rep 15(9):1901–1909. https://doi.org/10.1016/j.celrep.2016.04.080

McCracken JM, Chalise P, Briley SM et al (2017) C57BL/6 substrains exhibit different responses to acute carbon tetrachloride exposure: implications for work involving transgenic mice. Gene Expr 17(3):187–205. https://doi.org/10.3727/105221617X695050

Mogil JS, Bailey AL (2010) Sex and gender differences in pain and analgesia. Prog Brain Res 186:141–157. https://doi.org/10.1016/B978-0-444-53630-3.00009-9. Review. PMID: 21094890

Mogil JS, Chanda ML (2005) The case for the inclusion of female subjects in basic science studies of pain. Pain 117(1-2):1–5. Review. PMID: 16098670

Mora MS, Nestoriuc Y, Rief W (2011) Lessons learned from placebo groups in antidepressant trials. Philos Trans R Soc Lond B Biol Sci 366(1572):1879–1888. https://doi.org/10.1098/rstb.2010.0394

Mosedale M, Kim Y, Brock WJ, Roth SE, Wiltshire T, Eaddy JS, Keele GR, Corty RW, Xie Y, Valdar W, Watkins PB (2017) Candidate risk factors and mechanisms for Tolvaptan-induced liver injury are identified using a collaborative cross approach. Toxicol Sci 156(2):438–454. https://doi.org/10.1093/toxsci/kfw269

Nachshon A, Abu-Toamih Atamni HJ, Steuerman Y et al (2016) Dissecting the effect of genetic variation on the hepatic expression of drug disposition genes across the collaborative cross mouse strains. Front Genet 7:172. https://doi.org/10.3389/fgene.2016.00172

Nilson JH, Abbud RA, Keri RA, Quirk CC (2000) Chronic hypersecretion of luteinizing hormone in transgenic mice disrupts both ovarian and pituitary function, with some effects modified by the genetic background. Recent Prog Horm Res 55:69–89

Peer CJ, Strope JD, Beedie S, Ley AM, Holly A, Calis K, Farkas R, Parepally J, Men A, Fadiran EO, Scott P, Jenkins M, Theodore WH, Sissung TM (2016) Alcohol and aldehyde dehydrogenases contribute to sex-related differences in clearance of Zolpidem in rats. Front Pharmacol 7:260. https://doi.org/10.3389/fphar.2016.00260. eCollection 2016. PMID: 27574509

Prendergast BJ, Onishi KG, Zucker I (2014) Female mice liberated for inclusion in neuroscience and biomedical research. Neurosci Biobehav Rev 40:1–5. https://doi.org/10.1016/j.neubiorev.2014.01.001. Epub 2014 Jan 20. PMID: 24456941

Regitz-Zagrosek V (2012) Sex and gender differences in health: Science & Society Series on sex and science. EMBO Rep 13(7):596–603. https://doi.org/10.1038/embor.2012.87

Sagvolden T, Johansen EB, Wøien G, Walaas SI, Storm-Mathisen J, Bergersen LH, Hvalby Ø, Jensen V, Aase H, Russell VA, Killeen PR, DasBanerjee T, Middleton F, Faraone SV (2009) The spontaneously hypertensive rat model of ADHD – the importance of selecting the appropriate reference strain. Neuropharmacology 57(7-8):619–626

Shapiro BH, Agrawal AK, Pampori NA (1995) Gender differences in drug metabolism regulated by growth hormone. Int J Biochem Cell Biol 27(1):9–20. Review. PMID: 7757886

Simon MM, Greenaway S, White JK, Fuchs H, Gailus-Durner V, Wells S, Sorg T, Wong K, Bedu E, Cartwright EJ, Dacquin R, Djebali S, Estabel J, Graw J, Ingham NJ, Jackson IJ, Lengeling A, Mandillo S, Marvel J, Meziane H, Preitner F, Puk O, Roux M, Adams DJ, Atkins S, Ayadi A, Becker L, Blake A, Brooker D, Cater H, Champy MF, Combe R, Danecek P, di Fenza A, Gates H, Gerdin AK, Golini E, Hancock JM, Hans W, Hölter SM, Hough T, Jurdic P, Keane TM, Morgan H, Müller W, Neff F, Nicholson G, Pasche B, Roberson LA, Rozman J, Sanderson M, Santos L, Selloum M, Shannon C, Southwell A, Tocchini-Valentini GP, Vancollie VE, Westerberg H, Wurst W, Zi M, Yalcin B, Ramirez-Solis R, Steel KP, Mallon AM, de Angelis MH, Herault Y, Brown SD (2013) A comparative phenotypic and genomic analysis of C57BL/6J and C57BL/6N mouse strains. Genome Biol 14(7):R82. https://doi.org/10.1186/gb-2013-14-7-r82

Sorge RE, LaCroix-Fralish ML, Tuttle AH et al (2011) Spinal cord toll-like receptor 4 mediates inflammatory and neuropathic hypersensitivity in male but not female mice. J Neurosci 31 (43):15450–15454. https://doi.org/10.1523/JNEUROSCI.3859-11.2011

Srivastava A, Morgan AP, Najarian ML et al (2017) Genomes of the mouse collaborative cross. Genetics 206(2):537–556. https://doi.org/10.1534/genetics.116.198838

St Lezin E, Simonet L, Pravenec M, Kurtz TW (1992) Hypertensive strains and normotensive 'control' strains. How closely are they related? Hypertension 19:419–424

Summers deLuca LE, Pikor NB, O'Leary J et al (2010) Substrain differences reveal novel disease-modifying gene candidates that alter the clinical course of a rodent model of multiple sclerosis. J Immunol 184(6):3174–3185. https://doi.org/10.4049/jimmunol.0902881

Sundberg JP, Schofield PN (2010) Mouse genetic nomenclature: standardization of strain, gene, and protein symbols. Vet Pathol 47(6):1100–1104. https://doi.org/10.1177/0300985810374837

United States Food and Drug Agency (2018) Drug safety communication: risk of next-morning impairment after use of insomnia drugs; FDA requires lower recommended doses for certain drugs containing zolpidem (Ambien, Ambien CR, Edluar, and Zolpimist). http://www.fda.gov/downloads/Drugs/DrugSafety/UCM335007. Accessed 23 Sept 2018

Voelkl B, Vogt L, Sena ES, Würbel H (2018) Reproducibility of preclinical animal research improves with heterogeneity of study samples. PLoS Biol 16(2):e2003693. https://doi.org/10.1371/journal.pbio.2003693

Watson JL, Ryan L, Silverberg N, Cahan V, Bernard MA (2014) Obstacles and opportunities in Alzheimer's clinical trial recruitment. Health Aff (Millwood) 33(4):574–579. https://doi.org/10.1377/hlthaff.2013.1314

Wautheir V, Waxman DJ (2008) Sex-specific early growth hormone response genes in rat liver. Mol Endocrinol 22(8):1962–1974

Waxman DJ, Holloway MG (2009) Sex differences in the expression of hepatic drug metabolizing enzymes. Mol Pharmacol 76(2):215–228. https://doi.org/10.1124/mol.109.056705

Wolbold R, Klein K, Burk O, Nüssler AK, Neuhaus P, Eichelbaum M, Schwab M, Zanger UM (2003) Sex is a major determinant of CYP3A4 expression in human liver. Hepatology 38 (4):978–988. PMID: 14512885

Würbel H (2000) Behaviour and the standardization fallacy. Nat Genet 26:263

Würbel H (2002) Behavioral phenotyping enhanced—beyond (environmental standardization). Genes Brain Behav 1:3–8

Yang X, Schadt EE, Wang S, Wang H, Arnold AP, Ingram-Drake L, Drake TA, Lusis AJ (2006) Tissue-specific expression and regulation of sexually dimorphic genes in mice. Genome Res 16 (8):995–1004. Epub 2006 Jul 6. PMID: 16825664; PMCID: PMC1524872

Building Robustness into Translational Research

Betül R. Erdogan and Martin C. Michel

Contents

Abstract

Nonclinical studies form the basis for the decision whether to take a therapeutic candidate into the clinic. These studies need to exhibit translational robustness for both ethical and economic reasons. Key findings confirmed in multiple species have a greater chance to also occur in humans. Given the heterogeneity of patient populations, preclinical studies or at least programs comprising multiple studies need to reflect such heterogeneity, e.g., regarding strains, sex, age, and comorbidities of experimental animals. However, introducing such heterogeneity requires larger studies/programs to maintain statistical power in the face of greater variability. In addition to classic sources of bias, e.g., related to lack of randomization and concealment, translational studies face specific sources of potential

B. R. Erdogan
Department of Pharmacology, School of Pharmacy, Ankara University, Ankara, Turkey

M. C. Michel (✉)
Department of Pharmacology, Johannes Gutenberg University, Mainz, Germany
e-mail: marmiche@uni-mainz.de

A. Bespalov et al. (eds.), *Good Research Practice in Non-Clinical Pharmacology and Biomedicine*, Handbook of Experimental Pharmacology 257,
https://doi.org/10.1007/164_2019_283

bias such as that introduced by a model that may not reflect the full spectrum of underlying pathophysiology in patients, that defined by timing of treatment, or that implied in dosing decisions and interspecies differences in pharmacokinetic profiles. The balance of all these factors needs to be considered carefully for each study and program.

Keywords
Age · Comorbidity · Heterogeneity · Robustness · Sex · Translational research

1   Introduction

The testing of new medicines in humans, particularly in patients, requires a sound and robust body of nonclinical evidence for both ethical and economic reasons (Wieschowski et al. 2018). Ethically, participation of humans in a clinical study is based on the premise that a new drug may provide more efficacious and/or safer treatment of a disease. Economically, it requires an investment of around 20 million € to bring a drug candidate to clinical proof-of-concept evaluation, particularly for a first-in-class medicine. The supporting evidence for the ethical and the economical proposition is typically based on animal studies. Animal models can become particularly important if human material is difficult to obtain (Alini et al. 2008) or the condition to be treated is too complex or too poorly understood to solely extrapolate from in vitro studies (Michel and Korstanje 2016). While existing treatments show that this can be done successfully, numerous examples exist where experimental treatments looked promising in animal models but have failed in clinical studies due to lack of efficacy. Prominent examples of failed clinical programs include amyotrophic lateral sclerosis (Perrin 2014), anti-angiogenic treatment in oncology (Martić-Kehl et al. 2015), several cardiovascular diseases (Vatner 2016), sepsis (Shukla et al. 2014), and stroke-associated neuroprotection (Davis 2006). Therefore, the idea of enhancing robustness of nonclinical studies is not new and has been advocated for more than 20 years (Hsu 1993; Stroke Therapy Academic Industry Roundtable (STAIR) 1999). Nonetheless, poor technical quality and reporting issues remain abundant (Chang et al. 2015; Kilkenny et al. 2009), and clinical development programs continue to fail due to lack of efficacy despite promising findings in animals.

Generalizability shows how applicable the results from one model are for others. In the context of translational research, this translates into the question whether findings from experimental models are likely to also occur in patients. Generalizability of preclinical animal studies is possible, only if the studies are reproducible, replicable, and robust. This chapter discusses causes contributing to lack of robustness of translational studies and the cost/benefit in addressing them. In this context we define robustness as an outcome that can be confirmed in principle despite some modifications of experimental approach, e.g., different strains or species. Only robust findings in the nonclinical models are likely to predict those in clinical proof-of-concept studies. For obvious reasons, a translational study can only be robust if it is reproducible, i.e., if another investigator doing everything exactly as the original researchers will obtain a comparable result. General factors enhancing reproducibility such as randomization, blinding, choice of appropriate sample sizes and analytical techniques, and avoiding bias due to selective reporting of findings (Lapchak et al. 2013; Snyder et al. 2016) will not be covered here because they are discussed in depth in other chapters of this book. However, it should be noted that generally accepted measures to enhance reproducibility have not been adhered to in most studies intended to have translational value (Kilkenny et al. 2009) and reporting standards have often been poor (Chang et al. 2015).

Against this background, communities interested in various diseases have developed specific recommendation for the design, conduct, analysis, and reporting of animal studies in their field, e.g., Alzheimer's disease (Snyder et al. 2016), atherosclerosis (Daugherty et al. 2017), lung fibrosis (Jenkins et al. 2017), multiple sclerosis (Amor and Baker 2012; Baker and Amor 2012), rheumatology (Christensen et al. 2013), stroke (Lapchak et al. 2013; Stroke Therapy Academic Industry Roundtable (STAIR) 1999), or type 1 diabetes (Atkinson 2011; Graham and Schuurman 2015); disease-overarching guidelines for animal studies with greater translational value have also been proposed (Anders and Vielhauer 2007). We will also discuss these disease-overarching approaches.

2 Homogeneous vs. Heterogeneous Models

Homogeneous models, e.g., inbred strains or single sex of experimental animals, intrinsically exhibit less variability and, accordingly, have greater statistical power to find a difference with a given number of animals (sample size). In contrast, human populations to be treated tend to be more heterogeneous, e.g., regarding gender, ethnicity, age, comorbidities, and comedications. While heterogeneity often remains limited in phase III clinical studies due to strict inclusion and exclusion criteria, marketed drugs are used in even more heterogeneous populations. This creates a fundamental challenge for translational studies. More homogeneous models tend to need fewer animals to have statistical power but may have a smaller chance to reflect the broad patient population intended to use a drug. In contrast, more heterogeneous translational programs are likely to be costlier but, if consistently showing efficacy, should have a greater chance to predict efficacy in patients. The following discusses some frequent sources of heterogeneity. However, these are just examples, and investigators are well advised to systematically consider the costs and opportunities implied in selection of models and experimental conditions (Fig. 1).

2.1 Animal Species and Strain

While mammals share many regulatory systems, individual species may differ regarding the functional role of a certain pathway. For instance, β_3-adrenoceptors are a major regulator of lipolysis in rodents, particularly in brown adipose tissue; this

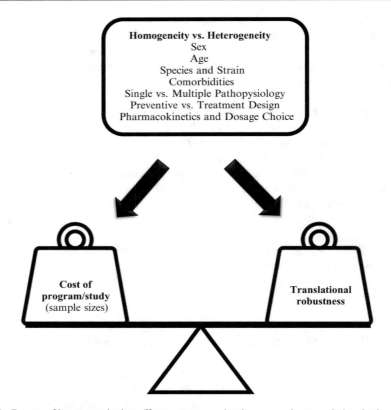

Fig. 1 Degree of heterogeneity has effects on program/study costs and on translational robustness. An appropriate balance must be defined on a project-specific basis

was the basis of clinical development of β_3-adrenoceptor agonists for the treatment of obesity and type 2 diabetes. However, metabolic β_3-adrenoceptor agonist programs of several pharmaceutical companies have failed in phase II trials (Michel and Korstanje 2016) because adult humans have little brown adipose tissue and lipolysis in human white adipose tissues is primarily driven by β_1-adrenoceptors (Barbe et al. 1996). Similarly, α_1-adrenoceptors are major regulators of inotropy in rat heart but have a limited role, if any, in the human heart (Brodde and Michel 1999). Thus, the targeted mechanism should be operative in the animal species used as preclinical model and in humans. When corresponding human data are lacking, multiple animal strains and species should be compared that are phylogenetically sufficiently distinct, i.e., the confirmation of rat studies should not be in mice but perhaps in dogs or nonhuman primates. This approach has long been standard practice in toxicology, where regulatory agencies require data in at least two species, one of which must be a nonrodent.

A variation of this theme are differences between strains within a species. For instance, rat strains can differ in the thymic atrophy in response to induction of experimental autoimmune encephalomyelitis (Nacka-Aleksić et al. 2018) or in the degree of urinary bladder hypertrophy in streptozotocin-induced type 1 diabetes (Arioglu Inan et al. 2018). Similarly, inbred strains may yield more homogeneous responses than outbred strains and, accordingly, may require smaller sample sizes; however, the traits selected in an inbred strain may be less informative. For example, inbred Wistar-Kyoto rats are frequently used as normotensive control in studies with spontaneously hypertensive rats. However, Wistar-Kyoto rats may share some phenotypes with spontaneously hypertensive rats such as increased frequency of micturition and amplitude of urinary bladder detrusor activity with the hypertensive animals that are not observed in other normotensive strains (Jin et al. 2010).

2.2 Sex of Animals

Except for a small number of gender-specific conditions such as endometriosis or benign prostatic hyperplasia, diseases typically affect men and women – although often with different prevalence. Thus, most drugs must work in both genders. Many drug classes are similarly effective in both genders, for instance, muscarinic antagonists in the treatment of overactive bladder syndrome (Witte et al. 2009), and preclinical data directly comparing both sexes had predicted this (Kories et al. 2003). On the other hand, men and women may exhibit differential responsiveness to a given drug, at least at the quantitative level; for instance, the vasopressin receptor agonist desmopressin reduced nocturia to a greater extent in women than in men (Weiss et al. 2012). Such findings can lead to failed studies in a mixed gender population. Therefore, robust preclinical data should demonstrate efficacy in both sexes. However, most preclinical studies do not account for sex as a variable and have largely been limited to male animals (Tierney et al. 2017; Pitkänen et al. 2014). For instance, only 12 out of 71 group comparisons of urinary bladder hypertrophy in the streptozotocin model of type 1 diabetes were reported for female rats (Arioglu Inan et al. 2018). In reaction, the NIH have published guidance on the consideration of sex as a biological variable (National Institutes of Health 2015). It requires to use both sexes in grant application unless the target disease predominantly affects one gender. For a more detailed discussion of the role of sex differences, see chap. 9 by Rizzo et al.

Generally performing preclinical studies in both sexes comes at a price. A study designed to look at drug effects vs. vehicle in male and female rats and compare the effect between sexes needs not only twice as many groups but also a greater number of animals per group to maintain statistical power when adjusting for multiple comparisons. This makes a given study more expensive (Tannenbaum and Day 2017), and lack of funding is seen as a main reason not to incorporate both sexes in study design (McCarthy 2015). An alternative approach could be to do a single study based on mixed sexes. This may be more robust to detect an efficacious treatment but also may have more false negatives if a drug is considerably less effective in one of the two sexes. As it may be useful to study multiple animal models for a given condition (see below), a third option could be to use males in one and females in the other model, targeting a balanced representation of sexes across a program.

This works well if the two studies yield similar results. However, if they show different results, one does not know whether such difference comes from that in sex of the experimental animal or from that of model, necessitating additional studies.

2.3 Age

Studies in various organ systems and pathologies show that older animals react differently than adolescent ones, for instance, in the brain (Scerri et al. 2012), blood vessels (Mukai et al. 2002), or urinary bladder (Frazier et al. 2006). Nonetheless, the most frequently used age group in rat experiments is about 12 weeks old at the start of the experiment, i.e., adolescent, whereas most diseases preferentially affect patients at a much higher age. Moreover, the elderly may be more sensitive to side effects, for instance, because they exhibit a leakier blood-brain barrier (Farrall and Wardlaw 2009) or are more prone to orthostasis (Mets 1995), thereby shifting the efficacy/tolerability balance in an unfavorable way. The same applies to the reduced renal function in the elderly, which can affect pharmacokinetics of an investigational drug in a way not predicted from studies in adolescent animals. While experiments in old animals are more expensive than those in adolescent ones (a 2-year-old rat may be ten times as expensive as a 12-week-old animal), it cannot necessarily be expected that young animals are highly predictive for conditions predominantly affecting the elderly. The conflict between a need for preclinical data in an age group comparable with that of the target population and the considerably higher cost of aged animals could be resolved by performing at least one key preclinical study in old animals.

2.4 Comorbidities

A condition targeted by a new medication frequently is associated with other diseases. This can reflect that two conditions with high prevalence in an age group often coincide based on chance, e.g., because both increase in prevalence with age. However, it can also result from two conditions sharing a root cause. An example of the former is that patients seeking treatment for the overactive bladder syndrome were reported to concomitantly suffer from arterial hypertension (37.8%), diabetes (15.4%), benign prostatic hyperplasia (37.3% of male patients), and depression and congestive heart failure (6.5% each) (Schneider et al. 2014), all of which are age-related conditions. An example of the latter are diseases involving atherosclerosis as a common risk factor, including hypertension, heart failure, and stroke. Regardless of the cause of association between two disease states, comorbidity may affect drug effects in the target disease. Thus, studies in experimental stroke treatment have largely been performed in otherwise healthy animals (Sena et al. 2007). That this may yield misleading data is highlighted by a presumed anti-stroke drug, the free radical scavenger NXY-059. While this drug had been tested in nine preclinical studies, it failed in a clinical proof-of-concept study; however, the preclinical package included only a single study involving animals with relevant comorbidity, and that study showed a considerably smaller efficacy than the others (MacLeod et al. 2008). Greater reliance on animal models with relevant comorbidities may have prevented this drug candidate advancing to clinical studies, thereby sparing study participants from an ineffective treatment and saving the sponsoring company a considerable amount of resources. Again, the balance between greater robustness and greater cost may be achieved by doing some studies in animal models with and some without relevant comorbidity.

In a more general vein, lack of consistency across species, sexes, age groups, or comorbidities is not only a hurdle but can also be informative as diverging findings in one group (if consistent within this group) may point to important new avenues of research (Bespalov et al. 2016a).

3 Translational Bias

Bias leading to poor reproducibility can occur at many levels, including selection of model and samples, execution of study, detection of findings, unequal attrition across experimental groups, and reporting (Green 2015; Hooijmans et al. 2014). Most of these are generic issues of poor reproducibility and are covered elsewhere in this book. We will focus on aspects with particular relevance for translational studies.

3.1 Single Versus Multiple Pathophysiologies

Choice of animal model is easier if the primary defect is known and can be recreated in an animal, e.g., mutation of a specific gene. However, even then it is not fully clear whether physiological response to mutation is the same in animals and humans. Many syndromes can be the result of multiple underlying pathophysiologies. For instance, arterial hypertension can be due to excessive catecholamine release as in pheochromocytoma, to increased glucocorticoid levels as in Cushing's disease, to an increased activity of the renin-angiotensin system such as in patients with renal disease, or to increased mineralocorticoid levels such as in patients with Conn's syndrome. Accordingly, angiotensin receptor antagonists effectively lower blood pressure in most animal models but have little effect in normotensive animals or animals hypertensive due to high salt intake or the diabetic Goto-Kakizaki rat (Michel et al. 2016). As a counterexample, β-adrenoceptor antagonists lower blood pressure in many hypertensive patients but may increase it in pheochromocytoma. Thus, reliance on a limited panel of animal models can yield misleading results if those models only reflect pathophysiologies relevant to a minor fraction of the patient population to be treated. While irrelevant models often yield positive findings (MacLeod et al. 2008), they do not advance candidates likely to become approved treatments. The choice of relevant animal models should involve careful consideration of the pathophysiology underlying the human disease to be treated (Green 2002).

However, this may be easier said than done, particularly in diseases that are multifactorial and/or for which limited insight into the underlying pathophysiology exists.

3.2 Timing of Intervention

Activation or inhibition of a given mechanism is not necessarily equally effective in the prevention and treatment of a disease or in early and late phases of it (Green 2002). An example of this is sepsis. A frequently used model of sepsis is the administration of lipopolysaccharide. Many agents found active in this model when used as pre-treatment or co-administered with the causative agent have failed to be of benefit in patients, presumably at least partly because the early pathophysiological cascade initiated by lipopolysaccharide differs from the mechanisms in later phases of sepsis. Patients typically need treatment for symptoms of sepsis when the condition has fully developed; therefore, a higher translational value is expected from studies with treatment starting several hours after onset of septic symptoms (Wang et al. 2015). Similarly, many conditions often are diagnosed in patients at an advanced stage, where partly irreversible processes may have taken place. One example of this is tissue fibrosis, which is more difficult to resolve than to prevent (Michel et al. 2016). Another example is oncology where growth of the primary tumor may involve different mechanisms than metastasis (Heger et al. 2014). Moreover, the assessment of outcomes must match a clinically relevant time point (Lapchak et al. 2013); for instance, β-adrenoceptor agonists acutely improve cardiac function in patients with heart failure, but their chronic use increased mortality (The German Austrian Xamoterol-Study Group 1988). Therefore, animal models can only be expected to be predictive if they reflect the clinical setting in which treatment is intended to be used.

3.3 Pharmacokinetics and Dosage Choice

Most drugs have only limited selectivity for their molecular target. If a drug is underdosed relative to what is needed to engage that target, false negative effects may occur (Green 2002), and a promising drug candidate may be wrongly abandoned. More often an experimental treatment is given to animals in doses higher than required, which may lead to off-target effects that can yield false positive data. Testing of multiple doses, preferably accompanied by pharmacokinetics of each dose, may help avoiding the false negative and false positive conclusions based on under- and overdosing, respectively. Moreover, a too high dose may cause adverse effects which shift the efficacy/tolerability ration in an unfavorable way, potentially leading to unjustified termination of a program. Careful comparison of pharmacokinetic and pharmacodynamic effects can improve interpretation of data from heterogeneous models (Snyder et al. 2016), as has been shown in QT_c prolongation (Gotta et al. 2015). QT_c prolongation can happen as a consequence of alterations of ion channel function, which can lead to impairing ventricular repolarization in the heart and may predispose to polymorphic ventricular tachycardia (torsade de pointes), which in turn may cause syncope and sudden death. This may be further aided by the use of biomarkers, particularly target engagement markers (Bespalov et al. 2016b). It is generally advisable to search for information on specific-specific pharmacokinetic data to identify most suitable doses prior to finalizing a study design (Kleiman and Ehlers 2016).

Moreover, desired and adverse drug effects may exhibit differential time profiles. An example of this are α_1-adrenoceptor antagonists intended for the treatment of symptoms of benign prostatic hyperplasia. Concomitant lowering of blood pressure may be an adverse effect in their use. Original animal studies have typically compared drug effects on intraurethral pressure as a proxy for efficacy and those on blood pressure as a proxy for tolerability (Witte et al. 2002). However, it became clear that some α_1-adrenoceptor antagonists reach higher concentrations in target tissue than in plasma at late time points; this allows dosing with smaller peak plasma levels and blood pressure effects which maintain therapeutic efficacy, thereby providing a basis for improving tolerability (Franco-Salinas et al. 2010).

4   Conclusions

Bias at all levels of planning, execution, analysis, interpretation, and reporting of studies is a general source of poor reproducibility (Green 2015; Hooijmans et al. 2014). Additional layers such as choice of animal model, stage of condition, and tested doses add to the potential bias in translational research. A major additional hurdle for translational research is related to the balancing between homo- and heterogeneity of models. While greater heterogeneity is more likely to be representative for patient groups to be treated, it also requires larger studies. These are not only more expensive but also have ethical implications balancing number of animals being used against statistical power and robustness of results (Hawkins et al. 2013). Conclusions on the appropriate trade-off may not only be specific for a disease or a treatment but also depend on the setting in which the research is performed, e.g., academia, biotech, and big pharma (Frye et al. 2015). As there is no universally applicable recipe to strike the optimal balance, translational investigators should carefully consider the implications of a chosen study design and apply the limitations implied in their interpretation of the resulting findings.

While considerations will need to be adapted to the disease and intervention of interest, some more general recommendations emerge. While we often have a good understanding which degree of symptom improvement in a patient is clinically meaningful, similar understanding in animal models if mostly missing. Nonetheless, we consider it advisable that investigators critically consider whether the effects they observe in an experimental model are of a magnitude that can be deemed to be clinically meaningful if confirmed in patients. It has been argued that the nonclinical studies building the final justification to take a new treatment into patients "need to apply the same attention to detail, experimental rigour and statistical power . . . as in the clinical trials themselves" (Howells et al. 2014). In this respect, it has long been the gold standard that pivotal clinical trials should be based on a multicenter design to limit site-specific biases, largely reflecting unknown unknowns. While most nonclinical studies are run within a single lab, examples of preclinical multicenter studies are emerging (Llovera et al. 2015; Maysami et al. 2016). While preclinical multicenter studies improve robustness, they are one component for improved quality but do not substitute for critical thinking.

Acknowledgments BRE is a Ph.D student supported by Turkish Scientific and Technical Research Council (TUBITAK – 2211/A). Work on reproducibility in the lab of MCM has been funded in part by the European Quality in Preclinical Data (EQIPD) consortium. This project has received funding from the Innovative Medicines Initiative 2 Joint Undertaking under grant agreement No 777364. This Joint Undertaking receives support from the European Union's Horizon 2020 research and innovation program and EFPIA.

References

Alini M, Eisenstein SM, Ito K, Little C, Kettler AA, Masuda K, Melrose J, Ralphs J, Stokes I, Wilke HJ (2008) Are animal models useful for studying human disc disorders/degeneration? Eur Spine J 17:2–19

Amor S, Baker D (2012) Checklist for reporting and reviewing studies of experimental animal models of multiple sclerosis and related disorders. Mult Scler Relat Disord 1:111–115

Anders H-J, Vielhauer V (2007) Identifying and validating novel targets with in vivo disease models: guidelines for study design. Drug Discov Today 12:446–451

Arioglu Inan E, Ellenbroek JH, Michel MC (2018) A systematic review of urinary bladder hypertrophy in experimental diabetes: part I. streptozotocin-induced rat models. Neurourol Urodyn 37:1212–1219

Atkinson MA (2011) Evaluating preclinical efficacy. Sci Transl Med 3:96cm22

Baker D, Amor S (2012) Publication guidelines for refereeing and reporting on animal use in experimental autoimmune encephalomyelitis. J Neuroimmunol 242:78–83

Barbe P, Millet L, Galitzky J, Lafontan M, Berlan M (1996) In situ assessment of the role of the β_1-, β_2- and β_3-adrenoceptors in the control of lipolysis and nutritive blood flow in human subcutaneous adipose tissue. Br J Pharmacol 117:907–913

Bespalov A, Emmerich CH, Gerlach B, Michel MC (2016a) Reproducibility of preclinical data: one man's poison is another man's meat. Adv Precision Med 1:1–10

Bespalov A, Steckler T, Altevogt B, Koustova E, Skolnick P, Deaver D, Millan MJ, Bastlund JF, Doller D, Witkin J, Moser P, O'Donnell P, Ebert U, Geyer MA, Prinssen E, Ballard T, MacLeod M (2016b) Failed trials for central nervous system disorders do not necessarily invalidate preclinical models and drug targets. Nat Rev Drug Discov 15:516–516

Brodde OE, Michel MC (1999) Adrenergic and muscarinic receptors in the human heart. Pharmacol Rev 51:651–689

Chang C-F, Cai L, Wang J (2015) Translational intracerebral hemorrhage: a need for transparent descriptions of fresh tissue sampling and preclinical model quality. Transl Stroke Res 6:384–389

Christensen R, Bliddal H, Henriksen M (2013) Enhancing the reporting and transparency of rheumatology research: a guide to reporting guidelines. Arthritis Res Ther 15:109

Daugherty A, Tall AR, Daemen MJAP, Falk E, Fisher EA, García-Cardeña G, Lusis AJ, Owens AP, Rosenfeld ME, Virmani R (2017) Recommendation on design, execution, and reporting of animal atherosclerosis studies: a scientific statement from the American Heart Association. Arterioscler Thromb Vasc Biol 37:e131–e157

Davis S (2006) Optimising clinical trial design for proof of neuroprotection in acute ischaemic stroke: the SAINT clinical trial programme. Cerebrovasc Dis 22(Suppl 1):18–24

Farrall AJ, Wardlaw JM (2009) Blood-brain-barrier: ageing and microvascular disease: systematic review and meta-analysis. Neurobiol Aging 30:337–352

Franco-Salinas G, de la Rosette JJMCH, Michel MC (2010) Pharmacokinetics and pharmacodynamics of tamsulosin in its modified-release and oral-controlled absorption system formulations. Clin Pharmacokinet 49:177–188

Frazier EP, Schneider T, Michel MC (2006) Effects of gender, age and hypertension on ß-adrenergic receptor function in rat urinary bladder. Naunyn Schmiedebergs Arch Pharmacol 373:300–309

Frye SV, Arkin MR, Arrowsmith CH, Conn PJ, glicksman MA, Hull-Ryde EA, Slusher BS (2015) Tackling reproducibility in academic preclinical drug discovery. Nat Rev Drug Discov 14:733–734

Gotta V, Cools F, van Ammel K, Gallacher DJ, Visser SAG, Sannajust F, Morissette P, Danhof M, van der Graaf PH (2015) Inter-study variability of preclinical in vivo safety studies and translational exposure-QTc relationships – a PKPD meta-analysis. Br J Pharmacol 172:4364–4379

Graham ML, Schuurman H-J (2015) Validity of animal models of type 1 diabetes, and strategies to enhance their utility in translational research. Eur J Pharmacol 759:221–230

Green AR (2002) Why do neuroprotective drugs that are so promising in animals fail in the clinic? An industry perspective. Clin Exp Pharmacol Physiol 29:1030–1034

Green SB (2015) Can animal data translate to innovations necessary for a new era of patient-centred and individualised healthcare? Bias in preclinical animal research. BMC Med Ethics 16:53

Hawkins D, Gallacher E, Gammell M (2013) Statistical power, effect size and animal welfare: recommendations for good practice. Anim Welf 22:339–344

Heger M, van Golen RF, Broekgaarden M, Michel MC (2014) The molecular basis for the pharmacokinetics and pharmacodynamics of curcumin and its metabolites in relation to cancer. Pharmacol Rev 66:222–307

Hooijmans CR, Rovers MM, de Vries RB, Leenaars M, Ritskes-Hoitinga M, Langendam MW (2014) SYRCLE's risk of bias tool for animal studies. BMC Med Res Methodol 14:43

Howells DW, Sena ES, Macleod MR (2014) Bringing rigour to translational medicine. Nat Rev Neurol 10:37

Hsu CY (1993) Criteria for valid preclinical trials using animal stroke models. Stroke 24:633–636

Jenkins RG, Moore BB, Chambers RC, Eickelberg O, Königshoff M, Kolb M, Laurent GJ, Nanthakumar CB, Olman MA, Pardo A, Selman M, Sheppard D, Sime PJ, Tager AM, Tatler AL, Thannickal VJ, White ES (2017) An official American Thoracic Society workshop report: use of animal models for the preclinical assessment of potential therapies for pulmonary fibrosis. Am J Respir Cell Mol Biol 56:667–679

Jin LH, Andersson KE, Kwon YH, Yoon SM, Lee T (2010) Selection of a control rat for conscious spontaneous hypertensive rats in studies of detrusor overactivity on the basis of measurement of intra-abdominal pressures. Neurourol Urodyn 29:1338–1343

Kilkenny C, Parsons N, Kadyszewski E, Festing ME, Cuthill IC, Fry D, Hutton J, Altman DG (2009) Survey of the quality of experimental design, statistical analysis and reporting of research using animals. PLoS One 4:e7824

Kleiman RJ, Ehlers MD (2016) Data gaps limit the translational potential of preclinical research. Sci Transl Med 8:320ps321–320ps321

Kories C, Czyborra C, Fetscher C, Schneider T, Krege S, Michel MC (2003) Gender comparison of muscarinic receptor expression and function in rat and human urinary bladder: differential regulation of M_2 and M3? Naunyn Schmiedebergs Arch Pharmacol 367:524–531

Lapchak PA, Zhang JH, Noble-Haeusslein LJ (2013) RIGOR guidelines: escalating STAIR and STEPS for effective translational research. Transl Stroke Res 4:279–285

Llovera G, Hofmann K, Roth S, Salas-Perdomo A, Ferrer-Ferrer M, Perego C, Zanier ER, Mamrak U, Rex A, Party H, Agin V, Frauchon C, Orset C, Haelewyn B, De Simoni MG, Dirnagl U, Grittner U, Planas AM, Plesnila N, Vivien D, Liesz A (2015) Results of a preclinical randomized controlled multicenter trial (pRCT): anti-CD49d treatment for acute brain ischemia. Sci Transl Med 7:299ra121

MacLeod MR, van der Worp HB, Sena ES, Howells DW, Dirnagl U, Donnan GA (2008) Evidence for the efficacy of NXY-059 in experimental focal cerebral ischaemia is confounded by study quality. Stroke 39:2824–2829

Martić-Kehl MI, Wernery J, Folkers G, Schubiger PA (2015) Quality of animal experiments in anti-angiogenic cancer drug development – a systematic review. PLoS One 10:e0137235

Maysami S, Wong R, Pradillo JM, Denes A, Dhungana H, Malm T, Koistinaho J, Orset C, Rahman M, Rubio M, Schwaninger M, Vivien D, Bath PM, Rothwell NJ, Allan SM (2016) A cross-laboratory preclinical study on the effectiveness of interleukin-1 receptor antagonist in stroke. J Cereb Blood Flow Metab 36:596–605

McCarthy MM (2015) Incorporating sex as a variable in preclinical neuropsychiatric research. Schizophr Bull 41:1016–1020

Mets TF (1995) Drug-induced orthostatic hypotension in older patients. Drugs Aging 6:219–228

Michel MC, Korstanje C (2016) ß$_3$-Adrenoceptor agonists for overactive bladder syndrome: role of translational pharmacology in a re-positioning drug development project. Pharmacol Ther 159:66–82

Michel MC, Brunner HR, Foster C, Huo Y (2016) Angiotensin II type 1 receptor antagonists in animal models of vascular, cardiac, metabolic and renal disease. Pharmacol Ther 164:1–81

Mukai Y, Shimokawa H, Higashi M, Morikawa K, Matoba T, Hiroki J, Kunihiro I, Talukder HMA, Takeshita A (2002) Inhibition of renin-angiotensin system ameliorates endothelial dysfunction associated with aging in rats. Arterioscler Thromb Vasc Biol 22:1445–1450

Nacka-Aleksić M, Stojanović M, Pilipović I, Stojić-Vukanić Z, Kosec D, Leposavić G (2018) Strain differences in thymic atrophy in rats immunized for EAE correlate with the clinical outcome of immunization. PLoS One 13:e0201848

National Institutes of Health (2015) Consideration of sex as a biological variable in NIH-funded research, NIH guideline for using male and female animals. NIH, Betheda. https://grants.nih.gov/grants/guide/notice-files/NOT-OD-15-102.html

Perrin S (2014) Preclinical research: make mouse studies work. Nature 507:423–425

Pitkänen A, Huusko N, Ndode-Ekane XE, Kyyriäinen J, Lipponen A, Lipsanen A, Sierra A, Bolkvadze T (2014) Gender issues in antiepileptogenic treatments. Neurobiol Dis 72:224–232

Scerri C, Stewart CA, Balfour DJK, Breen KC (2012) Nicotine modifies in vivo and in vitro rat hippocampal amyloid precursor protein processing in young but not old rats. Neurosci Lett 514:22–26

Schneider T, Arumi D, Crook TJ, Sun F, Michel MC (2014) An observational study of patient satisfaction with fesoterodine in the treatment of overactive bladder: effects of additional educational material. Int J Clin Pract 68:1074–1080

Sena E, van der Worp HB, Howells D, Macleod M (2007) How can we improve the pre-clinical development of drugs for stroke? Trends Neurosci 30:433–439

Shukla P, Rao GM, Pandey G, Sharma S, Mittapelly N, Shegokar R, Mishra PR (2014) Therapeutic interventions in sepsis: current and anticipated pharmacological agents. Br J Pharmacol 171:5011–5031

Snyder HM, Shineman DW, Friedman LG, Hendrix JA, Khachaturian A, Le Guillou I, Pickett J, Refolo L, Sancho RM, Ridley SH (2016) Guidelines to improve animal study design and reproducibility for Alzheimer's disease and related dementias: for funders and researchers. Alzheimers Dement 12:1177–1185

Stroke Therapy Academic Industry Roundtable (STAIR) (1999) Recommendations for standards regarding preclinical neuroprotective and restorative drug development. Stroke 30:2752–2758

Tannenbaum C, Day D (2017) Age and sex in drug development and testing for adults. Pharmacol Res 121:83–93

Tierney MC, Curtis AF, Chertkow H, Rylett RJ (2017) Integrating sex and gender into neurodegeneration research: a six-component strategy. Alzheimers Dement Transl Res Clin Interv 3:660–667

The German Austrian Xamoterol-Study Group (1988) Double-blind placebo-controlled comparison of digoxin and xamoterol in chronic heart failure. Lancet 1:489–493

Vatner SF (2016) Why so few new cardiovascular drugs translate to the clinics. Circ Res 119:714–717

Wang Z, Sims CR, Patil NK, Gokden N, Mayeux PR (2015) Pharmacological targeting of sphingosine-1-phosphate receptor 1 improves the renal microcirculation during sepsis in the mouse. J Pharmacol Exp Ther 352:61–66

Weiss JP, Zinner NR, Klein BM, Norgaard JP (2012) Desmopressin orally disintegrating tablet effectively reduces nocturia: results of a randomized, double-blind, placebo-controlled trial. Neurourol Urodyn 31:441–447

Wieschowski S, Chin WWL, Federico C, Sievers S, Kimmelman J, Strech D (2018) Preclinical efficacy studies in investigator brochures: do they enable risk–benefit assessment? PLoS Biol 16:e2004879

Witte DG, Brune ME, Katwala SP, Milicic I, Stolarik D, Hui YH, Marsh KC, Kerwin JF Jr, Meyer MD, Hancock AA (2002) Modeling of relationships between pharmacokinetics and blockade of agonist-induced elevation of intraurethral pressure and mean arterial pressure in conscious dogs treated with α_1-adrenoceptor antagonists. J Pharmacol Exp Ther 300:495–504

Witte LPW, Peschers U, Vogel M, de la Rosette JJMCH, Michel MC (2009) Does the number of previous vaginal deliveries affect overactive bladder symptoms or their response to treatment? LUTS 1:82–87

Minimum Information and Quality Standards for Conducting, Reporting, and Organizing In Vitro Research

Christoph H. Emmerich and Christopher M. Harris

Contents

C. H. Emmerich
PAASP GmbH, Heidelberg, Germany
e-mail: christoph.emmerich@paasp.net

C. M. Harris (✉)
Molecular and Cellular Pharmacology, AbbVie Immunology, AbbVie Bioresearch Center, Worcester, MA, USA
e-mail: christopher.harris@abbvie.com

© The Author(s) 2019
A. Bespalov et al. (eds.), *Good Research Practice in Non-Clinical Pharmacology and Biomedicine*, Handbook of Experimental Pharmacology 257,
https://doi.org/10.1007/164_2019_284

Abstract

Insufficient description of experimental practices can contribute to difficulties in reproducing research findings. In response to this, "minimum information" guidelines have been developed for different disciplines. These standards help ensure that the complete experiment is described, including both experimental protocols and data processing methods, allowing a critical evaluation of the whole process and the potential recreation of the work. Selected examples of minimum information checklists with relevance for in vitro research are presented here and are collected by and registered at the MIBBI/FAIRsharing Information Resource portal.

In addition, to support integrative research and to allow for comparisons and data sharing across studies, ontologies and vocabularies need to be defined and integrated across areas of in vitro research. As examples, this chapter addresses ontologies for cells and bioassays and discusses their importance for in vitro studies.

Finally, specific quality requirements for important in vitro research tools (like chemical probes, antibodies, and cell lines) are suggested, and remaining issues are discussed.

Keywords

In vitro research · MIAME guidelines · Minimum Information · Ontologies · Quality standards

1   Introduction: Why Details Matter

As laboratory workflows become increasingly diverse and complex, it has become more challenging to adequately describe the actual methodology followed. Efficient solutions that specify a minimum of information/items that clearly and transparently define all experimental reagents, procedures, data processing, and findings of a research study are required. This is important not only to fully understand the new information generated but also to provide sufficient details for other scientists to independently replicate and verify the results.

However, it can be very difficult to decide which parameters, settings, and experimental factors are critical and therefore need to be reported. Although the level of detail might differ, the need to define minimum information (MI) requirements to follow experiments in all different fields of life sciences is not a new phenomenon (Shapin and Schaffer 1985).

In 1657, Robert Boyle and his associate, Robert Hooke, designed an air pump in order to prove the existence of the vacuum, a space devoid of matter. At that time, Boyle's air pump was the first scientific apparatus that produced vacuum – a controversial concept that many distinguished philosophers considered impossible. Inspired by Boyle's success, the Dutch mathematician and scientist Christiaan Huygens built his own air pump in Amsterdam a few years later, which was the first machine built outside Boyle's direct supervision. Interestingly, Huygens produced a phenomenon where water appeared to levitate inside a glass jar within the air pump. He called it "anomalous suspension" of water, an effect never noticed by Boyle. Boyle and Hooke could not replicate the effect in their air pump, and so the Royal Society (and with it, all of England) consequently rejected Huygens' claims. After months of dispute, Huygens finally visited Boyle and Hooke in England in 1663 and managed to reproduce his results on Boyle's own air pump. Following this, the anomalous suspension of water was accepted as a matter of fact, and Huygens was elected a Foreign Member of the Royal Society. In this way, a new form of presenting scientific experiments and studies emerged, and it enabled the reproduction of published results, thereby establishing the credibility of the author's work. In this context, Robert Boyle is recognized as one of the first scientists to introduce the Materials and Methods section into scientific publications.

As science progressed, it became more and more obvious that further progress could only be achieved if new projects and hypotheses were built on results described by other scientists. Hence, new approaches were needed to report and standardize experimental procedures and to ensure the documentation of all essential and relevant information.

The example above speaks to the long-term value of thoroughly reporting materials and methods to the scientific process. In our own time, there has been ongoing discussion of a reproducibility crisis in science. When scientists were asked what contributes to irreproducible research, concerns were both behavioral and technical (Baker 2016). Importantly, the unavailability of methods, code, and raw data from the original lab was found to be "always or often" a factor for more than 40% and "sometimes" a factor for 80% of the 1,500 respondents. For in vitro research, low external validity can partially be explained by issues with incorrect cell lines that have become too common, including an example of a cell line that may never have existed as a unique entity (Lorsch et al. 2014). Additionally, the use of so-called big data requires sophisticated data organization if such data are to be meaningful to scientists other than the source laboratory. Deficiencies in annotation of such data restrict their utility, and capturing appropriate metadata is key to understanding how the data were generated and in facilitating novel analyses and interpretations.

2   Efforts to Standardize In Vitro Protocols

Today, most in vitro techniques not only require skilled execution and experimental implementation but also the handling of digital information and large, interlinked data files, the selection of the most appropriate protocol, and the integration of quality requirements to increase reproducibility and integrity of study results. Thus, the management and processing of data has become an integral part of the daily laboratory work. There is an increasing need to employ highly specialized techniques, but optimal standards may not be intuitive to scientists not experienced in a particular method or field.

This situation has led to a growing trend for communities of researchers to define "minimum information" (MI) checklists and guidelines for the description and contextualization of research studies. These MI standards facilitate sharing and publication of data, increase data quality, and provide guidance for researchers, publishers, and reviewers. In contrast to the recording/reporting of all the information generated during an experiment, MI specifications define specific information subsets, which need to be reported and should therefore be used to standardize the content of descriptions of protocols, materials, and methods.

2.1   The MIAME Guidelines

Microarrays have become a critical platform to compare different biological conditions or systems (e.g., organs, cell types, or individuals). Importantly, data obtained and published from such assays could only be understood by other scientists and analyzed in a meaningful manner if the biological properties of all samples (e.g., sample treatment and handling) and phenotypes were known. These accompanying microarray data, however, were initially deposited on authors' websites in different formats or were not accessible at all. To address this issue, and given the often complex experimental microarray settings and the amount of data produced in a single experiment, information needed to be recorded systematically (Brazma 2009).

Thus in 2001, the Minimum Information About a Microarray Experiment (MIAME) guidelines were developed and published by the Microarray Gene Expression Data (MGED) Society (now the "Functional Genomics Data [FGED] Society"). The MIAME guidelines describe the information that allows the interpretation of microarray-based experiments unambiguously and that enables independent reproduction of published results (Brazma et al. 2001). The six most critical elements identified by MIAME are:

- The raw data for each hybridization
- The final processed (normalized) data for the set of hybridizations
- The essential sample annotation, including experimental factors and their values
- The experimental design, including sample data relationships (e.g., which raw data file relates to which sample, which hybridizations are technical or biological replicates)
- Sufficient annotation of the array (e.g., the gene identifiers, genomic coordinates, probe oligonucleotide sequences, or reference commercial array catalogue number)
- The essential laboratory and data processing protocols (e.g., which normalization method was used to obtain the final processed data)

Since its publication, the MIAME position paper has been cited over 4,100 times (as of August 2018; source: Google Scholar), demonstrating the commitment from the microarray community to these standards. Most of the major scientific journals now require authors to comply with the MIAME principles (Taylor et al. 2008). In addition, MIAME-supportive public repositories have been established, which enable the deposition and accession of experimental data and provide a searchable index functionality, enabling results to be used for new analyses and interpretations. For annotating and communicating MIAME-compliant microarray data, the spreadsheet-based MAGE-TAB (MicroArray Gene Expression Tabular) format has been developed by the FGED Society. Documents in this format can be created, viewed, and edited using commonly available spreadsheet software (e.g., Microsoft Excel) and will support the collection as well as the exchange of data between tools and databases, including submissions to public repositories (Rayner et al. 2006).

2.2   The MIBBI Portal

The success of MIAME spurred the development of appropriate guidelines for many different in vitro disciplines, summarized and collected at the Minimum Information about Biological and Biomedical Investigations (MIBBI) portal. MIBBI was created as an open-access, online resource for MI checklist projects, thereby harmonizing the various checklist development efforts (Taylor et al. 2008). MIBBI is managed by representatives of its various participant communities, which is especially valuable since it combines standards and information from several distinct disciplines.

Since 2011, MIBBI has evolved into the FAIRsharing Information Resource (https://fairsharing.org/collection/MIBBI). Being an extension of MIBBI, FAIRsharing collects and curates reporting standards, catalogues bioscience data policies, and hosts a communication forum to maintain linkages between funders, journals, and leaders of the standardization efforts. Importantly, records in FAIRsharing are both manually curated by the FAIRsharing team and edited by the research community. The FAIRsharing platform also provides a historical overview to understand versions of guidelines and policies, as well as database updates (McQuilton et al. 2016). In summary, the MIBBI/FAIRsharing initiative aims to increase connectivity between minimum information checklist projects to unify the standardization community and to maximize visibility for guideline and database developers.

Selected examples of minimum information initiatives from different in vitro disciplines are given in Table 1. Many of these are registered at the MIBBI/FAIRsharing Information Resource. Some, like Encyclopedia of DNA Elements (ENCODE) and Standards for Reporting Enzymology Data (STRENDA), serve a similar mission as the MI projects but do not refer to their output by the MI name.

2.3   Protocol Repositories

The MI approach ensures the adequacy of reported information from each study. Increasingly, scientific data are organized in databases into which dispersed groups contribute data. Biopharma databases that capture assay data on drug candidate function and disposition serve as an example, although the concepts discussed here apply generally. Databases of bioassay results often focus on final results and

Table 1 Examples of minimum information checklists from different disciplines, to ensure the reproducibility and appropriate interpretability of experiments within their domains

Name	Scope/goal	Developer	Link/publication
ENCODE	Experimental guidelines, quality standards, uniform analysis pipeline, software tools, and ontologies for epigenetic experiments	ENCODE consortium	https://www.encodeproject.org/data-standards/ (Sloan et al. 2016)
MIABE	Descriptions of interacting entities: small molecules, therapeutic proteins, peptides, carbohydrates, food additives	EMBL-EBI industry program	http://www.psidev.info/miabe (Orchard et al. 2011)
MIAME	Specification of microarray experiments: raw data, processed data, sample annotation, experimental design, annotation of the array, laboratory and data processing protocols	FGED society	http://fged.org/projects/miame/ (Brazma et al. 2001)
MIAPE	Minimum set of information about a proteomics experiment	Human proteome organization (HUPO) proteomics Standards initiative	http://www.psidev.info/miape (Binz et al. 2008)
MIFlowCyt	Flow cytometry experimental overview, sample description, instrumentation, reagents, and data analysis	International Society for Analytical Cytology (ISAC)	http://flowcyt.sourceforge.net/miflowcyt/ (Lee et al. 2008)
MIMIx	Minimum information guidelines for molecular interaction experiments	HUPO proteomics Standards initiative	http://www.psidev.info/mimix (Orchard et al. 2007)
MIQE	Quantitative PCR assay checklist, including experimental design, sample, nucleic acid extraction, reverse transcription, target information, oligonucleotides, protocol, validation, and data analysis	Group of research-active scientists	http://www.rdml.org/miqe.php (Bustin et al. 2009)
MISFISHIE	Specifications for in situ hybridization and IHC experiments: experimental design, biomaterials and treatments, reporters, staining, imaging data, and image characterization	NIH/NIDDK stem cell genome anatomy projects consortium	http://mged.sourceforge.net/misfishie/ (Deutsch et al. 2008)
STRENDA	Reagents and conditions used for enzyme activity and enzyme inhibition studies	STRENDA consortium	http://www.beilstein-institut.de/en/projects/strenda (Tipton et al. 2014)

IHC immunohistochemistry, *NIH* National Institutes of Health, *NIDDK* National Institute of Diabetes and Digestive and Kidney Diseases, *PCR* polymerase chain reaction

key metadata, and methodology is typically shared in separate protocols. Protocol repositories are used to provide accessibility, transparency, and consistency. At its simplest, a protocol repository may consist of short prose descriptions, like those used in journal articles or patent applications, and they may be stored in a shared location. A set of minimal information is necessary, but the unstructured format makes manual curation essential and tedious. Greater functionality is provided by a spreadsheet or word processor file template with structured information. For this medium solution, files can be managed as database attachments or on a web-based platform (e.g., SharePoint or Knowledge Notebook) that supports filtering, searching, linking, version tracking, and the association of limited metadata. Curation is still manual, but the defined format facilitates completeness. The most sophisticated option is a protocol database with method information contained in structured database tables. Benefits include the ability to search, filter, sort, and change at the resolution of each of the contributing pieces of data and metadata, as opposed to managing the file as a whole. In addition, the protocol database can mandate the completion of all essential fields as a condition for completing protocol registration, thereby minimizing the burden on curation. These approaches build on each other, such that completion of a simple solution facilitates implementation of the next level of functionality.

3 The Role of Ontologies for In Vitro Studies

Ontologies are a set of concepts and categories that establish the properties and relationships within a subject area. Ontologies are imperative in organizing sets of data by enabling the assignment of like and unlike samples or conditions, a necessary prelude to drawing insights on similarities and differences between experimental groups. Insufficient ontology harmonization is a limiting factor for the full utilization of large data sets to compare data from different sources. In addition, ontologies can facilitate compliance with method-reporting standards by defining minimal information fields for a method, such as those in Table 1, whose completion can be set as a condition for data deposition. Perhaps the most fundamental ontology in the life sciences is the Gene Ontology (http://www.geneontology.org/) (Gene Ontology Consortium 2001), on which others build to categorize increasing levels of complexity from gene to transcript to protein. The Ontology for Biomedical Investigations (http://obi-ontology.org/) was established to span the medical and life sciences and provides a general scope (Bandrowski et al. 2016). We will discuss two specific ontologies that are particularly relevant to quality and reproducibility of in vitro experiments, those that address cells and bioassays.

3.1 Ontologies for Cells and Cell Lines

Nearly all of the in vitro methods discussed in Sect. 2 above start with a cell-derived sample, and that cell identity is crucial for reproducibility and complete data utilization. The Cell Ontology (CL; http://cellontology.org/) provides a structured and classified vocabulary for natural cell types (Bard et al. 2005; Diehl et al. 2016). The Cell Line Ontology (CLO; http://www.clo-ontology.org/) was created to categorize cell lines, defined as a "genetically stable and homogeneous population of cultured cells that share a common propagation history" (Sarntivijai et al. 2008, 2014). The CL and CLO enable unambiguous identification of the sample source and are thus critical for data quality. The Encyclopedia of DNA Elements (ENCODE) Project employed the CL to organize its database of genomic annotations for human and mouse with over 4,000 experiments in more than 350 cell types (Malladi et al. 2015). The FANTOM5 effort uses the CL to classify transcriptomic data from more than 1,000 human and mouse samples (Lizio et al. 2015). An appropriate cell ontology is now required as a metadata standard for large data sets within the transcriptomic and functional genomics fields, and it is anticipated that additional areas will mandate this (Diehl et al. 2016).

3.2   The BioAssay Ontology

Bioassay databases store information on drug candidates and represent very large and heterogeneous collections of data that can be challenging to organize and for which there is a need to continually draw novel insights. To address these and other challenges, the BioAssay Ontology (BAO; http://bioassayontology.org) was created as the set of concepts and categories that define bioassays and their interrelationships (Vempati and Schurer 2004; Visser et al. 2011). The BAO is organized into main hierarchies that describe assay format, assay design, the target of the assay or the metatarget, any perturbagen used, and the detection technology utilized. The BAO can also define if an assay has a confirmatory, counter-screen, or other relationship to another assay. Excellent ontologies already exist for several aspects used in the BAO, and so the BAO is integrated with the Gene Ontology (Ashburner et al. 2000), the Cell Line Ontology (Sarntivijai et al. 2008), protein names from UniProt (http://www.uniprot.org), and the Unit Ontology (http://bioportal.bioontology.org/visualize/45500/). Many terms exist beneath the hierarchy summarized above, and they use a defined vocabulary. These terms constitute a set of metadata that collectively describe a unique bioassay.

3.3   Applications of the BAO to Bioassay Databases

PubChem and ChEMBL are notable publicly accessible databases with screening results on the bioactivity of millions of molecules. But the manner in which these data are structured limits their utility. For example, PubChem lists reagent concentrations in column headers, rather than as a separate field. To address this and other data organization issues, the BioAssay Research Database (BARD; https://bard.nih.gov/) was developed by seven National Institutes of Health (NIH) and academic centers (Howe et al. 2015). The BARD utilizes the BAO to organize data in the NIH Molecular Libraries Program, with over 4,000 assay definitions. In addition, BARD provides a data deposition interface that captures the appropriate metadata to structure the bioassay data.

Data organization with the BAO enables new analyses and insights. Scientists at AstraZeneca used the BAO to annotate their collection of high-throughput screening data and compared their set of assays with those available in PubChem (Zander Balderud et al. 2015). They extracted metrics on the utilization of assay design and detection approaches and considered over- vs. underutilization of potential technologies. BAO terms were also used to identify similar assays from which hits were acknowledged as frequent false-positive results in high-throughput screening (Moberg et al. 2014; Schürer et al. 2011).

As an example of a BAO implementation, AbbVie's Platform Informatics and Knowledge Management organization negotiated with representatives from the relevant scientific functions to determine those categories from the BAO that minimally defined classes of their assays. New assays are created by choosing from within the defined vocabularies, and the assay names are generated by the concatenation of those terms, e.g., 2nd messenger__ADRB2__HEK293__Isoprenaline__Human__cAMP__Antagonist__IC50 for an assay designed to measure antagonism of the β_2-adrenergic receptor. Unforeseen assay variables are accommodated by adding new terms within the same categories. AbbVie has found that adopting the BAO has reduced the curation burden, accelerated the creation of new assays, and eased the integration of external data sources.

4 Specific Examples: Quality Requirements for In Vitro Research

4.1 Chemical Probes

Chemical probes have a central role in advancing our understanding of biological processes. They provide the opportunity to modulate a biologic system without the compensatory mechanisms that come with genetic approaches, and they mimic the manner in which a disease state can be treated with small-molecule therapeutics. In the field of receptor pharmacology, some families bear the name of the natural chemical probe that led to their identification: opiate, muscarinic, nicotinic, etc. However, not all chemical probes are equal. Davies et al. described the selectivity of 28 frequently used protein kinase inhibitors and found that none were uniquely active on their presumed target at relevant concentrations (Davies et al. 2000). A probe must also be available at the site of action at sufficient concentration to modulate the target. Unfortunately, peer-reviewed articles commonly use chemical probes without reference to selectivity, solubility, or in vivo concentration.

To promote the use of quality tools, the Chemical Probes Portal (www.chemicalprobes.org) was created as a community-driven Wiki that compiles characterization data, describes optimal working conditions, and grades probes on the appropriateness of their use (Arrowsmith et al. 2015). New probes are added to the Wiki only after curation. The portal captures characterization of potency, selectivity, pharmacokinetics, and tolerability data. It is hoped that use of this portal becomes widespread, both in terms of the evaluation of probes as well as their application.

4.2   Cell Line Authentication

Cells capable of proliferating under laboratory conditions are essential tools for the study of cellular mechanisms and to define disease molecular markers and evaluate therapeutic candidates. But cell line misidentification and cross-contamination have been recognized since the 1960s (Gartler 1967; Nelson-Rees and Flandermeyer 1977). Recent examples include esophageal cell lines, which were used in over 100 publications before it was shown that they actually originated from other parts of the body (Boonstra et al. 2010). The first gastric MALT lymphoma cell line (MA-1) was described as a model for this disease in 2011. Due to misidentification, MA-1 turned out to be the already known Pfeiffer cell line, derived from diffuse large B-cell lymphoma (Capes-Davis et al. 2013). The RGC-5 rat retinal ganglion cell line was used in at least 230 publications (On authentication of cell lines 2013) but was later identified by the lab in which it originated to actually be the same as the mouse 661 W cell line, derived from photoreceptor cells (Krishnamoorthy et al. 2013). A list of over 480 known misidentified cell lines (as of August 2018) is available from the International Cell Line Authentication Committee (ICLAC; http://iclac.org/). It shows that a large number of cell lines have been found to be contaminated – HeLa cells, the first established cancer cell line, are the most frequent contaminant. It is therefore critical to ensure that all cell lines used in in vitro studies are authentic. In fact, expectations for the proper identification of cell lines have been communicated both by journal editors (On authentication of cell lines 2013) and by the National Institutes of Health (Notice Regarding Authentication of Cultured Cell Lines, NOT-OD-08-017 2007).

Short tandem repeat (STR) profiling compares the genetic signature of a particular cell line with an established database and is the standard method for unambiguous authentication of cell lines. An 80% or higher match in profiled STR loci is recommended for cell line authentication following the ANSI/ATCC ASN-0002-2011 Authentication of Human Cell Lines: Standardization of STR Profiling (http://webstore.ansi.org). This standard was developed in 2011 by the American Type Culture Collection (ATCC) working group of scientists from academia, regulatory agencies, major cell repositories, government agencies, and industry.

To provide support for bench scientists working with cell lines and to establish principles for standardization, rationalization, and international harmonization of cell and tissue culture laboratory practices, minimal requirements for quality standards in cell and tissue culture were defined (Good Cell Culture Practice) (Coecke et al. 2005), and the "guidelines for the use of cell lines" were published (Geraghty et al. 2014).

4.3 Antibody Validation

There is growing attention to the specificity and sensitivity of commercial antibodies for research applications, with respect to intended application. For example, an antibody validated for an unfolded condition (Western blotting) may not work in native context assays (immunohistochemistry or immunoprecipitation) and vice versa. In addition, lot variation can be a concern, particularly for polyclonal antibodies and particularly when raised against an entire protein and undefined epitope. Therefore, validation steps are warranted for each new batch.

A specific example where differences in detection antibodies have provided conflicting results comes from the field of neutrophil extracellular traps (NETs), which are extended complexes of decondensed chromatin and intracellular proteins. Measurement of NETs commonly uses confocal microscopy to image extracellular chromatin bound to histones, whose arginine sidechains have undergone enzymatic deimidation (conversion of arginine to citrulline, a process termed citrullination). There are conflicting reports about the presence and nature of citrullinated histones in these NET structures (Li et al. 2010; Neeli and Radic 2013). A recent report compared samples from ten NETosis stimuli using six commercially available antibodies that recognize citrullinated histones (Neeli and Radic 2016), four of which are specific for citrullinated histone H3. The report found significant differences in the number and intensity of Western blot signals detected by these antibodies, some of which were dependent on the NETosis stimulus. Since each H3 citrullination site is adjacent to a lysine known to undergo epigenetic modification, changes in epitope structure may be confounding measurements of citrullination, particularly for antibodies raised against synthetic peptides that lack lysine modification.

Current efforts to increase the reproducibility and robustness of antibody-based methods involve information-sharing requirements from journals (Gilda et al. 2015), high-quality antibody databases (CiteAB 2017), and international frameworks for antibody validation standards (Bradbury and Pluckthun 2015; GBSI 2017; Uhlen et al. 2016). In 2016, the International Working Group on Antibody Validation (IWGAV) identified key criteria for conducting antibody validation and assessing performance (Table 2) (Uhlen et al. 2016): The IWGAV recommends the use of several of the procedures described in Table 2 to properly validate an antibody for a specific application.

4.4 Webtools Without Minimal Information Criteria

Many websites serve as helpful compendiums for a diverse range of biological disciplines. Some examples include BRENDA (http://www.brenda-enzymes.org) for enzyme kinetic data, SABIO-RK (http://sabio.villa-bosch.de) for enzymatic reactions, ToxNet (https://toxnet.nlm.nih.gov/) for chemical toxicology, and Medicalgenomics (http://www.medicalgenomics.org/) for gene expression and molecular associations. But a lack of minimal information about experimental

Table 2 IWGAV strategy for antibody validation

Strategy	Model	Number of antibodies	Analysis
Genetic	Knockout or knockdown cells/tissues	Antibody of interest	Antibody-based method of choice
Orthogonal	Several samples	Antibody of interest	Correlation between antibody-based and antibody-independent assays
Independent antibodies	Lysate or tissue with target protein	Several independent antibodies with different epitopes	Specificity analysis through comparative and quantitative analysis
Tagged proteins	Lysate or tissue containing tagged and native protein	Anti-tag antibody compared with antibody of interest	Correlating the signal from the tagged and non-tagged proteins
Immunocapture followed by MS	Lysate containing protein of interest	Antibody of interest	Target immunocapture and mass spectrometry of target and potential binding partners

conditions and nonstandard ontologies often prevent the use of scientific websites to answer questions about complex systems. For example, BRENDA uses the Enzyme Commission system as ontology, but multiple gene products fall within the same reaction category. Also, comparing enzyme-specific activities without greater detail on methods is challenging. It is important to note, however, that enabling systems-wide bioinformatics analyses is not the purpose of these sites and was not envisioned when these tools were developed. Nevertheless, these sites still serve an essential function by indicating the existence of and references to data not otherwise searchable and are an evolutionary step forward in the accessibility of biological data.

4.5 General Guidelines for Reporting In Vitro Research

The animal research community has provided a guidance document called Animal Research: Reporting In Vivo Experiments (ARRIVE) (Kilkenny et al. 2010), sponsored by the National Centre for the Replacement, Refinement, and Reduction of Animals in Research. Similar guidelines for in vitro studies have not been established. The content from ARRIVE regarding title, abstract, introduction, etc., applies equally to in vitro biology. In Table 3 we discuss some salient differences with the in vivo standards, since a complete set of guidelines on reporting in vitro experiments would be best done by a dedicated community of practicing scientists. Additional reporting needs for individual areas of in vitro biology will vary greatly. Where specific MI criteria are lacking, practitioners are encouraged to work together to establish such.

Table 3 Guidance for reporting of in vitro studies and methodologies

Ethical statement	An ethical statement should indicate the ethical review process and permissions for any materials derived from human volunteers, including appropriate privacy assurances
Experimental procedures	Experimental procedures should follow MI guidelines wherever such exist. Where nonexistent, the method details given must be sufficient to reproduce the work. Parameters to consider include buffer (e.g., cell culture medium) and lysis (e.g., for cell-based studies) conditions, sample preparation and handling, volumes, concentrations, temperatures, and incubation times. Complex procedures may require a flow diagram, and novel equipment may require a picture
Materials	Commercial materials (cells, antibodies, enzymes or other proteins, nucleic acids, chemicals) should include the vendor, catalogue number, and lot number. Non-commercially sourced materials should include the quality control analyses performed to validate their identity, purity, biological activity, etc. (e.g., sequencing to confirm the correct sequence of cDNA plasmids). Similar analyses should be performed on commercial material where not supplied by the vendor, in a manner appropriate to its intended purpose
Recombinant proteins	The source of producing recombinant proteins should be disclosed, including the sequence, expression system, purification, and analyses for purity and bioactivity. Proteins produced from bacteria should be measured for endotoxin prior to any use on live cells (or animals)
Inhibitors and compounds	For inhibitors and chemical compounds, it should be stated whether or not specificity screenings to identify potential off-target effects have been performed
Cell lines	The method for purifying or preparing primary cells should be stated clearly. Cell lines should have their identity verified as described in Sect. 4.2 above, and cross-contamination should be checked regularly. Furthermore, routine testing should be performed for successful control of mycoplasma contamination. The passage number should be given, as over-passaging of cells can potentially lead to experimental artifacts. Alternatively, purchasing fresh cells for a set of experiments from a recognized animal cell culture repository (such as the American Type Culture Collection, Manassas, Virginia, or the Leibniz Institute DSMZ-German Collection of Microorganisms and Cell Cultures) may be an attractive option from both a logistic and cost perspective. Where studies monitor functional changes in live cells, results should be interpreted in respect to parallel viability/cytotoxicity measurements, particularly where a loss of function is observed
Antibodies	The specificity and possible cross-reactivity of each antibody used need to be controlled. This applies to internally generated as well as commercially available antibodies. Relevant procedures to validate an antibody for a specific method or technique are given in Table 2. Details about performed experiments to investigate antibody specificity should be described
Study design	The size of experimental and control groups should be indicated, and it should distinguish between biological and technical replicates and between distinct experiments. It should be stated whether or not randomization steps have been used to control for the spatial arrangement of samples (e.g., to avoid technical artifacts when using multi-well microtiter plates) and the order of sample collection and processing (e.g., when there is a circadian rhythm or time-of-day effect)
Statistical analysis	The type of statistical analyses should be stated clearly, including the parameter represented by any error bars. In addition, an explicit statement of how many experiments/data points were excluded from analysis and how often experiments were repeated should be included

5   Open Questions and Remaining Issues

The measures and initiatives described above for the high-quality annotation of methods and data sets were designed to increase the total value derived from (in vitro) biomedical research. However, some open questions and issues remain.

5.1   Guidelines vs. Standards

MI checklists for in vitro research, such as MIAME, are reporting guidelines and cannot describe all factors that could potentially affect the outcome of an experiment and should therefore be considered when planning and reporting robust and repro-ducible studies. MI guidelines address the question "What information is required to reproduce experiments?" rather than "Which parameters are essential, and what are potential sources of variation?" However, answering both questions is necessary to increase data robustness and integrity. Reporting guidelines alone cannot prevent irreproducible research. A newly established initiative, the RIPOSTE framework (Masca et al. 2015), therefore aims to increase data reproducibility by encouraging early discussions of study design and planning within a multidisciplinary team (including statisticians), at the time when specific questions or hypotheses are proposed.

To avoid misunderstandings, it is essential to distinguish between "guidelines" and "standards." Broadly written, guidelines do not specify the minimum threshold for data recording and reporting, and for study interpretation, but rather serve as an important starting point for the development of community-supported and vetted standards. Standards, in contrast, must define their elements clearly, specifically, and unambiguously, including the information detail necessary to fulfill all standard requirements (Burgoon 2006). As an example, the MIAME guidelines are often referred to as a standard. However, all possible experimental settings and specifications (see Sect. 2.1) are not defined by MIAME, leading to potentially alternative interpretations and therefore heterogeneous levels of experimental detail. Consequently, different MIAME-compliant studies may not collect or provide the exact same information about an experiment, complicating true data sharing and communication (Burgoon 2006).

5.2   Compliance and Acceptance

To achieve the highest acceptance and compliance among scientists, journals, database curators, funding agencies, and all other stakeholders, MI guidelines need to maintain a compromise between detail requirements and practicality in reporting, so that compliance with the developed guidelines remains practical, efficient, and realistic to implement. An evaluation of 127 microarray articles published between July 2011 and April 2012 revealed that ~75% of these publications were not compliant with the MIAME guidelines (Witwer 2013). A survey of scientists attending the 2017 European Calcified Tissue Society meeting found that a majority were familiar with performing RT-qPCR experiments, but only 6% were aware of the MIQE guidelines (Bustin 2017). These examples show that the engagement of and comprehensive vetting by the scientific community is critical for the successful adoption of workable guidelines and standards. Additional challenges to compliance involve the publication process. Some journals impose space limitations but don't support supplemental material. Some journals may encourage MI approaches in their instructions to authors, but reviewers may not be sufficiently versed in all of the reported methodology to adequately critique them or to request compliance with MI approaches.

5.3   Coordinated Efforts

The Minimum Information checklists are usually developed independently from each other. Consequently, some guidelines can be partially redundant and overlapping. Although differences in wording/nomenclature and substructuring will complicate an integration, these overlaps need to be resolved through a coordinated effort and to the satisfaction of all concerned parties.

5.4   Format and Structured Data

The three basic components of a modern reporting structure are MI specifications (see Sect. 2), controlled vocabularies (see Sect. 3), and data formats (Taylor 2007). Most MI guidelines do not provide a standard format or structured templates for presenting experimental results and accompanying information, for transmitting information from data entry to analysis software, or for the storage of data in repositories. For some guidelines (e.g., MIAME and the MAGE-TAB format), data exchange formats were developed to support scientists, but finding the perfect compromise between ease of use and level of complexity so that a standard format for most guidelines is accepted by the research community still remains a challenge.

6   Concluding Remarks

Undoubtedly, requirements regarding reporting, ontologies, research tools, and data standards will improve robustness and reproducibility of in vitro research and will facilitate the exchange and analysis of future research. In the meantime, all different stakeholders and research communities need to be engaged to ensure that the various guideline development projects and initiatives are coordinated and harmonized in a meaningful way.

Acknowledgments The authors would like to thank Chris Butler of AbbVie for helpful discussions.

References

Arrowsmith CH, Audia JE, Austin C, Baell J, Bennett J, Blagg J, Bountra C, Brennan PE, Brown PJ, Bunnage ME, Buser-Doepner C, Campbell RM, Carter AJ, Cohen P, Copeland RA, Cravatt B, Dahlin JL, Dhanak D, Edwards AM, Frederiksen M, Frye SV, Gray N, Grimshaw CE, Hepworth D, Howe T, Huber KVM, Jin J, Knapp S, Kotz JD, Kruger RG, Lowe D, Mader MM, Marsden B, Mueller-Fahrnow A, Müller S, O'Hagan RC, Overington JP, Owen DR, Rosenberg SH, Ross R, Roth B, Schapira M, Schreiber SL, Shoichet B, Sundström M, Superti-Furga G, Taunton J, Toledo-Sherman L, Walpole C, Walters MA, Willson TM, Workman P, Young RN, Zuercher WJ (2015) The promise and peril of chemical probes. Nat Chem Biol 11:536. https://doi.org/10.1038/nchembio.1867

Ashburner M, Ball CA, Blake JA, Botstein D, Butler H, Cherry JM, Davis AP, Dolinski K, Dwight SS, Eppig JT, Harris MA, Hill DP, Issel-Tarver L, Kasarskis A, Lewis S, Matese JC, Richardson JE, Ringwald M, Rubin GM, Sherlock G (2000) Gene ontology: tool for the unification of biology. The gene ontology consortium. Nat Genet 25(1):25–29. https://doi.org/10.1038/75556

Baker M (2016) 1,500 scientists lift the lid on reproducibility. Nature 533(7604):452–454. https://doi.org/10.1038/533452a

Bandrowski A, Brinkman R, Brochhausen M, Brush MH, Bug B, Chibucos MC, Clancy K, Courtot M, Derom D, Dumontier M, Fan L, Fostel J, Fragoso G, Gibson F, Gonzalez-Beltran A, Haendel MA, He Y, Heiskanen M, Hernandez-Boussard T, Jensen M, Lin Y, Lister AL, Lord P, Malone J, Manduchi E, McGee M, Morrison N, Overton JA, Parkinson H, Peters B, Rocca-Serra P, Ruttenberg A, Sansone SA, Scheuermann RH, Schober D, Smith B, Soldatova LN, Stoeckert CJ Jr, Taylor CF, Torniai C, Turner JA, Vita R, Whetzel PL, Zheng J (2016) The Ontology for biomedical investigations. PLoS One 11(4):e0154556. https://doi.org/10.1371/journal.pone.0154556

Bard J, Rhee SY, Ashburner M (2005) An ontology for cell types. Genome Biol 6(2):R21–R21. https://doi.org/10.1186/gb-2005-6-2-r21

Binz PA, Barkovich R, Beavis RC, Creasy D, Horn DM, Julian RK Jr, Seymour SL, Taylor CF, Vandenbrouck Y (2008) Guidelines for reporting the use of mass spectrometry informatics in proteomics. Nat Biotechnol 26(8):862. https://doi.org/10.1038/nbt0808-862

Boonstra JJ, van Marion R, Beer DG, Lin L, Chaves P, Ribeiro C, Pereira AD, Roque L, Darnton SJ, Altorki NK, Schrump DS, Klimstra DS, Tang LH, Eshleman JR, Alvarez H, Shimada Y, van Dekken H, Tilanus HW, Dinjens WN (2010) Verification and unmasking of widely used human esophageal adenocarcinoma cell lines. J Natl Cancer Inst 102(4):271–274. djp499 [pii]. https://doi.org/10.1093/jnci/djp499

Bradbury A, Pluckthun A (2015) Reproducibility: standardize antibodies used in research. Nature 518(7537):27–29.. 518027a [pii]. https://doi.org/10.1038/518027a

Brazma A (2009) Minimum information about a microarray experiment (MIAME)--successes, failures, challenges. ScientificWorldJournal 9:420–423. https://doi.org/10.1100/tsw.2009.57

Brazma A, Hingamp P, Quackenbush J, Sherlock G, Spellman P, Stoeckert C, Aach J, Ansorge W, Ball CA, Causton HC, Gaasterland T, Glenisson P, Holstege FC, Kim IF, Markowitz V, Matese JC, Parkinson H, Robinson A, Sarkans U, Schulze-Kremer S, Stewart J, Taylor R, Vilo J, Vingron M (2001) Minimum information about a microarray experiment (MIAME)-toward standards for microarray data. Nat Genet 29(4):365–371. https://doi.org/10.1038/ng1201-365

Burgoon LD (2006) The need for standards, not guidelines, in biological data reporting and sharing. Nat Biotechnol 24(11):1369–1373. https://doi.org/10.1038/nbt1106-1369

Bustin S (2017) The continuing problem of poor transparency of reporting and use of inappropriate methods for RT-qPCR. Biomol Detect Quantif 12:7–9. https://doi.org/10.1016/j.bdq.2017.05.001

Bustin SA, Benes V, Garson JA, Hellemans J, Huggett J, Kubista M, Mueller R, Nolan T, Pfaffl MW, Shipley GL, Vandesompele J, Wittwer CT (2009) The MIQE guidelines: minimum information for publication of quantitative real-time PCR experiments. Clin Chem 55 (4):611–622.. clinchem.2008.112797 [pii]. https://doi.org/10.1373/clinchem.2008.112797

Capes-Davis A, Alston-Roberts C, Kerrigan L, Reid YA, Barrett T, Burnett EC, Cooper JR, Freshney RI, Healy L, Kohara A, Korch C, Masters JR, Nakamura Y, Nims RW, Storts DR, Dirks WG, MacLeod RA, Drexler HG (2013) Beware imposters: MA-1, a novel MALT lymphoma cell line, is misidentified and corresponds to Pfeiffer, a diffuse large B-cell lymphoma cell line. Genes Chromosomes Cancer 52(10):986–988. https://doi.org/10.1002/gcc.22094

CiteAB (2017) The antibody search engine. https://www.citeab.com/

Coecke S, Balls M, Bowe G, Davis J, Gstraunthaler G, Hartung T, Hay R, Merten OW, Price A, Schechtman L, Stacey G, Stokes W (2005) Guidance on good cell culture practice. A report of the second ECVAM task force on good cell culture practice. Altern Lab Anim 33(3):261–287

Davies SP, Reddy H, Caivano M, Cohen P (2000) Specificity and mechanism of action of some commonly used protein kinase inhibitors. Biochem J 351(Pt 1):95–105

Deutsch EW, Ball CA, Berman JJ, Bova GS, Brazma A, Bumgarner RE, Campbell D, Causton HC, Christiansen JH, Daian F, Dauga D, Davidson DR, Gimenez G, Goo YA, Grimmond S, Henrich T, Herrmann BG, Johnson MH, Korb M, Mills JC, Oudes AJ, Parkinson HE, Pascal LE, Pollet N, Quackenbush J, Ramialison M, Ringwald M, Salgado D, Sansone SA, Sherlock G, Stoeckert CJ Jr, Swedlow J, Taylor RC, Walashek L, Warford A, Wilkinson DG, Zhou Y, Zon LI, Liu AY, True LD (2008) Minimum information specification for in situ hybridization and immunohistochemistry experiments (MISFISHIE). Nat Biotechnol 26(3):305–312.. nbt1391 [pii]. https://doi.org/10.1038/nbt1391

Diehl AD, Meehan TF, Bradford YM, Brush MH, Dahdul WM, Dougall DS, He Y, Osumi-Sutherland D, Ruttenberg A, Sarntivijai S, van Slyke CE, Vasilevsky NA, Haendel MA, Blake JA, Mungall CJ (2016) The cell ontology 2016: enhanced content, modularization, and ontology interoperability. J Biomed Semant 7(1):44. https://doi.org/10.1186/s13326-016-0088-7

Gartler SM (1967) Genetic markers as tracers in cell culture. Natl Cancer Inst Monogr 26:167–195

GBSI (2017) Global biological standards institute. https://www.gbsi.org/

Gene Ontology Consortium (2001) Creating the gene ontology resource: design and implementation. Genome Res 11(8):1425–1433. https://doi.org/10.1101/gr.180801

Geraghty RJ, Capes-Davis A, Davis JM, Downward J, Freshney RI, Knezevic I, Lovell-Badge R, Masters JR, Meredith J, Stacey GN, Thraves P, Vias M (2014) Guidelines for the use of cell lines in biomedical research. Br J Cancer 111(6):1021–1046. bjc2014166 [pii]. https://doi.org/10.1038/bjc.2014.166

Gilda JE, Ghosh R, Cheah JX, West TM, Bodine SC, Gomes AV (2015) Western blotting inaccuracies with unverified antibodies: need for a Western blotting minimal reporting standard (WBMRS). PLoS One 10(8):e0135392. https://doi.org/10.1371/journal.pone.0135392. PONE-D-15-18879 [pii]

Howe EA, de Souza A, Lahr DL, Chatwin S, Montgomery P, Alexander BR, Nguyen DT, Cruz Y, Stonich DA, Walzer G, Rose JT, Picard SC, Liu Z, Rose JN, Xiang X, Asiedu J, Durkin D, Levine J, Yang JJ, Schurer SC, Braisted JC, Southall N, Southern MR, Chung TD, Brudz S, Tanega C, Schreiber SL, Bittker JA, Guha R, Clemons PA (2015) BioAssay research database (BARD): chemical biology and probe-development enabled by structured metadata and result types. Nucleic Acids Res 43:D1163–D1170. https://doi.org/10.1093/nar/gku1244

Kilkenny C, Browne WJ, Cuthill IC, Emerson M, Altman DG (2010) Improving bioscience research reporting: the ARRIVE guidelines for reporting animal research. PLoS Biol 8(6): e1000412. https://doi.org/10.1371/journal.pbio.1000412

Krishnamoorthy RR, Clark AF, Daudt D, Vishwanatha JK, Yorio T (2013) A forensic path to RGC-5 cell line identification: lessons learned. Invest Ophthalmol Vis Sci 54(8):5712–5719. https://doi.org/10.1167/iovs.13-12085

Lee JA, Spidlen J, Boyce K, Cai J, Crosbie N, Dalphin M, Furlong J, Gasparetto M, Goldberg M, Goralczyk EM, Hyun B, Jansen K, Kollmann T, Kong M, Leif R, McWeeney S, Moloshok TD, Moore W, Nolan G, Nolan J, Nikolich-Zugich J, Parrish D, Purcell B, Qian Y, Selvaraj B, Smith C, Tchuvatkina O, Wertheimer A, Wilkinson P, Wilson C, Wood J, Zigon R, Scheuermann RH, Brinkman RR (2008) MIFlowCyt: the minimum information about a flow cytometry experiment. Cytometry A 73(10):926–930. https://doi.org/10.1002/cyto.a.20623

Li P, Li M, Lindberg MR, Kennett MJ, Xiong N, Wang Y (2010) PAD4 is essential for antibacterial innate immunity mediated by neutrophil extracellular traps. J Exp Med 207(9):1853–1862. https://doi.org/10.1084/jem.20100239

Lizio M, Harshbarger J, Shimoji H, Severin J, Kasukawa T, Sahin S, Abugessaisa I, Fukuda S, Hori F, Ishikawa-Kato S, Mungall CJ, Arner E, Baillie JK, Bertin N, Bono H, de Hoon M, Diehl AD, Dimont E, Freeman TC, Fujieda K, Hide W, Kaliyaperumal R, Katayama T, Lassmann T, Meehan TF, Nishikata K, Ono H, Rehli M, Sandelin A, Schultes EA, 't Hoen PA, Tatum Z, Thompson M, Toyoda T, Wright DW, Daub CO, Itoh M, Carninci P, Hayashizaki Y, Forrest AR, Kawaji H (2015) Gateways to the FANTOM5 promoter level mammalian expression atlas. Genome Biol 16(1):22. https://doi.org/10.1186/s13059-014-0560-6

Lorsch JR, Collins FS, Lippincott-Schwartz J (2014) Fixing problems with cell lines. Science 346 (6216):1452–1453. https://doi.org/10.1126/science.1259110

Malladi VS, Erickson DT, Podduturi NR, Rowe LD, Chan ET, Davidson JM, Hitz BC, Ho M, Lee BT, Miyasato S, Roe GR, Simison M, Sloan CA, Strattan JS, Tanaka F, Kent WJ, Cherry JM, Hong EL (2015) Ontology application and use at the ENCODE DCC. Database 2015:bav010. https://doi.org/10.1093/database/bav010

Masca NG, Hensor EM, Cornelius VR, Buffa FM, Marriott HM, Eales JM, Messenger MP, Anderson AE, Boot C, Bunce C, Goldin RD, Harris J, Hinchliffe RF, Junaid H, Kingston S, Martin-Ruiz C, Nelson CP, Peacock J, Seed PT, Shinkins B, Staples KJ, Toombs J, Wright AK, Teare MD (2015) RIPOSTE: a framework for improving the design and analysis of laboratory-based research. eLife 4. https://doi.org/10.7554/eLife.05519

McQuilton P, Gonzalez-Beltran A, Rocca-Serra P, Thurston M, Lister A, Maguire E, Sansone SA (2016) BioSharing: curated and crowd-sourced metadata standards, databases and data policies in the life sciences. Database (Oxford) 2016:baw075. https://doi.org/10.1093/database/baw075

Moberg A, Zander Balderud L, Hansson E, Boyd H (2014) Assessing HTS performance using BioAssay Ontology: screening and analysis of a bacterial phospho-N-acetylmuramoyl-penta-peptide translocase campaign. Assay Drug Dev Technol 12(9–10):506–513. https://doi.org/10.1089/adt.2014.595

Neeli I, Radic M (2013) Opposition between PKC isoforms regulates histone deimination and neutrophil extracellular chromatin release. Front Immunol 4:38. https://doi.org/10.3389/fimmu.2013.00038

Neeli I, Radic M (2016) Current challenges and limitations in antibody-based detection of citrullinated histones. Front Immunol 7:528. https://doi.org/10.3389/fimmu.2016.00528

Nelson-Rees WA, Flandermeyer RR (1977) Inter- and intraspecies contamination of human breast tumor cell lines HBC and BrCa5 and other cell cultures. Science 195(4284):1343–1344

Notice Regarding Authentication of Cultured Cell Lines, NOT-OD-08-017 (2007). https://grants.nih.gov/grants/guide/notice-files/not-od-08-017.html. Accessed 28 Nov 2007

On Authentication of Cell Lines (2013) Mol Vis 19:1848–1851

Orchard S, Salwinski L, Kerrien S, Montecchi-Palazzi L, Oesterheld M, Stumpflen V, Ceol A, Chatr-aryamontri A, Armstrong J, Woollard P, Salama JJ, Moore S, Wojcik J, Bader GD, Vidal M, Cusick ME, Gerstein M, Gavin AC, Superti-Furga G, Greenblatt J, Bader J, Uetz P, Tyers M, Legrain P, Fields S, Mulder N, Gilson M, Niepmann M, Burgoon L, De Las Rivas J, Prieto C, Perreau VM, Hogue C, Mewes HW, Apweiler R, Xenarios I, Eisenberg D, Cesareni G, Hermjakob H (2007) The minimum information required for reporting a molecular interaction experiment (MIMIx). Nat Biotechnol 25(8):894–898. nbt1324 [pii]. https://doi.org/10.1038/nbt1324

Orchard S, Al-Lazikani B, Bryant S, Clark D, Calder E, Dix I, Engkvist O, Forster M, Gaulton A, Gilson M, Glen R, Grigorov M, Hammond-Kosack K, Harland L, Hopkins A, Larminie C, Lynch N, Mann RK, Murray-Rust P, Lo Piparo E, Southan C, Steinbeck C, Wishart D, Hermjakob H, Overington J, Thornton J (2011) Minimum information about a bioactive entity (MIABE). Nat Rev Drug Discov 10(9):661–669. https://doi.org/10.1038/nrd3503

Rayner TF, Rocca-Serra P, Spellman PT, Causton HC, Farne A, Holloway E, Irizarry RA, Liu J, Maier DS, Miller M, Petersen K, Quackenbush J, Sherlock G, Stoeckert CJ Jr, White J, Whetzel PL, Wymore F, Parkinson H, Sarkans U, Ball CA, Brazma A (2006) A simple spreadsheet-based, MIAME-supportive format for microarray data: MAGE-TAB. BMC Bioinf 7:489. https://doi.org/10.1186/1471-2105-7-489

Sarntivijai S, Ade AS, Athey BD, States DJ (2008) A bioinformatics analysis of the cell line nomenclature. Bioinformatics 24(23):2760–2766. https://doi.org/10.1093/bioinformatics/btn502

Sarntivijai S, Lin Y, Xiang Z, Meehan TF, Diehl AD, Vempati UD, Schurer SC, Pang C, Malone J, Parkinson H, Liu Y, Takatsuki T, Saijo K, Masuya H, Nakamura Y, Brush MH, Haendel MA, Zheng J, Stoeckert CJ, Peters B, Mungall CJ, Carey TE, States DJ, Athey BD, He Y (2014) CLO: the cell line ontology. J Biomed Semant 5:37. https://doi.org/10.1186/2041-1480-5-37

Schürer SC, Vempati U, Smith R, Southern M, Lemmon V (2011) BioAssay Ontology annotations facilitate cross-analysis of diverse high-throughput screening data sets. J Biomol Screen 16 (4):415–426. https://doi.org/10.1177/1087057111400191

Shapin S, Schaffer S (1985) Leviathan and the air-pump. Princeton University Press, Princeton

Sloan CA, Chan ET, Davidson JM, Malladi VS, Strattan JS, Hitz BC, Gabdank I, Narayanan AK, Ho M, Lee BT, Rowe LD, Dreszer TR, Roe G, Podduturi NR, Tanaka F, Hong EL, Cherry JM (2016) ENCODE data at the ENCODE portal. Nucleic Acids Res 44(D1):D726–D732. https://doi.org/10.1093/nar/gkv1160

Taylor CF (2007) Standards for reporting bioscience data: a forward look. Drug Discov Today 12 (13–14):527–533. https://doi.org/10.1016/j.drudis.2007.05.006

Taylor CF, Field D, Sansone SA, Aerts J, Apweiler R, Ashburner M, Ball CA, Binz PA, Bogue M, Booth T, Brazma A, Brinkman RR, Michael Clark A, Deutsch EW, Fiehn O, Fostel J, Ghazal P, Gibson F, Gray T, Grimes G, Hancock JM, Hardy NW, Hermjakob H, Julian RK Jr, Kane M, Kettner C, Kinsinger C, Kolker E, Kuiper M, Le Novere N, Leebens-Mack J, Lewis SE, Lord P, Mallon AM, Marthandan N, Masuya H, McNally R, Mehrle A, Morrison N, Orchard S, Quackenbush J, Reecy JM, Robertson DG, Rocca-Serra P, Rodriguez H, Rosenfelder H, Santoyo-Lopez J, Scheuermann RH, Schober D, Smith B, Snape J, Stoeckert CJ Jr, Tipton K, Sterk P, Untergasser A, Vandesompele J, Wiemann S (2008) Promoting coherent minimum reporting guidelines for biological and biomedical investigations: the MIBBI project. Nat Biotechnol 26(8):889–896. nbt.1411 [pii]. https://doi.org/10.1038/nbt.1411

Tipton KF, Armstrong RN, Bakker BM, Bairoch A, Cornish-Bowden A, Halling PJ, Hofmeyr J-H, Leyh TS, Kettner C, Raushel FM, Rohwer J, Schomburg D, Steinbeck C (2014) Standards for reporting enzyme data: the STRENDA consortium: what it aims to do and why it should be helpful. Perspect Sci 1(1):131–137. https://doi.org/10.1016/j.pisc.2014.02.012

Uhlen M, Bandrowski A, Carr S, Edwards A, Ellenberg J, Lundberg E, Rimm DL, Rodriguez H, Hiltke T, Snyder M, Yamamoto T (2016) A proposal for validation of antibodies. Nat Methods 13(10):823–827.. nmeth.3995 [pii]. https://doi.org/10.1038/nmeth.3995

Vempati UD, Schurer SC (2004) Development and applications of the Bioassay Ontology (BAO) to describe and categorize high-throughput assays. In: Sittampalam GS, Coussens NP, Brimacombe K et al (eds) Assay guidance manual. US National Library of Medicine, Bethesda

Visser U, Abeyruwan S, Vempati U, Smith RP, Lemmon V, Schurer SC (2011) BioAssay Ontology
    (BAO): a semantic description of bioassays and high-throughput screening results. BMC Bioinf
    12:257. https://doi.org/10.1186/1471-2105-12-257
Witwer KW (2013) Data submission and quality in microarray-based microRNA profiling. Clin
    Chem 59(2):392–400. https://doi.org/10.1373/clinchem.2012.193813
Zander Balderud L, Murray D, Larsson N, Vempati U, Schürer SC, Bjäreland M, Engkvist O (2015)
    Using the BioAssay Ontology for analyzing high-throughput screening data. J Biomol Screen
    20(3):402–415. https://doi.org/10.1177/1087057114563493

Minimum Information in In Vivo Research

Patrizia Voehringer and Janet R. Nicholson

Contents

Abstract

Data quality, reproducibility and reliability are a matter of concern in many scientific fields including biomedical research. Robust, reproducible data and scientific rigour form the foundation on which future studies are built and determine the pace of knowledge gain and the time needed to develop new and innovative drugs that provide benefit to patients. Critical to the attainment of this is the precise and transparent reporting of data. In the current chapter, we will describe literature highlighting factors that constitute the minimum information that is needed to be included in the reporting of in vivo research. The main part of the chapter will focus on the minimum information that is essential for reporting in a scientific publication. In addition, we will present a table distinguishing information necessary to be recorded in a laboratory notebook or another form of internal protocols versus information that should be reported in a paper. We will use examples from the behavioural literature, in vivo studies where the use of

P. Voehringer · J. R. Nicholson (✉)
Boehringer Ingelheim Pharma GmbH & Co. KG, Biberach an der Riss, Germany
e-mail: patrizia.voehringer@boehringer-ingelheim.com;
janet.nicholson@boehringer-ingelheim.com

© The Author(s) 2019
A. Bespalov et al. (eds.), *Good Research Practice in Non-Clinical Pharmacology
and Biomedicine*, Handbook of Experimental Pharmacology 257,
https://doi.org/10.1007/164_2019_285

anaesthetics and analgesics are used and finally ex vivo studies including histological evaluations and biochemical assays.

Keywords

Behavior · Data quality · In vivo · Publication · Reporting · Reproducibility · Standards

1   Introduction

Data quality, reproducibility and reliability are a matter of concern in many scientific fields including biomedical research. Robust, reproducible data and scientific rigour form the foundation on which future studies are built and determine the pace of knowledge gain and the time needed to develop new and innovative drugs that provide benefit to patients (Freedman and Gibson 2015). In particular, research involving animals is essential for the progression of biomedical science, assuming that experiments are well designed, performed, analysed, interpreted as well as reported.

However, it has been described many times over the last few years that in preclinical research – particularly preclinical animal research – many findings presented in high-profile journals are not reliable and cannot be replicated (Begley and Ellis 2012; Peers et al. 2012; Prinz et al. 2011). This has led to the so-called reproducibility crisis which, according to some, may largely be due to the failure to adhere to good scientific and research practices and the neglect of rigorous and careful application of scientific methods (Begley and Ioannidis 2015; Collins and Tabak 2014). In this context, various reasons have been suggested to contribute to and perhaps explain the lack of reliability and reproducibility in preclinical research including inadequacies in the design, execution and statistical analysis of experiments as well as deficiencies in their reporting (Glasziou et al. 2014; Ioannidis et al. 2014; Jarvis and Williams 2016).

It has been reported that only a minority of animal studies described in the scientific literature use critical experimental design features such as randomisation and blinding despite these components being essential to the production of robust results with minimal risk of experimental bias (Hirst et al. 2014; Macleod et al. 2015). Furthermore, in a study by Bebarta et al., it was described that studies, which did not utilise randomisation and blinding, were more likely to display differences between control and treatment groups, leading to an overestimation of the magnitude of the treatment effects (Bebarta et al. 2003). Another kind of bias that may compromise the validity of preclinical research is reporting bias, consisting of publication bias as well as selective analysis and outcome reporting bias. In many cases, animal studies with negative, neutral or inconclusive results are not reported at all (publication bias), or only the analysis yielding the best statistically significant effect is selectively presented from a host of outcomes that were measured (selective analysis and outcome reporting bias) (Tsilidis et al. 2013). This under-representation of negative research findings can be misleading concerning the interpretation of presented data, often associated with falsely inflated efficacy estimates of an intervention (Korevaar et al. 2011). Furthermore, unnecessary repetitions of similar studies by investigators unaware of earlier efforts may result.

In 2005, Ioannidis stated that it can be proven that most published research findings are irreproducible or even false due to the incorrect and inadequate use of statistics for their quantification. Specifically, underlying factors such as flexible study designs, flexible statistical analyses and the conductance of small studies with low statistical power were described (Button et al. 2013; Ioannidis 2005). Along these lines, Marino expressed the view that poor understanding of statistical concepts is a main contributory factor to why so few research findings can be reproduced (Marino 2014). Thus, it is urgently required that best practices in statistical design and analysis are incorporated into the framework of the scientific purpose, thereby increasing confidence in research findings.

Additionally, transparent, clear and consistent reporting of research involving animals has become a further substantial issue. Systematic analysis has revealed that a significant proportion of publications reporting in vivo research lack information on study planning, study execution and/or statistical analysis (Avey et al. 2016; Kilkenny et al. 2009; Landis et al. 2012). This failure in reporting makes it difficult to identify potential drawbacks in the experimental design and/or data analysis of the underlying experiment, limiting the benefit and impact of the findings. Moreover, when many of these factors are intertwined, this can lead to negative consequences such as higher failure rates and poor translation between preclinical and clinical phases (Hooijmans and Ritskes-Hoitinga 2013).

Importantly, from an ethical perspective, laboratory animals should be used responsibly. In this context, it is of utmost importance to implement Russell and Burch's 3Rs (reduction, refinement, replacement) principle in the planning and execution of animal studies (Carlsson et al. 2004; Tannenbaum and Bennett 2015; Wuerbel 2017) as well as more efficient study designs, improved research methods including experimental practice, animal husbandry and care. Also the availability of sufficient information and detailed descriptions of animal studies may help to improve animal welfare and to avoid unnecessary animal experiments and wasting animals on inconclusive research.

In the past decade, several guidelines and frameworks have been released in order to improve the scientific quality, transparency and reproducibility of animal experiments (Hooijmans et al. 2010; Kilkenny et al. 2010a; Nature 2013; NIH, Principles and Guidelines for Reporting Preclinical Research). The ARRIVE (Animal Research: Reporting In Vivo Experiments) guidelines focus on the clear and transparent reporting of the minimum information that all scientific publications reporting preclinical animal research should include such as study design, experimental procedures and specific characteristics of the animals used (Kilkenny et al. 2010b). Similarly, the Gold Standard Publication Checklist (GSPC) also aims at improving the planning, design and execution of animal experiments (Hooijmans et al. 2011). The ARRIVE guidelines were launched in 2010 by a team led by the UK National Centre for the Replacement, Refinement and Reduction of Animals in Research (NC3Rs) and have steadily gained credence over the past years. Endorsed by more than 1,000 biomedical journals, the ARRIVE guidelines are now the most widely accepted key reporting recommendations for animal research (NC3Rs, ARRIVE: Animal Research: Reporting In Vivo Experiments). In addition, various leading scientific journals have begun to change their review practices and place greater emphasis on experimental details prompting authors to report all relevant information on how the study was designed, conducted and analysed (Curtis and Abernethy 2015; Curtis et al. 2015; McGrath and Lilley 2015; McNutt 2014a, b; Nature 2013). Such initiatives may help to ensure transparency and reproducibility of preclinical animal research, thereby improving its reliability and predictive value as well as maximising a successful translation into clinically-relevant applications. However, the compliance with these guidelines remains low several years later. An evaluation of papers published in *Nature* and *PLOS* journals in the 2 years before and after the ARRIVE guidelines were communicated suggests that there has been only little improvement in reporting standards and that authors, referees and editors generally are ignoring the guidelines (Baker et al. 2014). Quite recently, a similar analysis by Leung et al. has shown that the reporting quality in animal research continues to be low and that supporting the ARRIVE guidelines by several journals has not resulted in a considerable improvement of reporting standards (Leung et al. 2018). Obviously, despite the widespread endorsement of the guiding principles by multiple journals in various research areas, the impact of this endorsement on the quality of reporting standards of animal studies is only modest (Avey et al. 2016; Delgado-Ruiz et al. 2015; Liu et al. 2016; Schwarz et al. 2012; Ting et al. 2015). In part, this may be caused by the fact that the recommendations have limitations regarding feasibility and applicability across the diversity of scientific fields that comprise biomedical research making them impractical for some kind of studies. Moreover, researchers may not be convinced that it is necessary to apply effort in order to achieve maximum transparency and reproducibility of animal-based research. It is crucial to increase the awareness of the existence of animal research reporting guidelines as well as the importance of their implementation. A serious problem of guiding principles in general and the ARRIVE guidelines in particular is that most biomedical research journals endorse them but do not rigorously enforce them by urgently requiring comprehensive and detailed reporting of the performed research. A direct consequence of enforced compliance may be increased time and financial burdens making an balanced weighting between what is ideal and what is feasible and practical absolutely essential (Leung et al. 2018).

Nevertheless, the scientific community needs effective, practical and simple tools, maybe in the form of guidelines or checklists, to promote the quality of reporting preclinical animal research. Ideally, such guiding principles should be used as references earlier in the research process before performing the study, helping scientists to focus on key methodological and analytical principles and to avoid errors in the design, execution and analysis of the experiments.

A recent study by Han et al. showed that the mandatory application of a checklist improved the reporting of crucial methodological details, such as randomisation, blinding and sample size estimation, in preclinical in vivo animal studies (Han et al. 2017). Such positive examples support optimism that when reporting is distinctly required, important improvements will be achieved (Macleod 2017). Accordingly, the strict adherence to reporting guidelines will become useful to address the concerns about data reproducibility and reliability that are widely recognised in the scientific community.

In the present chapter, we discuss the minimum information that should be provided for an adequate description of in vivo experiments, in order to allow others to interpret, evaluate and eventually reproduce the study. The main part of the chapter will focus on the minimum information that is essential for the reporting in a scientific publication. In addition, a table will be presented distinguishing information necessary to be recorded in a laboratory notebook or another form of internal record versus information that should be reported in a paper. Examples of specific research areas such as behavioural experiments, anaesthesia and analgesia and their possible interference with experimental outcomes as well as ex vivo biochemical and histological analysis will be described.

2   General Aspects

Over the last decade, several guiding principles, such as the GSPC and the ARRIVE guidelines, have been developed in order to improve the quality of designing, conducting, analysing and particularly reporting preclinical animal research. These recommendations have in common that all major components of animal studies that can affect experimental outcomes, including conditions of animal housing, husbandry and care, have to be efficiently reported. In the following section, the most important aspects mentioned in these guidelines are summarised (Hooijmans et al. 2010; Kilkenny et al. 2010a). Finally, a table will be presented comparing information that is necessary to be recorded in a laboratory notebook or another form of internal protocols versus information that should be reported in a scientific publication (Table 1).

At the beginning of a preclinical animal study report, readers should be introduced to the research topic within the context of the scientific area as well as the motivation for performing the current study and the focus of the research question, specific aims and objectives. Primarily, it should be explained why the specific animal species and strain have been chosen and how this animal model can address the scientific hypotheses, particularly with regard to the clinical relevance of the project.

Any studies involving the use of laboratory animals must be formally approved by national regulatory authorities. Therefore, it is necessary to provide information indicating that the protocol used in the study has been ethically reviewed and approved. Additionally, any compliance to national or institutional guidelines and recommendations for the care and use of animals that cover the research should be stated (Jones-Bolin 2012).

In order to allow the replication of a reported study, a detailed description of the experimental animals has to be provided, including species, strain (exact genetic code/nomenclature), gender, age (at the beginning and the end of the experiment), weight (at the start of the experiment) and the origin and source of the animals.

Table 1 Necessary information for including in a publication and recording in a laboratory notebook

	Publication	Laboratory notebook
(A) *Experimental model*		
Model name	✓	✓
Background and purpose of test	✓	✓
Species	✓	✓
Sex of animals used	✓	✓
Genetic background		
• Standard name	✓	✓
• Original and current parental strain		✓
• Number of backcrosses from original to current		✓
Genetic manipulation		
• Knockout/transgenic	✓	✓
• Constitutive/inducible	✓	✓
• Cre line	✓	✓
• Doxycycline/tamoxifen	✓	✓
Previous use in an experiment		
• Drug and test naïve Y/N	✓	✓
• Description of previous procedures including details of drug washout period/return to baseline values	✓	✓
Source of animals		
• Name of commercial vendor/collaborator/in-house breeding	✓	✓
• Age and weight when received	✓	✓
• Health/immune status	✓	✓
In-house colony		
• Breeding/husbandry	✓	✓
• Breeding scheme (state female genotype first)		✓
• Duos/trios		✓
• Are all animals littermates Y/N (if Y then how)		✓
• How many cohorts are planned for each study		✓
• How far apart are the cohorts		✓
• Are all experimental groups equally represented in all cohorts (Y/N)		✓
• How often are breeders changed		✓
• Birth check frequency		✓
• Sexing age		✓
• Age at weaning		✓
• Are litters mixed at weaning		✓
Age		✓
• At start/end of experiment	✓	✓
(B) *Experimental details*		
Habituation to vivarium period (days) (if sourced from external) prior to experimental procedure	✓	✓
Assignment to experimental groups		

(continued)

Table 1 (continued)

	Publication	Laboratory notebook
• Randomisation method	✓	✓
• Matching for group assignment (name of variable matched)		✓
• Procedures to minimise bias (e.g. litter, cohort, cage, treatment order)		✓
• SOPs available (Y/N)		✓
Experimenter blinding procedures		
• Procedures to keep treatments blind	✓	✓
• Procedures to keep experimenter blind	✓	✓
• Blinding code and decoding timeline		✓
• SOPs available		✓
Training of experimenters		
• Are experimenters trained and certified in each procedure?	✓	✓
• Method of training and certification		✓
• How often is certification renewed?		✓
Sample		
• Sample size	✓	✓
• Power analysis conducted for each measure for each test	✓	✓
Experimental protocols for each test	✓	✓
• Description	✓	✓
• Tests order and rationale	✓	✓
• Duration of habituation to testing room	✓	✓
• SOPs available (Y/N)		✓
Food/water access during experiment (description)	✓	✓
• Ad libitum or restricted access to food and water during experiment	✓	✓
Adverse/noteworthy events during test	✓	✓
Exclusion criteria	✓	✓
Data processing and analysis	✓	✓
• QC methods		✓
• Primary and secondary measures for each test	✓	✓
• Analysis for each measure for each test	✓	✓
• Check to see if data meets statistical test assumptions	✓	✓
• Treatment of outliers	✓	✓
• Experimental units of analysis (animal/cage/litter)	✓	✓
• Notebooks and data storage		✓
Drug	✓	✓
Name of drug used	✓	✓
Source of drug	✓	✓
Drug batch/sample number	✓	✓
Storage prior to preparation	✓	✓
Drug preparation		✓
• Vehicle name and details of preparation	✓	✓

(continued)

Table 1 (continued)

	Publication	Laboratory notebook
• Doses and rational	✓	✓
• Dose volume	✓	✓
• Route of administration	✓	✓
• Time of administration and pretreatment time	✓	✓
• Drug storage (cold/dark/duration)	✓	✓
Blood sampling time and method (for bioanalysis)		
• Blood sampling method	✓	✓
• Blood sample volume	✓	✓
• Type of collection tube	✓	✓
• Plasma/serum preparation method	✓	✓
• Plasma/serum freezing and storage	✓	✓
Anaesthesia method and monitoring	✓	✓
Euthanasia method and monitoring	✓	✓
Genotyping tissue collection		
• Age at genotyping	✓	✓
• Method of genotyping	✓	✓
• Is genotyping repeated at end of study? (Y/N)	✓	✓
Tail samples kept (Y/N)		✓
Animal ID		
• Method used to ID animals, frequency of checking		✓
(C) *Animal facility*		
Microbial/pathogen status (if specific pathogen-free (SPF) specify pathogens)	✓	✓
Housing		
• Housing room used		✓
• Experimental rooms used		✓
• Species/sex of animals housed in same room		✓
• Caging type	✓	✓
• Controls in place for position of cages? (e.g. light differences, proximity to door)		✓
• Use of ventilated racks	✓	✓
• Number of animals per cage	✓	✓
• Are cages homogeneous for genotype	✓	✓
• Are animals regrouped at any time? If so, at what age?	✓	✓
Enrichment	✓	
• Type of bedding	✓	✓
• Toys in cage? Running wheel?	✓	✓
• Shredded paper?	✓	✓
• Igloos? Other?	✓	✓
Light/dark cycle	✓	
• Time of lights on/off	✓	✓
• Light/dark change with dawn and dusk light gradient? If Y, over what time frame?	✓	✓

(continued)

Table 1 (continued)

	Publication	Laboratory notebook
Music/sound used. If so, specify details	✓	✓
Humidity	✓	✓
Type of chow	✓	✓
Water (acidified/tap/distilled/autoclaved/filtered/other?)	✓	✓
Air exchange frequency		✓
Handling		
Frequency and duration of handling	✓	✓
Husbandry		
• No. cage changes/week		✓
• No. health checks/week		✓
Health reports from facility		✓
Personal protective equipment, description		✓
(D) *Approvals and authorisation*		
For example, IACUC or AAALAC approval number and date	✓	✓
Ethical approval statement/animal license application	✓	✓
(E) *Equipment*		
Description of equipment used	✓	✓
• Model number	✓	✓
• Vendor	✓	✓
Calibration		
• Method	✓	✓
• Frequency	✓	✓

Adapted from Brunner et al. (2016)

These biological variables are scientifically important since they often represent critical factors affecting health or disease of the animals and therefore may influence research outcomes (GV-SOLAS 1985; Oebrink and Rehbinder 2000). For the same reason, it is also essential to comment on the animals' experience and to state if they are drug naïve or if they have received any previous procedures or treatments. Additionally, information about the health, microbiological and immune status of the animals can be of high relevance for study outcomes and the ability to replicate findings and therefore should be given (GV-SOLAS 1999). This means, e.g. to depict that the animals are kept under specific pathogen-free (SPF) conditions (accompanied by a list of pathogens excluded) and that their health and microbiological status is checked and monitored according to the FELASA recommendations (Nicklas et al. 2002). When using genetically modified animals, it is important to describe their genetic background, how these animals were generated and which control animals were selected.

There is increasing evidence that elements of the laboratory environment as well as housing and husbandry practices can significantly affect the animals' biology and ultimately research outcomes (Hogan et al. 2018; Reardon 2016). This implicates an exact specification of the environmental conditions in which the animals were housed and where the experiments were conducted. The animal facility should be described concerning temperature, relative humidity, ventilation, lighting (light/dark cycle, light intensity) and noise (Baldwin et al. 2007; Speakman and Keijer 2012; Swoap et al. 2004; Van der Meer et al. 2004). In more detail, the specific housing conditions of the animals should be represented including type and size of the cages, bedding material, availability and type of environmental enrichment, number of animals per cage (and reasons for individual housing when applicable) as well as frequency of cage changes and handling procedures (Balcombe et al. 2004; Nicholson et al. 2009; Perez et al. 1997; Rock et al. 1997; van Praag et al. 2000). In addition, the reporting of nutrition and water regimes needs to be specified regarding the type (composition, special diets, purification) as well as access to food and water (ad libitum, restricted, amount of food/water and frequency and time of feeding or water supply).

When describing the procedures carried out in animal studies, several aspects require thorough consideration and need to be presented for each experiment and each experimental group, including controls. When has the experiment been performed (day and time of intervention and time interval between intervention and data sampling or processing)? Where has the experiment been performed (home cage, laboratory, special device/equipment for investigation)? What kind of intervention has been carried out? Here, details about the methodological techniques such as surgical procedures or sampling methods (including specialist equipment and suppliers) should be provided. Importantly, drugs and compounds used in the experiments need to be specified concerning name, manufacturer and concentration as well as the formulation protocol, dosage, application volume, frequency and route of administration. Additionally, when anaesthetics and analgesics are required for animal welfare reasons, it is crucial to include information about the name of these agents, administered doses, route and frequency of application as well as monitoring procedures of the animals' physiological signs that are used to guarantee a sufficient level of anaesthesia and analgesia (Flecknell 2018; Gaertner et al. 2008). Similarly, the method of euthanasia applied at the end of the study should be described (Sivula and Suckow 2018).

To ensure the quality and validity of preclinical animal research, it is crucial to indicate if the performed study is a confirmatory or hypothesis-testing one and to implement appropriate experimental study designs (Johnson and Besselsen 2002). This comprises a clear definition of the experimental unit (individual animal or group of animals in one cage) as well as the number of treatment and control (positive, negative, vehicle) groups. In this context, the reporting of animal numbers (total number per experiment as well as per experimental group) is essential to assess biological and statistical significance of the results and to re-analyse the data. Additionally, any power and sample size calculations used for the determination of adequate animal numbers that allow the generation of statistically meaningful results should be reported (Button et al. 2013). Moreover, any actions undertaken to minimise the effects of subjective bias when allocating animals to experimental groups (e.g. randomisation) and when assessing results (e.g. blinding) should be stated (Bello et al. 2014; Hirst et al. 2014; Moser 2019). Randomisation is the best method to achieve balance between treatment and control groups, whereas blinded assessment of outcomes (assessing, measuring or quantifying) improves qualitative scoring of subjective experimental observations and promotes comparable handling of data. Both strategies enhance the rigour of the experimental procedure and the scientific robustness of the results.

When reporting the results of the experiments, statistics needs to be fully described including the statistical method/test used to analyse the primary and secondary outcomes of the study (Marino 2014). The exact number of analysed animals and a measure of precision (mean, median, standard deviation, standard error of the mean, confidence interval) should be presented. This is of high relevance for interpreting the results and for evaluating the reliability of the findings. Importantly, the number of excluded animals as well as reasons and criteria to exclude them from the experiment, and hence analysis, should be well documented. Furthermore, the description of outcomes should comprise the full spectrum of positive and negative results as well as whether there were attempts to repeat or confirm the data. Equally, all relevant adverse events and any modifications that were made to the experimental protocol in order to reduce these unwanted effects should be reported.

Finally, when discussing and interpreting the findings, it is important to take into account the objectives and hypotheses of the study as predetermined in the experimental study design. Additionally, a comment on the overall scientific relevance of the outcomes as well as their potential to translate into clinical significance should be included. In order to demonstrate how animal welfare issues have been addressed in the current study, any implications of the experimental methods or results for the replacement, refinement or reduction of the use of laboratory animals in research need to be described (Taylor 2010).

In conclusion, the meaningful and accurate reporting of preclinical animal studies encompasses a plethora of aspects, ranging from a detailed description of the experimental animal to a complete documentation of the statistical analysis. Creating transparency in this way can help to evaluate studies in terms of their planning, methodology, statistical verification and reproducibility. It is highly recommended to make all raw data, analyses and protocols available to the whole research community in order to provide insight into the full workflow of the scientific project.

3 Behavioural Experiments

Behavioural animal studies are of great importance to increase the scientific knowledge about the complex processes underlying animal behaviour in general as well as to investigate potential drug effects on behavioural outcomes. Furthermore, translational research aims to identify disease-relevant endpoints in behavioural animal studies that are robust, reliable and reproducible and ultimately can be used to assess the potential of novel therapeutic agents to treat human diseases (Sukoff Rizzo and Silverman 2016).

However, performing behavioural experiments in animals is largely challenging for scientists since studies of this nature are extremely sensitive to external and environmental factors (Crabbe et al. 1999). Specific housing conditions, e.g. the lack of environmental stimulation, can interfere with brain development and selectively alter brain functions, thereby affecting the expression of certain behaviour (Wuerbel 2001). Resulting stereotypies and other abnormal repetitive behaviours can be severely confounding in behavioural experiments and have an impact on the validity, reliability and reproducibility of scientific outcomes (Garner 2005).

Additionally, when measuring behaviour in animals, there are multiple other factors that may influence the generation of a behavioural response which can be classified as 'trait', 'state' and 'technical' factors (Sousa et al. 2006). 'Trait' factors include genetic (e.g. genetic background, gender) as well as developmental characteristics (e.g. stress experience, handling, housing conditions, social hierarchy) of the animals. 'State' factors comprise the time of the experiment, the experience and training status of the investigator, characteristics of the animal (e.g. age, health status, pharmacological treatment) as well as features of the experimental setup (e.g. construction, illumination, test environment, cleansing). 'Technical' factors encompass data acquisition (e.g. automated vs. manual observation, calibration, choice of behavioural parameters) as well as data analysis (e.g. distribution, normalisation of data).

In preclinical research settings, it is difficult to standardise all such factors, which may contribute to the poor reproducibility of behavioural observations in animals across different laboratories (Wahlsten 2001). Standardisation is assumed to minimise the variability of results and to increase sensitivity and precision of the experimental procedure. However, contrary to the assumption that rigorous standardisation of animal experiments may help to ensure their reproducibility, it has been proposed that rather, systematic variation of experimental conditions (heterogenisation) can lead to the generation of robust and generalisable results across behavioural animal studies since the external validity is enhanced, thereby improving reproducibility (Richter et al. 2010; Voelkl et al. 2018; Wuerbel 2000). Nevertheless, considering that a strict and universal standardisation of laboratory environmental and experimental conditions is exceptionally unlikely, it is of major importance to take into account any possible determinants that might exert an effect on animals' performance when designing, conducting and analysing behavioural experiments and to report these factors accurately and transparently.

As mentioned above, there is increasing evidence that the laboratory environment and distinct husbandry and housing conditions may influence animal welfare and hence behaviour. Moreover, test outcomes of behavioural animal studies are highly dependent on small but important details regarding these conditions that are usually poorly reported. One such example is light conditions: light is a fundamental environmental factor regulating animal activity and physiology, and it has been found in rats that intense light conditions can lead to retinal damage, suppression of social play behaviour and locomotion as well as dissociation of circadian rhythms

(Castelhano-Carlos and Baumans 2009). Similarly, environmental sounds that are inevitably present in animal research facilities also exert considerable effects on animals' physiology and behaviour influencing sleeping patterns, locomotor activity, learning and anxiety reactions. Provision of a stable and controlled light and noise environment for the animals will contribute to their wellbeing and to the reproducibility of experimental outcomes, making a clear reporting of light and noise conditions obligatory.

Standard husbandry practices such as regularly performed cage-changing as well as commonly-used experimental procedures such as injections can significantly affect behavioural parameters in rodents, as measured by increased arousal behaviour and locomotor activity (Duke et al. 2001; Gerdin et al. 2012). These stress-related responses may have a considerable influence on the validity and quality of experimental outcomes and should be considered by researchers when designing study protocols and comparing data. Similarly, it has been shown that a change in housing conditions, including a combination of standard vs. individually ventilated cages and single vs. social housing, has a major impact on several physiological parameters and behavioural features of mice such as body weight, locomotor activity and anxiety-related behaviour (Pasquarelli et al. 2017). Thus, it is mandatory to clearly state as well as maintain a well-defined housing protocol during the experiment in order to ensure better comparison, reliability and reproducibility of experimental results across research facilities.

Environmental cage enrichment, which should be transparently reported when describing animals' housing conditions, is strongly recommended by various guidelines regulating laboratory animal care and accommodation, as it is reported to enhance animal welfare, to protect against the development of stereotypies, to reduce anxiety and to positively influence brain development as well as learning and memory behaviour (Simpson and Kelly 2011). And indeed, it has been shown in rats and mice that environmental enrichment does not result in enhanced individual data variability nor generate inconsistent data in replicate studies between multiple laboratories, indicating that housing conditions can be improved without impacting the quality or reproducibility of behavioural results (Baumans et al. 2010; Wolfer et al. 2004).

Much evidence concerning the reproducibility of behavioural animal studies comes from the area of rodent phenotyping (Kafkafi et al. 2018). Some behavioural phenotypes, such as locomotor activity, can be highly reproducible across several laboratories, suggesting high stability and therefore better reproducibility (Wahlsten et al. 2006). In contrast, other behavioural phenotypes, such as anxiety-like behaviour, are more problematic to measure since they show increased susceptibility to a multitude of environmental factors that can affect the animals' performance. Indeed, it has been reported that animal handling procedures, particularly the specific handling method itself, can elicit profound effects on animals' anxiety levels and stress responses, indicating that the use of handling methods that will not induce strong anxiety responses will minimise confounding effects during experiments (Hurst and West 2010).

One of the most commonly used methods to investigate anxiety behaviour in rodents is the elevated plus maze (EPM) test (Lister 1987; Pellow et al. 1985). Besides strain, gender and age differences, it has been shown that the manipulation of the animals prior to the experiment (e.g. exposure to stressors, housing, handling procedures) and the averseness of the test conditions themselves (e.g. increased light levels) as well as repeated testing in the EPM can strongly influence the manifestation of anxiety behaviour (Bessa et al. 2005; File 2001; Hogg 1996). These crucial factors should not be excluded from experimental descriptions when reporting. Additionally, illumination of the EPM is a critical aspect that needs to be clearly specified. In fact, Pereira et al. concluded that it is not the absolute level of luminosity upon the arms, but the relative luminosity between the open and closed arms that predicts the behavioural performance of rats in the maze (Pereira et al. 2005).

Overall, it has been suggested that animal behaviour that is more closely linked to sensory input and motor output will probably be less affected by minimal modifications within the laboratory environment, whereas behaviour that is associated with emotional and social processes will be more sensitive (Wahlsten et al. 2006).

4   Anaesthesia and Analgesia

For numerous animal experiments such as surgeries or imaging studies, the use of anaesthetics and analgesics in order to reduce animal suffering from pain and distress is an ethical obligation and crucial to the 3Rs concept (Carbone 2011). However, it is known that these drugs (as well as untreated pain itself) can severely affect the animals' biology and physiology, thereby influencing experimental data and introducing variability into research outcomes. Focusing on animal pain management means both an issue of generating high-quality, reproducible data and a substantial animal welfare concern. Dealing with this ethical and methodological conflict can pose a challenging task for scientists.

The ARRIVE guidelines recommend the reporting of anaesthesia and analgesia in order to achieve a full and detailed description of the experimental procedures performed in preclinical animal studies and to allow the critical evaluation and reproduction of published data. However, there is evidence that the current scientific literature lacks important details concerning the use of animal anaesthetics and analgesics, underestimating their potential interference with experimental results (Carbone and Austin 2016; Uhlig et al. 2015). In many cases, it is not clear whether scientists actively withhold treatment of animals with anaesthetic or analgesic drugs or just fail to include this information in the reporting, perhaps due to assumed insignificance to the experimental outcome. This creates the false impression that the selection of appropriate anaesthetic and analgesic regimens is not considered as a crucial methodological concern for generating high-quality research data. Furthermore, under-reporting of anaesthesia and pain management may also shape ongoing practice among researchers and encourage under-treatment of animals, which represents a serious problem concerning animal welfare.

Surgical pain and insufficient analgesia act as stressors and can elicit various effects on the animals' immune system, food and water consumption, social behaviour, locomotor activity as well as metabolic and hormone state, among others, which may all influence the experimental outcomes of animal studies (Leach et al. 2012; Liles and Flecknell 1993). The use of anaesthetics and analgesics relieves surgical pain, thus contributing to the refinement of the experimental methods. Additionally, following the surgical procedure, an appropriate long-term pain management, which could last for several days, is required to ensure animal wellbeing. However, anaesthetic and analgesic drugs themselves may also confound experimental results, e.g. by regulating inflammatory pathways or exerting immunomodulatory effects (Al-Hashimi et al. 2013; Fuentes et al. 2006; Galley et al. 2000; Martucci et al. 2004). In cancer studies on tumour metastasis in rats, it has been shown that analgesic drugs such as tramadol are able to prevent the effect of experimental surgery on natural killer cell activity and on the enhancement of metastatic diffusion, which needs to be taken into account when using this kind of animal model (Gaspani et al. 2002). Furthermore, as demonstrated for inhalation anaesthesia using sevoflurane in rats, the expression of circadian genes may be severely influenced, which needs to be borne in mind in the design of animal studies analysing gene expression (Kobayashi et al. 2007).

As indicated in these few examples, the selection of appropriate anaesthetic and analgesic procedures is a key factor in preclinical animal studies and has to be carefully considered in the context of the specific research question and study protocol (Gargiulo et al. 2012). Scientists need to know which particular anaesthetic and analgesic drugs were used, including name, dose, application frequency and route of administration. Importantly, concerning long-term pain management after surgery, it is recommended to specify the duration of the analgesic treatment. Moreover, when it is decided to withhold analgesics because of interference with the research project, it is essential to include the reasons for this decision when reporting the study so that this information is available to those who may subsequently wish to replicate and extend such studies (Stokes et al. 2009).

Hypothermia, hypotension, hypoxemia and respiratory depression are frequently observed side effects during animal anaesthesia that can develop to serious health problems culminating in unexpected death (Davis 2008). These risks need to be incorporated when planning and performing experiments and highlight the importance of adequate animal monitoring procedures to eliminate the incidence of complications during anaesthesia. Additionally, the reporting of such events and their practical management (e.g. the use of warming pads) is crucial for scientists trying to reproduce and evaluate research data.

Animal imaging studies have specific requirements concerning anaesthesia that are related to the use of particular methodological techniques and the duration of the experiments. The primary reason for general anaesthesia in imaging studies is the need for the restraint and immobility of the animals in order to avoid movement artefacts and to obtain signals with maximal reproducibility (Gargiulo et al. 2012).

However, anaesthetic agents can unintentionally affect physiological parameters of animals and confound the outcomes of different imaging modalities (Hildebrandt et al. 2008). As shown for positron emission tomography (PET) neuroimaging studies, the use of anaesthetics such as ketamine or isoflurane may alter neuromolecular mechanisms in animal brains, thereby leading to an incorrect representation of normal properties of the awake brain (Alstrup and Smith 2013). Moreover, repeated anaesthesia procedures and the preparation of the animals for the study may influence the processes under investigation. Physical restraint stress before the experiment can increase the anaesthetic induction doses and negatively influence the quality of some molecular imaging procedures such as PET due to altered kinetics and biodistribution of radiotracers (Hildebrandt et al. 2008). The latter effect has also been observed to be dependent on the choice of anaesthetics, the duration of fasting periods as well as to result from hypothermia observed as an adverse event from anaesthesia (Fueger et al. 2006).

As for surgical procedures, the careful selection of the most appropriate anaesthesia method addressing all the needs and goals of the specific research project and imaging modality is important (Vesce et al. 2017). Since anaesthetics can influence various physiological and pharmacological functions of the animals, monitoring of anaesthetic levels and of vital functions during imaging studies has proven useful. In order to achieve reproducible experimental conditions in imaging studies, a clear and consistent reporting of methodological details concerning the animals, fasting conditions, anaesthesia regimens and monitoring is absolutely essential.

5 Ex Vivo Biochemical and Histological Analysis

Numerous ex vivo methods, including biochemical and histological analyses, are used routinely to complement in vivo studies to add additional information or to address scientific questions which are difficult to address in an in vivo setting. The starting point for such studies is a living organism, and as such, many of the previously described considerations in, e.g. the ARRIVE guidelines are entirely applicable and should be included when reporting data from such studies. In the following section, we will highlight examples of studies where specific methodological details have been evinced to be important for outcome and as such should be included in any reporting of data from studies where similar ex vivo analyses have been carried out.

6 Histology

Histology is the microscopic study of animal and plant cells and tissues. It comprises a multistage process of cell or tissue collection and processing, sectioning, staining and examining under a microscope to finally quantification. Various methods are routinely applied in numerous cell and tissue types. The field of histology has been as affected as others by the lack of reproducibility of data across labs. In a recent report, Dukkipati et al. made the observation that conflicting data on the presence of pathological changes in cholinergic synaptic inputs (C-boutons) exists in the field of amyotrophic lateral sclerosis (ALS), thus making it difficult to assess roles of these synaptic inputs in the pathophysiology of the disease (Dukkipati et al. 2017). The authors sought to determine whether or not the reported changes described in the scientific literature are indeed truly statistically and biologically significant and to evaluate the possible reasons for why reproducibility has proven problematic. Thus, histological analyses were conducted using several variations on experimental design and data analysis and indeed, it was shown that factors including the grouping unit, sampling strategy and lack of blinding could all be contributors to the failure in replication of results. Furthermore, the lack of power analysis and effect size made the assessment of biological significance difficult. Experimental design has also been the focus of a report by Torlakovic et al. who have highlighted the importance of inclusion of appropriate and standardised controls in immunohistochemistry studies so that data can be reproduced from one test to another and indeed from one lab to another (Torlakovic et al. 2015). Lai et al. point to the difficulty in standardising complex methods in their report of the development of the OPTIClear method using fresh and archived human brain tissue (Lai et al. 2018).

A comparison of different quantification methods has been described by Wang et al. to determine hippocampal damage after cerebral ischemia (Wang et al. 2015). The authors start with the comment that multiple techniques are used to evaluate histological damage following ischemic insult although the sensitivity and repro-ducibility of these techniques is poorly characterised. Nonetheless, their output has a pivotal impact on results and conclusions drawn therefrom. In this study, two factors emerged as being important methodological aspects. Firstly, since neuronal cell death does not occur homogeneously within the CA1 region of the hippocampus, it is critical that the time post ischemic insult is accurately reported. Secondly, in terms of analysis regarding counting strategy, window size and position were both shown to have a major impact on study results and should therefore be clearly reported. Ward et al. make the point that in order to reproduce histopathological results from, e.g. the mouse, the pathology protocol, including necropsy methods and slide preparation, should be followed by interpretation of the slides by a pathologist familiar with reading mouse slides and familiar with the consensus medical nomenclature used in mouse pathology (Ward et al. 2017). Additionally, for the peer review of manuscripts where histopathology is a key part of the investigation, pathologists should be consulted.

The importance of such studies to the field is further acknowledged by the existence of numerous initiatives to improve reproducibility. For in situ hybridisation (ISH) and immunohistochemistry (IHC) biomarkers, the minimum information specification for ISH and IHC experiments (MISFISHIE) guidelines has been developed by the Stem Cell Genome Anatomy Projects consortium, and it is anticipated that compliance should enable researchers at different laboratories to fully evaluate data and reproduce experiments (Deutsch et al. 2008). The MISFISHIE checklist includes six aspects of information to be provided in the reporting of experiments ranging from experimental design, biomaterials and treatments, reporter (probe or antibody) information, staining protocols and parameters, imaging data and parameters and imaging characterisations. The use of statistics and any guidance on interpretation of results is, however, not included. The authors stress that the implementation of MISFISHIE should not remove variability in data but rather facilitate the identification of specific sources of variability. A similarly intended study describes a checklist of 20 items that are recommended to be included when reporting histopathology studies (Knijn et al. 2015). Thus, while reproducibility in histological analyses has been a problem and has perhaps hindered scientific progress, the field has adapted and adherence to new tools and guidelines that are now available offer hope that we are moving rapidly in a positive direction.

7   Ex Vivo Biochemical Analysis

Biochemical assessments can be performed in numerous ex vivo biological materials ranging from CSF to organoids and are routinely used to assess mRNA and proteins such as hormones.

Flow cytometry of ventricular myocytes is an emerging technology in cardiac research. Cellular variability and cytometer flow cell size are known to affect cytometer performance, and these two factors of variance are considered to limit assay validity and reproducibility across laboratories. In a study by Lopez et al., the authors hypothesised that washing and filtering create a bias towards sampling smaller cells than actually exist in the adult heart and they performed a study to test this (Lopez et al. 2017). The study results revealed that there was indeed a significant impact of washing and filtering on the experimental outcome and thus proposed a no-wash step in the protocol that could become part of a standard experimental design to minimise variability across labs.

Deckardt et al. have investigated the effect of a range of commonly used anaesthetics on clinical pathology measures including glucose, serum proteins, hormones and cholinesterase (Deckardt et al. 2007). The authors demonstrated differential effects of the different anaesthetics with regard to some of the measured parameters and differences across the sex and species used, thus demonstrating the importance of understanding the impact that an anaesthetic can have – even on ex vivo readouts – and to include appropriate controls. A similar study was conducted by Lelovas et al. which further highlights the importance of concise and accurate reporting of the use of anaesthetics in the collection of biological samples for biochemical readouts since their use can have a significant impact on outcome (Lelovas et al. 2017). Watters and Goodman published a comparison of basic methods in clinical studies and in in vitro tissue and cell culture studies reported in three anaesthesia journals (Watters and Goodman 1999). The authors identified 16 in vitro articles, and although they were not able to identify anything inherently wrong with the studies, they noted the small sample sizes and the lack of reporting on failures (only 2 of 53) and describe anecdotal evidence of experimenters only reporting on the experiments that work. The authors conclude with a call for all investigators to give reasons for sample size, to use randomisation and blinding wherever possible and to report exclusions and withdrawals, thus enabling an improvement in robustness and general applicability of the data published.

Antibodies are commonly used tools in research, particularly in ex vivo analyses. A common cause for the lack of reproducibility of data using antibodies could be due to the lack of thorough validation (Drucker 2016). The importance of standardised reagents has been highlighted by Venkataraman et al. who have described the establishment of a toolbox of immunoprecipation-grade monoclonal antibodies to human transcription factors with the aim of improving quality and reproducibility across labs (Venkataraman et al. 2018). This work was conducted as part of the NIH protein capture reagents programme (PCRP) which has generated over 1,500 reagents that can be used by the scientific community.

8 Perspective

An improvement in quality in preclinical research and particularly where animals are used is urgently needed. To achieve this, it is of fundamental importance to change the way experimental results are reported in the scientific literature so that data can be more easily reproduced across labs. This should enable more rapid scientific progress and reduce waste. Scientists are encouraged to adopt the existing guidelines by defining all relevant information that has to be included in publications and study reports, with the aim of enhancing the transparency, reproducibility and reliability of scientific work. Ensuring that preclinical research proceeds along structured guidelines will strengthen the robustness, rigour and validity of scientific data and ultimately the suitability of animal studies for translation into clinical trials.

We have described several important factors relating to behavioural experiments that may influence the outcomes of some selected behavioural animal studies. Obviously, this represents only a small part of the various possible variations of the laboratory environment, equipment and methodological procedures that can affect animal behaviour. However, we have indicated the importance of considering and reporting all relevant details regarding behavioural experiments, which will help to resolve the common problem of poor reproducibility of certain findings across different laboratories and to ensure high quality of behaviour animal studies.

We have highlighted the use of anaesthesia and analgesia as factors that can have a significant impact on experimental data, and it is therefore of utmost importance that their use is reported comprehensively. High animal welfare standards require the use of anaesthetics and analgesics when performing painful and stress-inducing experiments. However, since these drugs may severely influence research outcomes, it is necessary to carefully select the most suitable procedures for the scientific question under investigation and to evaluate the importance of the scientific needs in the context of animal wellbeing and existing guidelines for the description of experimental animal research should be applied. The complete reporting of anaesthesia procedures as well as pain management could significantly improve the quality and reproducibility of preclinical animal studies and enhance animal welfare.

Ex vivo measures including histological analysis and biochemical readouts are seemingly just as prone to poor reproducibility as in vivo experiments. Clearly, the precise details of the in-life part of the study should not be overlooked in the reporting of such studies since this aspect can have a significant impact on overall experimental outcome and conclusions.

The field has reached a point where something needs to be done to improve standards, and indeed, to this end, numerous initiatives are ongoing. One such initiative is the Innovative Medicines Initiative consortium project "European Quality in Preclinical Research" (EQIPD). The EQIPD project aims to identify ways to enable a smoother, faster and safer transition from preclinical to clinical testing by establishing common guidelines to strengthen the robustness, rigour and validity of research data. Numerous academic and industrial partners are involved in this initiative, which should have a significant and positive impact in the next few years. Nevertheless, the output of EQIPD and similar efforts need to be embraced and for that, the entire scientific community has an important role to play.

References

Al-Hashimi M, Scott SW, Thompson JP, Lambert DG (2013) Opioids and immune modulation: more questions than answers. Br J Anaesth 111:80–88

Alstrup AK, Smith DF (2013) Anaesthesia for positron emission tomography scanning of animal brains. Lab Anim 47:12–18

Avey MT, Moher D, Sullivan KJ, Fergusson D, Griffin G, Grimshaw JM, Hutton B, Lalu MM, Macleod M, Marshall J, Mei SH, Rudnicki M, Stewart DJ, Turgeon AF, McIntyre L, Group CCCTB (2016) The devil is in the details: incomplete reporting in preclinical animal research. PLoS One 11:e0166733

Baker D, Lidster K, Sottomayor A, Amor S (2014) Two years later: journals are not yet enforcing the ARRIVE guidelines on reporting standards for pre-clinical animal studies. PLoS Biol 12: e1001756

Balcombe JP, Barnard ND, Sandusky C (2004) Laboratory routines cause animal stress. Contemp Top Lab Anim Sci 43:42–51

Baldwin AL, Schwartz GE, Hopp DH (2007) Are investigators aware of environmental noise in animal facilities and that this noise may affect experimental data? J Am Assoc Lab Anim Sci 46:45–51

Baumans V, Van Loo P, Pham TM (2010) Standardisation of environmental enrichment for laboratory mice and rats: utilisation, practicality and variation in experimental results. Scand J Lab Anim Sci 37:101–114

Bebarta V, Luyten D, Heard K (2003) Emergency medicine animal research: does use of randomization and blinding affect the results? Acad Emerg Med 10:684–687

Begley CG, Ellis LM (2012) Drug development: raise standards for preclinical cancer research. Nature 483:531–533

Begley CG, Ioannidis JP (2015) Reproducibility in science: improving the standard for basic and preclinical research. Circ Res 116:116–126

Bello S, Krogsboll LT, Gruber J, Zhao ZJ, Fischer D, Hrobjartsson A (2014) Lack of blinding of outcome assessors in animal model experiments implies risk of observer bias. J Clin Epidemiol 67:973–983

Bessa JM, Oliveira M, Cerqueira JJ, Almeida OF, Sousa N (2005) Age-related qualitative shift in emotional behaviour: paradoxical findings after re-exposure of rats in the elevated-plus maze. Behav Brain Res 162:135–142

Brunner D, Balci B, Kabitzke P, Hill H (2016) Consensus preclinical checklist (PRECHECK): experimental conditions – rodent disclosure checklist. Int J Comp Psychol 29(1):1–5

Button KS, Ioannidis JP, Mokrysz C, Nosek BA, Flint J, Robinson ES, Munafo MR (2013) Power failure: why small sample size undermines the reliability of neuroscience. Nat Rev Neurosci 14:365–376

Carbone L (2011) Pain in laboratory animals: the ethical and regulatory imperatives. PLoS One 6: e21578

Carbone L, Austin J (2016) Pain and laboratory animals: publication practices for better data reproducibility and better animal welfare. PLoS One 11:e0155001

Carlsson HE, Hagelin J, Hau J (2004) Implementation of the 'three Rs' in biomedical research. Vet Rec 154:467–470

Castelhano-Carlos MJ, Baumans V (2009) The impact of light, noise, cage cleaning and in-house transport on welfare and stress of laboratory rats. Lab Anim 43:311–327

Collins FS, Tabak LA (2014) Policy: NIH plans to enhance reproducibility. Nature 505:612–613

Crabbe JC, Wahlsten D, Dudek BC (1999) Genetics of mouse behavior: interactions with laboratory environment. Science 284:1670–1672

Curtis MJ, Abernethy DR (2015) Revision of instructions to authors for pharmacology research and perspectives: enhancing the quality and transparency of published work. Pharmacol Res Perspect 3:e00106

Curtis MJ, Bond RA, Spina D, Ahluwalia A, Alexander SP, Giembycz MA, Gilchrist A, Hoyer D, Insel PA, Izzo AA, Lawrence AJ, MacEwan DJ, Moon LD, Wonnacott S, Weston AH, McGrath JC (2015) Experimental design and analysis and their reporting: new guidance for publication in BJP. Br J Pharmacol 172:3461–3471

Davis JA (2008) Mouse and rat anesthesia and analgesia. Curr Protoc Neurosci. Appendix 4: Appendix 4B. https://doi.org/10.1002/0471142301.nsa04bs42

Deckardt K, Weber I, Kaspers U, Hellwig J, Tennekes H, van Ravenzwaay B (2007) The effects of inhalation anaesthetics on common clinical pathology parameters in laboratory rats. Food Chem Toxicol 45:1709–1718

Delgado-Ruiz RA, Calvo-Guirado JL, Romanos GE (2015) Critical size defects for bone regeneration experiments in rabbit calvariae: systematic review and quality evaluation using ARRIVE guidelines. Clin Oral Implants Res 26:915–930

Deutsch EW, Ball CA, Berman JJ, Bova GS, Brazma A, Bumgarner RE, Campbell D, Causton HC, Christiansen JH, Daian F, Dauga D, Davidson DR, Gimenez G, Goo YA, Grimmond S, Henrich T, Herrmann BG, Johnson MH, Korb M, Mills JC, Oudes AJ, Parkinson HE, Pascal LE, Pollet N, Quackenbush J, Ramialison M, Ringwald M, Salgado D, Sansone SA, Sherlock G, Stoeckert CJ Jr, Swedlow J, Taylor RC, Walashek L, Warford A, Wilkinson DG, Zhou Y, Zon LI, Liu AY, True LD (2008) Minimum information specification for in situ hybridization and immunohistochemistry experiments (MISFISHIE). Nat Biotechnol 26:305–312

Drucker DJ (2016) Never waste a good crisis: confronting reproducibility in translational research. Cell Metab 24:348–360

Duke JL, Zammit TG, Lawson DM (2001) The effects of routine cage-changing on cardiovascular and behavioral parameters in male Sprague-Dawley rats. Contemp Top Lab Anim Sci 40:17–20

Dukkipati SS, Chihi A, Wang Y, Elbasiouny SM (2017) Experimental design and data analysis issues contribute to inconsistent results of C-bouton changes in amyotrophic lateral sclerosis. eNeuro 4:0281

File SE (2001) Factors controlling measures of anxiety and responses to novelty in the mouse. Behav Brain Res 125:151–157

Flecknell P (2018) Rodent analgesia: assessment and therapeutics. Vet J 232:70–77

Freedman LP, Gibson MC (2015) The impact of preclinical irreproducibility on drug development. Clin Pharmacol Ther 97:16–18

Fueger BJ, Czernin J, Hildebrandt I, Tran C, Halpern BS, Stout D, Phelps ME, Weber WA (2006) Impact of animal handling on the results of 18F-FDG PET studies in mice. J Nucl Med 47:999–1006

Fuentes JM, Talamini MA, Fulton WB, Hanly EJ, Aurora AR, De Maio A (2006) General anesthesia delays the inflammatory response and increases survival for mice with endotoxic shock. Clin Vaccine Immunol 13:281–288

Gaertner DJ, Hallman TM, Hankenson FC, Batchelder MA (2008) Anesthesia and analgesia in laboratory animals, 2nd edn. Academic Press, London

Galley HF, DiMatteo MA, Webster NR (2000) Immunomodulation by anaesthetic, sedative and analgesic agents: does it matter? Intensive Care Med 26:267–274

Gargiulo S, Greco A, Gramanzini M, Esposito S, Affuso A, Brunetti A, Vesce G (2012) Mice anesthesia, analgesia, and care, part I: anesthetic considerations in preclinical research. ILAR J 53:E55–E69

Garner JP (2005) Stereotypies and other abnormal repetitive behaviors: potential impact on validity, reliability, and replicability of scientific outcomes. ILAR J 46:106–117

Gaspani L, Bianchi M, Limiroli E, Panerai AE, Sacerdote P (2002) The analgesic drug tramadol prevents the effect of surgery on natural killer cell activity and metastatic colonization in rats. J Neuroimmunol 129:18–24

Gerdin AK, Igosheva N, Roberson LA, Ismail O, Karp N, Sanderson M, Cambridge E, Shannon C, Sunter D, Ramirez-Solis R, Bussell J, White JK (2012) Experimental and husbandry procedures as potential modifiers of the results of phenotyping tests. Physiol Behav 106:602–611

Glasziou P, Altman DG, Bossuyt P, Boutron I, Clarke M, Julious S, Michie S, Moher D, Wager E (2014) Reducing waste from incomplete or unusable reports of biomedical research. Lancet 383:267–276

GV-SOLAS (1985) Guidelines for specification of animals and husbandry methods when reporting the results of animal experiments. Working Committee for the Biological Characterization of Laboratory Animals/GV-SOLAS. Lab Anim 19:106–108

GV-SOLAS (1999) Implications of infectious agents on results of animal experiments. Report of the Working Group on Hygiene of the Gesellschaft fur Versuchstierkunde--Society for Laboratory Animal Science (GV-SOLAS). Lab Anim 33(Suppl 1):S39–S87

Han S, Olonisakin TF, Pribis JP, Zupetic J, Yoon JH, Holleran KM, Jeong K, Shaikh N, Rubio DM, Lee JS (2017) A checklist is associated with increased quality of reporting preclinical biomedical research: a systematic review. PLoS One 12:e0183591

Hildebrandt IJ, Su H, Weber WA (2008) Anesthesia and other considerations for in vivo imaging of small animals. ILAR J 49:17–26

Hirst JA, Howick J, Aronson JK, Roberts N, Perera R, Koshiaris C, Heneghan C (2014) The need for randomization in animal trials: an overview of systematic reviews. PLoS One 9:e98856

Hogan MC, Norton JN, Reynolds RP (2018) Environmental factors: macroenvironment versus microenvironment. In: Weichbrod RH, GAH T, Norton JN (eds) Management of animal care and use programs in research, education, and testing. CRC Press, Boca Raton, pp 461–478

Hogg S (1996) A review of the validity and variability of the elevated plus-maze as an animal model of anxiety. Pharmacol Biochem Behav 54:21–30

Hooijmans CR, Ritskes-Hoitinga M (2013) Progress in using systematic reviews of animal studies to improve translational research. PLoS Med 10:e1001482

Hooijmans CR, Leenaars M, Ritskes-Hoitinga M (2010) A gold standard publication checklist to improve the quality of animal studies, to fully integrate the Three Rs, and to make systematic reviews more feasible. Altern Lab Anim 38:167–182

Hooijmans CR, de Vries R, Leenaars M, Curfs J, Ritskes-Hoitinga M (2011) Improving planning, design, reporting and scientific quality of animal experiments by using the Gold Standard Publication Checklist, in addition to the ARRIVE guidelines. Br J Pharmacol 162:1259–1260

Hurst JL, West RS (2010) Taming anxiety in laboratory mice. Nat Methods 7:825–826

Ioannidis JP (2005) Why most published research findings are false. PLoS Med 2:e124

Ioannidis JP, Greenland S, Hlatky MA, Khoury MJ, Macleod MR, Moher D, Schulz KF, Tibshirani R (2014) Increasing value and reducing waste in research design, conduct, and analysis. Lancet 383:166–175

Jarvis MF, Williams M (2016) Irreproducibility in preclinical biomedical research: perceptions, uncertainties, and knowledge gaps. Trends Pharmacol Sci 37:290–302

Johnson PD, Besselsen DG (2002) Practical aspects of experimental design in animal research. ILAR J 43:202–206

Jones-Bolin S (2012) Guidelines for the care and use of laboratory animals in biomedical research. Curr Protoc Pharmacol. Appendix 4: Appendix 4B. https://doi.org/10.1002/0471141755. pha04bs59

Kafkafi N, Agassi J, Chesler EJ, Crabbe JC, Crusio WE, Eilam D, Gerlai R, Golani I, Gomez-Marin A, Heller R, Iraqi F, Jaljuli I, Karp NA, Morgan H, Nicholson G, Pfaff DW, Richter SH, Stark PB, Stiedl O, Stodden V, Tarantino LM, Tucci V, Valdar W, Williams RW, Wuerbel H, Benjamini Y (2018) Reproducibility and replicability of rodent phenotyping in preclinical studies. Neurosci Biobehav Rev 87:218–232

Kilkenny C, Parsons N, Kadyszewski E, Festing MF, Cuthill IC, Fry D, Hutton J, Altman DG (2009) Survey of the quality of experimental design, statistical analysis and reporting of research using animals. PLoS One 4:e7824

Kilkenny C, Browne W, Cuthill IC, Emerson M, Altman DG, Group NCRRGW (2010a) Animal research: reporting in vivo experiments: the ARRIVE guidelines. Br J Pharmacol 160:1577–1579

Kilkenny C, Browne WJ, Cuthill IC, Emerson M, Altman DG (2010b) Improving bioscience research reporting: the ARRIVE guidelines for reporting animal research. PLoS Biol 8: e1000412

Knijn N, Simmer F, Nagtegaal ID (2015) Recommendations for reporting histopathology studies: a proposal. Virchows Arch 466:611–615

Kobayashi K, Takemori K, Sakamoto A (2007) Circadian gene expression is suppressed during sevoflurane anesthesia and the suppression persists after awakening. Brain Res 1185:1–7

Korevaar DA, Hooft L, ter Riet G (2011) Systematic reviews and meta-analyses of preclinical studies: publication bias in laboratory animal experiments. Lab Anim 45:225–230

Lai HM, Liu AKL, Ng HHM, Goldfinger MH, Chau TW, DeFelice J, Tilley BS, Wong WM, Wu W, Gentleman SM (2018) Next generation histology methods for three-dimensional imaging of fresh and archival human brain tissues. Nat Commun 9:1066

Landis SC, Amara SG, Asadullah K, Austin CP, Blumenstein R, Bradley EW, Crystal RG, Darnell RB, Ferrante RJ, Fillit H, Finkelstein R, Fisher M, Gendelman HE, Golub RM, Goudreau JL, Gross RA, Gubitz AK, Hesterlee SE, Howells DW, Huguenard J, Kelner K, Koroshetz W, Krainc D, Lazic SE, Levine MS, Macleod MR, McCall JM, Moxley RT 3rd, Narasimhan K, Noble LJ, Perrin S, Porter JD, Steward O, Unger E, Utz U, Silberberg SD (2012) A call for transparent reporting to optimize the predictive value of preclinical research. Nature 490:187–191

Leach MC, Klaus K, Miller AL, Scotto di Perrotolo M, Sotocinal SG, Flecknell PA (2012) The assessment of post-vasectomy pain in mice using behaviour and the Mouse Grimace Scale. PLoS One 7:e35656

Lelovas PP, Stasinopoulou MS, Balafas EG, Nikita MA, Sikos NT, Kostomitsopoulos NG (2017) Valuation of three different anaesthetic protocols on complete blood count and biochemical parameters on Wistar rats. J Hellenic Vet Med Soc 68:587–598

Leung V, Rousseau-Blass F, Beauchamp G, Pang DSJ (2018) ARRIVE has not ARRIVEd: support for the ARRIVE (animal research: reporting of in vivo experiments) guidelines does not improve the reporting quality of papers in animal welfare, analgesia or anesthesia. PLoS One 13:e0197882

Liles JH, Flecknell PA (1993) The effects of surgical stimulus on the rat and the influence of analgesic treatment. Br Vet J 149:515–525

Lister RG (1987) The use of a plus-maze to measure anxiety in the mouse. Psychopharmacology 92:180–185

Liu Y, Zhao X, Mai Y, Li X, Wang J, Chen L, Mu J, Jin G, Gou H, Sun W, Feng Y (2016) Adherence to ARRIVE guidelines in Chinese journal reports on neoplasms in animals. PLoS One 11:e0154657

Lopez JE, Jaradeh K, Silva E, Aminololama-Shakeri S, Simpson PC (2017) A method to increase reproducibility in adult ventricular myocyte sizing and flow cytometry: avoiding cell size bias in single cell preparations. PLoS One 12:e0186792

Macleod M (2017) Findings of a retrospective, controlled cohort study of the impact of a change in Nature journals' editorial policy for life sciences research on the completeness of reporting study design and execution. bioRxiv. https://doi.org/10.1101/187245

Macleod MR, Lawson McLean A, Kyriakopoulou A, Serghiou S, de Wilde A, Sherratt N, Hirst T, Hemblade R, Bahor Z, Nunes-Fonseca C, Potluru A, Thomson A, Baginskaite J, Egan K, Vesterinen H, Currie GL, Churilov L, Howells DW, Sena ES (2015) Risk of bias in reports of in vivo research: a focus for improvement. PLoS Biol 13:e1002273

Marino MJ (2014) The use and misuse of statistical methodologies in pharmacology research. Biochem Pharmacol 87:78–92

Martucci C, Panerai AE, Sacerdote P (2004) Chronic fentanyl or buprenorphine infusion in the mouse: similar analgesic profile but different effects on immune responses. Pain 110:385–392

McGrath JC, Lilley E (2015) Implementing guidelines on reporting research using animals (ARRIVE etc.): new requirements for publication in BJP. Br J Pharmacol 172:3189–3193

McNutt M (2014a) Journals unite for reproducibility. Science 346:679

McNutt M (2014b) Reproducibility. Science 343:229

Moser P (2019) Out of control? Managing baseline variability in experimental studies with control groups. In: Handbook of experimental pharmacology, good research practice in pharmacology/experimental life sciences (in press)

National Institutes of Health (NIH) Principles and guidelines for reporting preclinical research [Internet]. Available: https://www.nih.gov/research-training/rigor-reproducibility/principles-guidelines-reporting-preclinical-research. Accessed 18 Aug 2018

Nature (2013) Reducing our irreproducibility. Nature 496:398

NC3Rs ARRIVE: animal research: reporting in vivo experiments [Internet]. Available: https://www.nc3rs.org.uk/arrive-animal-research-reporting-vivo-experiments. Accessed 23 Aug 2018

Nicholson A, Malcolm RD, Russ PL, Cough K, Touma C, Palme R, Wiles MV (2009) The response of C57BL/6J and BALB/cJ mice to increased housing density. J Am Assoc Lab Anim Sci 48:740–753

Nicklas W, Baneux P, Boot R, Decelle T, Deeny AA, Fumanelli M, Illgen-Wilcke B, FELASA (2002) Recommendations for the health monitoring of rodent and rabbit colonies in breeding and experimental units. Lab Anim 36:20–42

Oebrink KJ, Rehbinder C (2000) Animal definition: a necessity for the validity of animal experiments? Lab Anim 34:121–130

Pasquarelli N, Voehringer P, Henke J, Ferger B (2017) Effect of a change in housing conditions on body weight, behavior and brain neurotransmitters in male C57BL/6J mice. Behav Brain Res 333:35–42

Peers IS, Ceuppens PR, Harbron C (2012) In search of preclinical robustness. Nat Rev Drug Discov 11:733–734

Pellow S, Chopin P, File SE, Briley M (1985) Validation of open:closed arm entries in an elevated plus-maze as a measure of anxiety in the rat. J Neurosci Methods 14:149–167

Pereira LO, da Cunha IC, Neto JM, Paschoalini MA, Faria MS (2005) The gradient of luminosity between open/enclosed arms, and not the absolute level of Lux, predicts the behaviour of rats in the plus maze. Behav Brain Res 159:55–61

Perez C, Canal JR, Dominguez E, Campillo JE, Guillen M, Torres MD (1997) Individual housing influences certain biochemical parameters in the rat. Lab Anim 31:357–361

Prinz F, Schlange T, Asadullah K (2011) Believe it or not: how much can we rely on published data on potential drug targets? Nat Rev Drug Discov 10:712

Reardon S (2016) A mouse's house may ruin experiments. Nature 530:264

Richter SH, Garner JP, Auer C, Kunert J, Wuerbel H (2010) Systematic variation improves reproducibility of animal experiments. Nat Methods 7:167–168

Rock FM, Landi MS, Hughes HC, Gagnon RC (1997) Effects of caging type and group size on selected physiologic variables in rats. Contemp Top Lab Anim Sci 36:69–72

Schwarz F, Iglhaut G, Becker J (2012) Quality assessment of reporting of animal studies on pathogenesis and treatment of peri-implant mucositis and peri-implantitis. A systematic review using the ARRIVE guidelines. J Clin Periodontol 39(Suppl 12):63–72

Simpson J, Kelly JP (2011) The impact of environmental enrichment in laboratory rats--behavioural and neurochemical aspects. Behav Brain Res 222:246–264

Sivula CP, Suckow MA (2018) Euthanasia. In: Weichbrod RH, Thompson GAH, Norton JN (eds) Management of animal care and use programs in research, education, and testing. CRC Press, Boca Raton, pp 827–840

Sousa N, Almeida OF, Wotjak CT (2006) A hitchhiker's guide to behavioral analysis in laboratory rodents. Genes Brain Behav 5(Suppl 2):5–24

Speakman JR, Keijer J (2012) Not so hot: optimal housing temperatures for mice to mimic the thermal environment of humans. Mol Metab 2:5–9

Stokes EL, Flecknell PA, Richardson CA (2009) Reported analgesic and anaesthetic administration to rodents undergoing experimental surgical procedures. Lab Anim 43:149–154

Sukoff Rizzo SJ, Silverman JL (2016) Methodological considerations for optimizing and validating behavioral assays. Curr Protoc Mouse Biol 6:364–379

Swoap SJ, Overton JM, Garber G (2004) Effect of ambient temperature on cardiovascular parameters in rats and mice: a comparative approach. Am J Physiol Regul Integr Comp Physiol 287:R391–R396

Tannenbaum J, Bennett BT (2015) Russell and Burch's 3Rs then and now: the need for clarity in definition and purpose. J Am Assoc Lab Anim Sci 54:120–132

Taylor K (2010) Reporting the implementation of the Three Rs in European primate and mouse research papers: are we making progress? Altern Lab Anim 38:495–517

Ting KH, Hill CL, Whittle SL (2015) Quality of reporting of interventional animal studies in rheumatology: a systematic review using the ARRIVE guidelines. Int J Rheum Dis 18:488–494

Torlakovic EE, Nielsen S, Vyberg M, Taylor CR (2015) Getting controls under control: the time is now for immunohistochemistry. J Clin Pathol 68:879–882

Tsilidis KK, Panagiotou OA, Sena ES, Aretouli E, Evangelou E, Howells DW, Al-Shahi Salman R, Macleod MR, Ioannidis JP (2013) Evaluation of excess significance bias in animal studies of neurological diseases. PLoS Biol 11:e1001609

Uhlig C, Krause H, Koch T, Gama de Abreu M, Spieth PM (2015) Anesthesia and monitoring in small laboratory mammals used in anesthesiology, respiratory and critical care research: a systematic review on the current reporting in top-10 impact factor ranked journals. PLoS One 10:e0134205

Van der Meer E, Van Loo PL, Baumans V (2004) Short-term effects of a disturbed light-dark cycle and environmental enrichment on aggression and stress-related parameters in male mice. Lab Anim 38:376–383

van Praag H, Kempermann G, Gage FH (2000) Neural consequences of environmental enrichment. Nat Rev Neurosci 1:191–198

Venkataraman A, Yang K, Irizarry J, Mackiewicz M, Mita P, Kuang Z, Xue L, Ghosh D, Liu S, Ramos P, Hu S, Bayron Kain D, Keegan S, Saul R, Colantonio S, Zhang H, Behn FP, Song G, Albino E, Asencio L, Ramos L, Lugo L, Morell G, Rivera J, Ruiz K, Almodovar R, Nazario L, Murphy K, Vargas I, Rivera-Pacheco ZA, Rosa C, Vargas M, McDade J, Clark BS, Yoo S, Khambadkone SG, de Melo J, Stevanovic M, Jiang L, Li Y, Yap WY, Jones B, Tandon A, Campbell E, Montelione GT, Anderson S, Myers RM, Boeke JD, Fenyo D, Whiteley G, Bader JS, Pino I, Eichinger DJ, Zhu H, Blackshaw S (2018) A toolbox of immunoprecipitation-grade monoclonal antibodies to human transcription factors. Nat Methods 15:330–338

Vesce G, Micieli F, Chiavaccini L (2017) Preclinical imaging anesthesia in rodents. Q J Nucl Med Mol Imaging 61:1–18

Voelkl B, Vogt L, Sena ES, Wuerbel H (2018) Reproducibility of preclinical animal research improves with heterogeneity of study samples. PLoS Biol 16:e2003693

Wahlsten D (2001) Standardizing tests of mouse behavior: reasons, recommendations, and reality. Physiol Behav 73:695–704

Wahlsten D, Bachmanov A, Finn DA, Crabbe JC (2006) Stability of inbred mouse strain differences in behavior and brain size between laboratories and across decades. Proc Natl Acad Sci U S A 103:16364–16369

Wang J, Jahn-Eimermacher A, Bruckner M, Werner C, Engelhard K, Thal SC (2015) Comparison of different quantification methods to determine hippocampal damage after cerebral ischemia. J Neurosci Methods 240:67–76

Ward JM, Schofield PN, Sundberg JP (2017) Reproducibility of histopathological findings in experimental pathology of the mouse: a sorry tail. Lab Anim 46:146–151

Watters MP, Goodman NW (1999) Comparison of basic methods in clinical studies and in vitro tissue and cell culture studies reported in three anaesthesia journals. Br J Anaesth 82:295–298

Wolfer DP, Litvin O, Morf S, Nitsch RM, Lipp HP, Wuerbel H (2004) Laboratory animal welfare: cage enrichment and mouse behaviour. Nature 432:821–822

Wuerbel H (2000) Behaviour and the standardization fallacy. Nat Genet 26:263

Wuerbel H (2001) Ideal homes? Housing effects on rodent brain and behaviour. Trends Neurosci 24:207–211

Wuerbel H (2017) More than 3Rs: the importance of scientific validity for harm-benefit analysis of animal research. Lab Anim 46:164–166

A Reckless Guide to *P*-values

Local Evidence, Global Errors

Michael J. Lew

Contents

Abstract

This chapter demystifies *P*-values, hypothesis tests and significance tests and introduces the concepts of local evidence and global error rates. The local evidence is embodied in *this* data and concerns the hypotheses of interest for *this* experiment, whereas the global error rate is a property of the statistical analysis and sampling procedure. It is shown using simple examples that local evidence and global error rates can be, and should be, considered together when making inferences. Power analysis for experimental design for hypothesis testing is explained, along with the

M. J. Lew (✉)
Department of Pharmacology and Therapeutics, University of Melbourne, Parkville, VIC, Australia
e-mail: michaell@unimelb.edu.au

© The Author(s) 2019
A. Bespalov et al. (eds.), *Good Research Practice in Non-Clinical Pharmacology and Biomedicine*, Handbook of Experimental Pharmacology 257,
https://doi.org/10.1007/164_2019_286

more locally focussed expected *P*-values. Issues relating to multiple testing, HARKing and P-hacking are explained, and it is shown that, in many situations, their effects on local evidence and global error rates are in conflict, a conflict that can always be overcome by a fresh dataset from replication of key experiments. Statistics is complicated, and so is science. There is no singular right way to do either, and universally acceptable compromises may not exist. Statistics offers a wide array of tools for assisting with scientific inference by calibrating uncertainty, but statistical inference is not a substitute for scientific inference. *P*-values are useful indices of evidence and deserve their place in the statistical toolbox of basic pharmacologists.

Keywords
Evidence · Hypothesis test · Multiple testing · P-hacking · *P*-values · Scientific inference · Significance filter · Significance test · Statistical inference

1 Introduction

There is a widespread consensus that we are in the midst of a 'reproducibility crisis' and that inappropriate application of statistical methods facilitates, or even causes, irreproducibility (Ioannidis 2005; Nuzzo 2014; Colquhoun 2014; George et al. 2017; Wagenmakers et al. 2018). *P*-values are a "pervasive problem" (Wagenmakers 2007) because they are misunderstood, misapplied, and answer a question that no one asks (Royall 1997; Halsey et al. 2015; Colquhoun 2014). They exaggerate evidence (Johnson 2013; Benjamin et al. 2018) or they are irreconcilable with evidence (Berger and Sellke 1987). What's worse, 'P-hacking' amplifies their intrinsic shortcomings (Fraser et al. 2018). The inescapable conclusion, it would seem, is that *P*-values should be eliminated by replacement with Bayes factors (Goodman 2001; Wagenmakers 2007) or confidence intervals (Cumming 2008), or by simply doing without (Trafimow and Marks 2015). However, much of the blame for irreproducibility that is apportioned to *P*-values is based on pervasive and pernicious misunderstandings.

This chapter is an attempt to resolve those misunderstandings. Some might say it is a reckless attempt because history suggests that it is doomed to failure, and reckless also because it goes against much of the conventional wisdom regarding *P*-values and will therefore be seen by some as promoting inappropriate statistical practices. That's OK though, because the conventional wisdom regarding *P*-values is mistaken in important ways, and those mistakes fuel false suppositions regarding what practices are appropriate.

1.1 On the Role of Statistics

Statistics is complicated[1] but is usually presented simplistically in the statistics textbooks and courses studied by pharmacologists. Readers of those books and

---

[1]Even its grammatical form is complicated: "statistics" looks like a plural noun, but it is both plural when referring to values calculated from data and singular when referring to the discipline or approaches to data analysis.

graduates of those course should therefore be forgiven if they make the false assumption that statistics is a set of rules to be applied in order to obtain a statistically valid statistically significant. The instructions say that you match the data to the recipe, turn the crank, and bingo: it's significant, or not. If you do it right, then you might be rewarded with a star! No matter how explicable that simplistic view of statistics might be, it is far too limiting. It leads to thoughtless use of a limited set of methods and to over-reliance on the familiar but misunderstood *P*-value. It prevents the full utilisation of statistical thinking within scientific inference, and allows bad statistics to license false inferences. We have to aim for more than the rote-learning of recipes in statistics courses because while statistics is not simple, good science is harder. I therefore take as a working assumption the notion that good scientists are capable of dealing with the intricacies of statistical thinking.

I will admit upfront that it is not essential to have a statistical inference in order to make a scientific inference. For example, there is little need for a formal statistical analysis if results can be dealt with using the inter-ocular impact test.[2] However, scientific inferences can be made more securely with statistics because it offers a rich set of tools for calibrating uncertainty. Statistical analysis is particularly helpful in the penumbral 'maybe zone' where the uncertainty is relatively evenly balanced – the zone where scientists are most likely to be swayed by biases into over-interpretation of random deviations within the noise. The extra insight from a well-implemented statistical analysis can protect from the desire to find something notable, and thereby reduce the number of false claims made.

> Most people need all the help they can get to prevent them making fools of themselves by claiming that their favourite theory is substantiated by observations that do nothing of the sort.
>
> – Colquhoun (1971, p. 1)

Improved utilisation of statistical approaches would indeed help to minimise the number of times that pharmacologists make fools of themselves by reducing the number of false positive results in pharmacological journals and, consequently, reduce the number of faulty leads that fail to translate into a therapeutic (Begley and Ellis 2012). However, even ideal application of the most appropriate statistical methods would not improve the replicability of published results quite as much as might be assumed because not every result that fails to be replicated is a false positive and not every mistaken conclusion would be prevented by better statistical inferences.

Basic pharmacological studies are typically performed using biological models such as cell lines, tissue samples, or laboratory animals and so even if the original results are not false positives a replication might fail when it is conducted using different models (Drucker 2016). Replications might also fail when the original results are critically dependent on unrecognised methodological details, or on

---

[2]In other words, results that hit you right between the eyes. In the Australian vernacular the inter-ocular impact test is the bloody obvious test.

reagents such as antibodies that have properties that can vary over time or between sources (Berglund et al. 2008; Baker and Dolgin 2017; Voelkl et al. 2018). It is those types of irreproducibility rather than false positives that are responsible for many failures of published leads to translate into clinical targets or therapeutics (see also chapter "Building Robustness intro Translational Research"). The distinction being made here is between false positive inferences which lack 'internal validity' and failures of generalisability which lack 'external validity' even though correct in themselves. It is an important distinction because the former can be reduced by more appropriate use of statistical methods but the latter cannot.

The inherent objectivity of statistics can minimise the number of times that we make fools of ourselves, but just *doing statistics* is not enough, because it is not a set of rules for scientists to follow to make automated scientific inferences. To get from calibrated statistical inferences to reliable inferences about the real world, the statistical analyses have to be interpreted; thoughtfully and in the full knowledge of the properties of the tool and the nature of the real world system being probed. Some researchers might be disconcerted by the fact that statistics cannot provide such certainty, because they just want to be told whether their latest result is "real". No matter how attractive it might be to fob off onto statistics the responsibility for inferences, the answers that scientists seek cannot be answered by statistics alone.

2   All About *P*-values

P-values are not everything, and they are certainly not nothing. There are many, many useful procedures and tools in statistics that do not involve or provide *P*-values, but *P*-values are by far the most widely used inferential statistic in basic pharmacological research papers.

> P-values are a practical success but a critical failure. Scientists the world over use them, but scarcely a statistician can be found to defend them.
>
> – Senn (2001, p. 193)

Not only are *P*-values rarely defended, they are frequently derided (e.g. Berger and Sellke 1987; Lecoutre et al. 2001; Goodman 2001; Wagenmakers 2007). Even so, support for the continued use of *P*-values for at least some purposes with some caveats can be found (e.g. Nickerson 2000; Senn 2001; García-Pérez 2016; Krueger and Heck 2017). One crucial caveat is that a clear distinction has to be drawn between the dichotomisation of *P*-values into 'significant' or 'not significant' (typically on the basis of a threshold set at 0.05) and the evidential meaning of the actual numerically specified *P*-value. The former comes from a *hypothesis test* and the latter from a *significance test*. Contrary to what many readers will think and have been taught, they are not the same things. It might be argued that the battle to retain a clear distinction between significance tests and hypothesis tests has long been lost, but I have to continue that battle here because that distinction is critical for understanding the uses and misuses of *P*-values. Detailed accounts can also be found elsewhere (Huberty 1993; Senn 2001; Hubbard et al. 2003; Lenhard 2006; Hurlbert and Lombardi 2009; Lew 2012).

2.1   Hypothesis Test and Significance Test

When comparing significance tests and hypothesis tests it is conventional to note that the former are 'Fisherian' (or, perhaps, "neoFisherian" (Hurlbert and Lombardi 2009)) and the latter are 'Neyman–Pearsonian'. R.A. Fisher did not invent significance tests per se – Gossett published what became Student's *t*-test before Fisher's career had begun (Student 1908) and even that is not the first example – but Fisher did effectively popularise their use with his book *Statistical Methods for Research Workers* (1925), and he is credited with (or blamed for!) the convention of $P < 0.05$ as a criterion for 'significance'. It is important to note that Fisher's 'significant' denoted something along the lines of worthy of further consideration or investigation, which is different from what is denoted by the same word applied to the results of a hypothesis test. Hypothesis tests came later, with the 1933 paper by Neyman and Pearson that set out the workings of dichotomising hypothesis tests and also introduced of the ideas "errors of the first kind" (false positive errors; type I errors) and "errors of the second kind" (false negative errors; type II errors) and a formalisation of the concept of statistical power.

    A Neyman–Pearsonian hypothesis test is more than a simple statistical calculation. It is a method that properly encompasses experimental planning and experimenter behaviour as well. Before an experiment is conducted, the experimenter chooses α, the size of the critical region in the distribution of the test statistic, on the basis of the acceptable false positive (i.e. type I) error rate and sets the sample size on the basis of an acceptable false negative (i.e. type II) error rate. In effect the sample size, power,[3] and α are traded off against each other to obtain an experimental design with the appropriate mix of cost and error rates. In order for the error rates of the procedure to be well calibrated, the sample size and α have to be set in advance of the experiment being performed, a detail that is often overlooked by pharmacologists. After the experiment has been run and the data are in hand, the mechanics of the test involves a determination of whether the observed value of the test statistic lies within a predetermined critical region under the sampling distribution provided by a statistical model and the null hypothesis. When the observed value of the test statistic falls within the critical range, the result is 'significant' and the analyst discards the null hypothesis. When the observed test statistic falls outside the critical range, the result is 'not significant' and the null hypothesis is not discarded.

    In current practice, dichotomisation of results into significant and not significant is most often made on the basis of the observed *P*-value being less than or greater

---

[3]The 'power' of the experiment is one minus the false positive error rate, but it is a function of the true effect size, as explained later.

than a conventional threshold of 0.05, so we have the familiar $P < 0.05$ for $\alpha = 0.05$. The one-to-one relationship between the test statistic being within the critical range and the P-value being less than α means that such practice is not intrinsically problematical, but using a P-value as an intermediate in a hypothesis test obscures the nature of the test and contributes to the conflation of significance tests and hypothesis tests.

The classical Neyman–Pearsonian hypothesis test is an acceptance procedure, or a decision theory procedure (Birnbaum 1977; Hurlbert and Lombardi 2009) that does not require, or provide, a P-value. Its output is a binary decision: either reject the null hypothesis or fail to reject the null hypothesis. In contrast, a Fisherian significance test yields a P-value that encodes the evidence in the data against the null hypothesis, but not, directly, a decision. The P-value is the probability of observing data as extreme as that observed, or more extreme, when the null hypothesis is true. That probability is generated or determined by a statistical model of some sort, and so we should really include the phrase 'according to the statistical model' into the definition. In the Fisherian tradition[4] a P-value is interpreted evidentially: the smaller the P-value, the stronger the evidence against the null hypothesis and the more implausible the null hypothesis is, according to the statistical model. No behavioural or inferential consequences attach to the observed P-value and no threshold need to be applied because the P-value is a continuous index.

In practice, the probabilistic nature of P-values has proved difficult to use because people tend to mistakenly assume that the P-value measures the probability of the null hypothesis or the probability of an erroneous decision – it seems that they prefer any probability that is more noteworthy or less of a mouthful than the probability according to a statistical model of observing data as extreme or more extreme when the null hypothesis is true. Happily, there are no ordinary uses of P-values that require them to be interpreted as probabilities. My advice is to forget that P-values can be defined as probabilities and instead look at them as indices of surprisingness or unusualness of data: the smaller the P-value, the more surprising are the data compared to what the statistical model predicts when the null hypothesis is true.

---

[4]It has been argued that because Fisher regularly described experimental results as 'significant' or 'not significant' he was treating P-values dichotomously and that he used a fixed threshold for that dichotomisation (e.g. Lehmann 2011, pp. 51–53). However, Fisher meant the word 'significant' to denote only a result that is worthy of attention and follow-up, and he quoted P-values as being less than 0.05, 0.02, and 0.01 because he was working from tables of critical values of test statistics rather than laboriously calculating exact P-values manually. He wrote about the issue on several occasions, for example:

> Convenient as it is to note that a hypothesis is contradicted at some familiar level of significance such as 5% or 2% or 1% we do not, in Inductive Inference, ever need to lose sight of the exact strength which the evidence has in fact reached, or to ignore the fact that with further trial it might come to be stronger, or weaker.

> – Fisher (1960, p. 25)

Conflation of significance tests and hypothesis tests may be encouraged by their apparently equivalent outputs (significance and *P*-values), but the conflation is too often encouraged by textbook authors, even to the extent of presenting a hybrid approach containing features of both. The problem has deep roots: when Neyman and Pearson published their hypothesis test in 1933 it was immediately assumed that their test was an extension of Fisher's significance tests. Substantive differences in the philosophical and theoretical underpinnings soon became apparent to the protagonists and a long-lasting and bitter personal enmity developed between Fisher and Neyman (Lenhard 2006; Lehmann 2011). That feud seems likely to be one of the causes of the confusion that we have today as it has been suggested that authors of statistics textbooks avoided taking sides in the feud – an understandable response given vehemence and the forceful personalities of the protagonists – either by presenting only one of the approaches without mention of the other or by presenting a mixture of both (Cowles 1989; Huberty 1993; Halpin and Stam 2006).

Whatever the origin of the confusion, the fact that significance tests and hypothesis tests are rarely explained as distinct alternatives in textbooks encourages many to mistakenly assume that 'significance test' and 'hypothesis test' are synonyms. It also encourages to use a hybrid of the two which is commonly called NHST (null hypothesis significance test). NHST has been derided, for example, as an "inconsistent mishmash" (Gigerenzer 1998) and as a "jerry-built framework" (Krueger and Heck 2017, p. 1) but versions of NHST are nonetheless more common than well-constructed hypothesis tests and significance tests together. Users of NHST almost universally assume that they are 'doing it right' and the assumption that *P*-value equals NHST persists, largely unnoticed, particularly in the commentaries of those clamouring for the elimination of *P*-values. I therefore feel compelled to add to the list of derogatory epithets: NHST is like a reverso-platypus. The platypus was at one time derided as a fake[5] – a composite creature consisting of parts of several animals – but is a real animal, rare but beautiful, and perfectly adapted to its ecological niche. The common NHST is assumed by its many users to be a proper statistical procedure but is, in fact, an ugly composite, maladapted for almost all analytic purposes.

2.2   Contradictory Instructions

No one should be using NHST, but should we use hypothesis testing or significance testing? The answer should depend on what your analytical objectives are, but in practice it more often depends on who you ask. Not all advice is good advice, and not even the experts agree. Responses to the American Statistical Association's official

---

[5]Well, that's the conventional wisdom, but it may be an exaggeration. The first scientific description of the "duck-billed platypus" was done in England by Shaw and Nodder (1789), who wrote "Of all Mammalia yet known it seems the most extraordinary in its conformation; exhibiting the perfect resemblance of the beak of a Duck engrafted on the head of a quadruped. So accurate is the similitude that, at first view, it naturally excites the idea of some deceptive preparation by artificial means". If Shaw and Nodder really thought it a fake, they did not do so for long.

statement on *P*-values provide a case in point. In response to the widespread expressions of concern over the misuse and misunderstanding of *P*-values, the ASA convened a group of experts to consider the issues and to collaborate on drafting an official statement on *P*-values (Wasserstein and Lazar 2016). Invited commentaries were published alongside the final statement, and even a brief reading of those commentaries on the statement will turn up misgivings and disagreements. Given that most of the commentaries were written by participants in the expert group, such disquiet and dissent confirms the difficulty of this topic. It should also signal to readers that their practical familiarity with *P*-values does not ensure that they understand *P*-values.

The official ASA statement on *P*-values sets out six numbered principles concerning *P*-values and scientific inference:

1. *P*-values can indicate how incompatible the data are with a specified statistical model.
2. *P*-values do not measure the probability that the studied hypothesis is true, or the chance that the data were produced by random chance.
3. Scientific conclusions and business or policy decisions should not be based only on whether a *P*-value passes a specific threshold.
4. Proper inference requires full reporting and transparency.
5. A *P*-value, or statistical significance, does not measure the size of an effect or the importance of a result.
6. By itself, a *P*-value does not provide a good measure of evidence regarding a model or hypothesis.

Those principles are all sound – some derive directly from the definition of *P*-values and some are self-evidently good advice about the formation and reporting of scientific conclusions – but hypothesis tests and significance tests are not even mentioned in the statement and so it does not directly answer the question about whether we should use significance tests or hypothesis tests that I asked at the start of this section. Nevertheless, the statement offers a useful perspective and is not entirely neutral on the question. It urges against the use of a threshold in Principle 3 which says "Scientific conclusions and business or policy decisions should not be based only on whether a p-value passes a specific threshold". Without a threshold we cannot use a hypothesis test. Lest any reader think that the intention is that *P*-values should not be used, I point out that the explanatory note for that principle in the ASA document begins thus:

> Practices that reduce data analysis or scientific inference to mechanical "bright-line" rules (such as "$p < 0.05$") for justifying scientific claims or conclusions can lead to erroneous beliefs and poor decision making.
>
> – Wasserstein and Lazar (2016, p. 131)

"Bright-line rule" is an American legal phrase denoting an approach to simplifying ambiguous or complex legal issues by establishment of a clear,

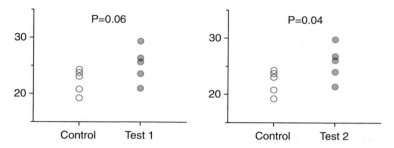

Fig. 1 P = 0.04 is not very different from P = 0.06. Pseudo-data devised to yield one-tailed $P = 0.06$ (left) and $P = 0.04$ (right) from a Student's *t*-test for independent samples, $n = 5$ per group. The *y*-axis is an arbitrarily scaled measure

consistent ruling on the basis of objective factors. In other words, subtleties of circumstance and subjective factors are ignored in favour of consistency and simplicity. Such a rule might be useful in the legal setting, but it does not sound like an approach well-suited to the considerations that should underlie scientific inference. It is unfortunate, therefore, that a mechanical bright-line rule is so often used in basic pharmacological research, and even worse that it is demanded by the instructions to authors of the *British Journal of Pharmacology*:

> When comparing groups, a level of probability (*P*) deemed to constitute the threshold for statistical significance should be defined in Methods, and not varied later in Results (by presentation of multiple levels of significance). Thus, ordinarily $P < 0.05$ should be used throughout a paper to denote statistically significant differences between groups. – Curtis et al. (2015)

An updated version of the guidelines retains those instructions (Curtis et al. 2018), but because it is a bad instruction I present three objections. The first is that routine use of an arbitrary *P*-value threshold for declaring a result significantly ignores almost all of the evidential content of the *P*-value by forcing an all-or-none distinction between a *P*-value small enough and one not small enough. The arbitrariness of a threshold for significance is well known and flows from the fact that there is no natural cutoff point or inflection point in the scale of *P*-values. Anyone who is unconvinced that it matters should note that the evidence in a result of $P = 0.06$ is not so different from that in a result of $P = 0.04$ as to support an opposite conclusion (Fig. 1).

The second objection to the instruction to use a threshold of $P < 0.05$ is that exclusive focus on whether the result is above or below the threshold blinds analysts to information beyond the sample in question. If the statistical procedure says discard the null hypothesis (or don't discard it), then that statistical decision seems to override and make redundant any further considerations of evidence, theory or scientific merit. That is quite dangerous, because all relevant material should be considered and integrated into scientific inferences.

The third objection refers to the strength of evidence needed to reach the threshold: the *British Journal of Pharmacology* instruction licenses claims on the basis of relatively weak evidence.[6] The evidential disfavouring of the null hypothesis in a P-value close to 0.05 is surprisingly weak when viewed as a likelihood ratio or Bayes factor (Goodman and Royall 1988; Johnson 2013; Benjamin et al. 2018), a weakness that can be confirmed by simply 'eyeballing' (Fig. 1).

A fixed threshold corresponding to weak evidence might sometimes be reasonable, but often it is not. As Carl Sagan said: "Extraordinary claims require extraordinary evidence".[7] It would be possible to overcome this last objection by setting a lower threshold whenever an extraordinary claim is to be made, but the *British Journal of Pharmacology* instructions preclude such a choice by insisting that the same threshold be applied to all tests within the whole study.

There has been a serious proposal that a lower default threshold of $P < 0.005$ be adopted as the default (Johnson 2013; Benjamin et al. 2018), but even if that would ameliorate the weakness of evidence objection, it doesn't address all of the problems posed by dichotomising results into significant and not significant, as is acknowledged by the many authors of that proposal.

Should the *British Journal of Pharmacology* enforce its guideline on the use of Neyman–Pearsonian hypothesis testing with a fixed threshold for statistical significance? Definitely not. Laboratory pharmacologists should usually avoid them because those tests are ill-suited to the reality of basic pharmacological studies.

The shortcoming of hypothesis testing is that it offers an all-or-none outcome and it engenders a one-and-done response to an experiment. All-or-none in that the significant or not significant outcome is dichotomous. One-and-done because once a decision has been made to reject the null hypothesis there is little apparent reason to re-test that null hypothesis the same way, or differently. There is no mechanism within the classical Neyman–Pearsonian hypothesis testing framework for a result to be treated as provisional. That is not particularly problematical in the context of a classical randomised clinical trial (RCT) because an RCT is usually conducted only after preclinical studies have addressed the relevant biological questions. That allows the scientific aims of the study to be simple – they are designed to provide a definitive answer to the primary question. An all-or-none one-and-done hypothesis test is therefore appropriate for an RCT.[8] But the majority of basic pharmacological laboratory studies do not have much in common with an RCT because they consist of a series of interlinked and inter-related experiments contributing variously to the primary inference. For example, a basic pharmacological study will often include experiments that validate experimental methods and reagents, concentration-

---

[6]Accepting $P = 0.05$ as a sufficient reason to suppose that a treatment is effective is akin to accepting 50% as a passing grade: it is traditional in many settings, but it is far from reassuring.

[7]That phrase comes from the television series *Cosmos*, 1980, but may derive from Laplace (1812), who wrote "The weight of evidence for an extraordinary claim must be proportioned to its strangeness". [translated, the original is in French].

[8]Clinical trials are sometimes aggregated in meta-analyses, but the substrate for meta-analytical combination is the observed effect sizes and sample sizes of the individual trials, not the dichotomised significant or not significant outcomes.

response curves for one or more of drugs, positive and negative controls, and other experiments subsidiary to the main purpose of the study. The design of the 'headline' experiment (assuming there is one) and interpretation of its results is dependent on the results of those subsidiary experiments, and even when there is a singular scientific hypothesis, it might be tested in several ways using observations within the study. It is the aggregate of all of the experiments that inform the scientific inferences. The all-or-none one-and-done outcome of a hypothesis test is less appropriate to basic research than it is to a clinical trial.

Pharmacological laboratory experiments also differ from RCTs in other ways that are relevant to the choice of statistical methodologies. Compared to an RCT, basic pharmacological research is very cheap, the experiments can be completed very quickly, with the results available for analysis almost immediately. Those advantages mean that a pharmacologist might design some of the experiments within a study in response to results obtained in that same study,[9] and so a basic pharmacological study will often contain preliminary or exploratory research. Basic research and clinical trials also differ in the consequences of erroneous inference. A false positive in an RCT might prove very damaging by encouraging the adoption of an ineffective therapy, but in the much more preliminary world of basic pharmacological research a false positive result might have relatively little influence on the wider world. It could be argued that statistical protections against false positive outcomes that are appropriate in the realm of clinical trials can be inappropriate in the realm of basic research. This idea is illustrated in a later section of this chapter.

The multi-faceted nature of the basic pharmacological study means that statistical approaches yielding dichotomous yes or no outcomes are less relevant than they are to the archetypical RCT. The scientific conclusions drawn from basic pharmacological experiments should be based on thoughtful consideration of the entire suite of results in conjunction with any other relevant information, including both pre-existing evidence and theory. The dichotomous all-or-none, one-and-done hypothesis test is poorly adapted to the needs of basic pharmacological experiments, and is probably poorly adapted to the needs of most basic scientific studies. Scientific studies depend on a detailed evaluation of evidence but a hypothesis test does not fully support such an evaluation.

2.3 Evidence Is Local; Error Rates Are Global

A way to understand difference between the Fisherian significance test and the Neyman–Pearsonian hypothesis test is to recognise that the former supports 'local' inference, whereas the latter is designed to protect against 'global' long-run error. The *P*-value of a significance test is local because it is an index of the evidence in *this* data against *this* null hypothesis. In contrast, the hypothesis test decision regarding

---

[9]Yes, that is also done in 'adaptive' clinical trials, but they are not the archetypical RCT that is the comparator here.

rejection of the null hypothesis is global because it is based on a parameter, α, which is set without reference to the observed data. The long run performance of the hypothesis test is a property of the procedure itself and is independent of any particular data, and so it is global. Local evidence; global errors. This is not an ahistoric imputation, because Neyman and Pearson were clear about their preference for global error protection rather than local evidence and their objectives in devising hypothesis tests:

> We are inclined to think that as far as a particular hypothesis is concerned, no test based upon the theory of probability can by itself provide any valuable evidence of the truth or falsehood of that hypothesis.
>
> But we may look at the purpose of tests from another view-point. Without hoping to know whether each separate hypothesis is true or false, we may search for rules to govern our behaviour with regard to them, in following which we insure that, in the long run of experience, we shall not be too often wrong.
>
> – Neyman and Pearson (1933)

The distinction between local and global properties or information is relatively little known, but Liu and Meng (2016) offer a much more technical and complete discussion of the local/global distinction, using the descriptors 'individualised' and 'relevant' for the local and the 'robust' for the global. They demonstrate a trade-off between relevance and robustness that requires judgement on the part of the analyst. In short, the desirability of methods that have good long-run error properties is undeniable, but paying attention exclusively to the global blinds us to the local information that is relevant to inferences. The instructions of the *British Journal of Pharmacology* are inappropriate because they attend entirely to the global and because the dichotomising of each experimental result into significant and not significant hinders thoughtful inference.

Many of the battles and controversies regarding statistical tests swirl around issues that might be clarified using the local versus global distinction, and so it will be referred to repeatedly in what follows.

2.4   On the Scaling of *P*-values

In order to be able to safely interpret the local, evidential, meaning of a P-value, a pharmacologist should understand its scaling. Just like the EC_{50}s with which pharmacologists are so familiar, P-values have a bounded scale, and just as is the case with EC_{50}s it makes sense to scale P-values geometrically (or logarithmically). The non-linear relationship between P-values and an intuitive scaling of evidence against the null hypothesis can be gleaned from Fig. 2. Of course, a geometric scaling of the evidential meaning of P-values implies that the descriptors of evidence should be similarly scaled and so such a scale is proposed in Fig. 3, with P-values around 0.05 being called 'trivial' in recognition of the relatively unimpressive evidence for a real difference between condition A and control in Fig. 2.

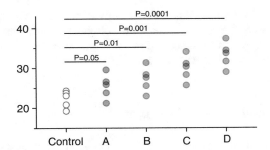

Fig. 2 What simple evidence looks like. Pseudo-data devised to yield one-tailed *P*-values from 0.05 to 0.0001 from a Student's *t*-test for independent samples, $n = 5$ per group. The left-most group of values is the control against which each of the other sets is compared, and the pseudo-datasets A, B, C, and D were generated by arithmetic adjustment of a single dataset to obtain the indicated *P*-values. The *y*-axis is an arbitrarily scaled measure

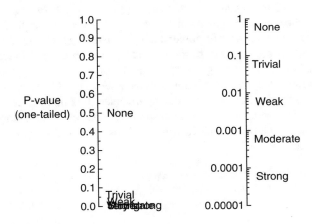

Fig. 3 Evidential descriptors for *P*-values. Strength of evidence against the null hypothesis scales semi-geometrically with the smallness of the *P*-value. Note that the descriptors for strength of evidence are illustrative only, and it would be a mistake to assume, for example, that a *P*-value of 0.001 indicates moderately strong evidence against the null hypothesis in every circumstance

Attentive readers will have noticed that the *P*-values in Figs. 1, 2 and 3 are all one-tailed. The number of tails that published *P*-values have is inconsistent, is often unspecified, and the number of tails that a *P*-value *should* have is controversial (e.g. see Dubey 1991; Bland and Bland 1994; Kobayashi 1997; Freedman 2008; Lombardi and Hurlbert 2009; Ruxton and Neuhaeuser 2010). Arguments about *P*-value tails are regularly confounded by differences between local and global considerations. The most compelling reasons to favour two tails relate to global error rates, which means that they apply only to *P*-values that are dichotomised into significant and not significant in a hypothesis test. Those arguments can safely be ignored when *P*-values are used as indices of evidence and I therefore recommend one-tailed *P*-values for general use in pharmacological experiments – as long as the *P*-values are interpreted as evidence and not as a surrogate for decision. (Either way, the number of tails should always be specified.)

2.5   Power and Expected *P*-values

The Neyman–Pearsonian hypothesis test is a decision procedure that, with a few assumptions, can be an optimal procedure. Optimal only in the restricted sense that the smallest sample gives the highest power to reject the null hypothesis when it is false, for any specified rate of false positive errors. To achieve that optimality the experimental sample size and α are selected prior to the experiment using a power analysis and with consideration of the costs of the two specified types of error and the benefits of potentially correct decisions. In other words, there is a loss function built into the design of experiments. However, outside of the clinical trials arena, few pharmacologists seem to design experiments in that way. For example, a study of 22 basic biomedical research papers published in *Nature Medicine* found that none of them included any mention of a power analysis for setting the sample size (Strasak et al. 2007), and a simple survey of the research papers in the most recent issue of *British Journal of Pharmacology* (2018, issue 17 of volume 175) gives a similar picture with power analyses specified in only one out of the 11 research papers that used $P < 0.05$ as a criterion for statistical significance. It is notable that all of those *BJP* papers included statements in their methods sections claiming compliance with the guidelines for experimental design and analysis, guidelines that include this as the first key point:

> Experimental design should be subjected to 'a priori power analysis' so as to ensure that the size of treatment and control groups is adequate[...]
>
> – Curtis et al. (2015)

The most recent issue of *Journal of Pharmacology and Experimental Therapeutics* (2018, issue 3 of volume 366) similarly contains no mention of power of sample size determination in any of its 9 research papers, although none of its authors had to pay lip service to guidelines requiring it.

In reality, power analyses are not always necessary or helpful. They have no clear role in the design of a preliminary or exploratory experiment that is concerned more with hypothesis generation than hypothesis testing, and a large fraction of the experiments published in basic pharmacological journals are exploratory or preliminary in nature. Nonetheless, they are described here in detail because experience suggests they are mysterious to many pharmacologists and they are very useful for planning confirmatory experiments.

For a simple test like Student's *t*-test a pre-experiment power analysis for determination of sample size is easily performed. The power of a Student's *t*-test is dependent on: (1) the predetermined acceptable false positive error rate, α (bigger

α gives more power); (2) the true effect size, which we will denote as δ (more power when δ is larger); (3) the population standard deviation, σ (smaller σ gives more power); and (4) the sample size (larger n for more power). The common approach to a power test is to specify an effect size of interest and the minimum desired power, so say we wish to detect a true effect of $\delta = 3$ in a system where we expect the standard deviation to be $\sigma = 2$. The free software[10] called R has the function power.t.test() that gives this result:

```
> power.t.test(delta=3, sd=2, power=0.8, sig.level = 0.05,
alternative ='one.sided', n=NULL)

    Two-sample t test power calculation

          n = 6.298691
      delta = 3
         sd = 2
  sig.level = 0.05
      power = 0.8
alternative = one.sided

NOTE: n is number in *each* group
```

It is conventional to round the sample size up to the next integer so the sample size would be 7 per group.

While a single point power analysis like that is straightforward, it provides relatively little information compared to the information supplied by the analyst, and its output is specific to the particular effect size specified, an effect size that more often than not has to be 'guesstimated' instead of estimated because it is the unknown that is the object of study. A plot of power versus effect size is far more informative than the point value supplied by the conventional power test (Fig. 4). Those graphical power functions show clearly the three-way relationship between sample size, effect size and the risk of a false negative outcome (i.e. one minus the power).

Some experimenters are tempted to perform a post-experiment power analysis when their observed *P*-value is unsatisfyingly large. They aim to answer the question of how large the sample *should* have been, and proceed to plug in the observed effect size and standard deviation and pulling out a larger sample size – always larger – that might have given them the desired small *P*-value. Their interpretation is then that the result *would have been significant* but for the fact that the experiment was under-powered. That interpretation ignores the fact that the observed effect size might be an exaggeration, or the observed standard deviation might be an underestimation and

---

[10]www.r-project.org.

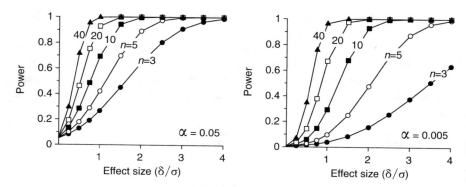

Fig. 4 Power functions for $\alpha = 0.05$ and 0.005. Power of one-sided Student's t-test for independent samples expressed as a function of standardised true effect size δ/σ for sample sizes (per group) from $n = 3$ to $n = 40$. Note that $\delta = \mu_1 - \mu_2$ and σ are population parameters rather than sample estimates

the null hypothesis might be true! Such a procedure is generally inappropriate and dangerous (Hoenig and Heisey 2001). There is a one-to-one correspondence of observed P-value and post-experiment power and no matter what the sample size, a larger than desired P-value *always* corresponds to a low power at the observed effect size, whether the null hypothesis is true or false. Power analyses are useful in the design of experiments, not for the interpretation of experimental results.

Power analyses are tied closely to dichotomising Neyman–Pearsonian hypothesis tests, even when expanded to provide full power functions as in Fig. 4. However, there is an alternative more closely tied to Fisherian significance testing – an approach better aligned to the objectives of evidence gathering. That alternative is a plot of average expected P-values as functions of effect size and sample size (Sackrowitz and Samuel-Cahn 1999; Bhattacharya and Habtzghi 2002). The median is more relevant than the mean, both because the distribution of expected P-values is very skewed and because the median value offers a convenient interpretation of there being a 50:50 bet that and observed P-value will be either side of it. An equivalent plot showing the 90th percentile of expected P-values gives another option for experiment sample size planning purposes (Fig. 5).

Should the *British Journal of Pharmacology* enforce its power guideline? In general no, but pharmacologists should use power curves or expected P-value curves for designing some of their experiments, and ought to say so when they do. Power analyses for sample size are very important for experiments that are intended to be definitive and decisive, and that's why sample size considerations are dealt with in detail when planning clinical trials. Even though the majority of experiments in basic pharmacological research papers are not like that, as discussed above, even preliminary experiments should be planned to a degree, and power curves and expected P-value curves are both useful in that role.

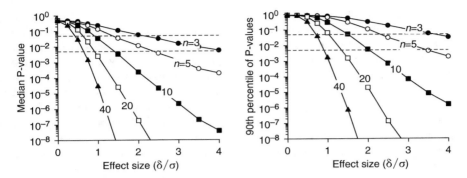

Fig. 5 Expected *P*-value functions. *P*-values expected from Student's *t*-test for independent samples expressed as a function of standardised true effect size δ/σ for sample sizes (per group) from $n = 3$ to $n = 40$. The graph on the left shows the median of expected *P*-values (i.e. the 50th percentile) and the graph on the right shows the 90th percentile. It can be expected that 50% of observed *P*-values will lie below the median lines and 90% will lie below the 90th percentile lines for corresponding sample sizes and effect sizes. The dashed lines indicate $P = 0.05$ and 0.005

3   Practical Problems with *P*-values

The sections above deal with the most basic misconceptions regarding the nature of *P*-values, but critics of *P*-values usually focus on other important issues. In this section I will deal with the significance filter, multiple comparisons and some forms of P-hacking, and I need to point out immediately that most of the issues are not specific to *P*-values even if some of them are enabled by the unfortunate dichotomisation of *P*-values into significant and not significant. In other words, the practical problems with *P*-values are largely the practical problems associated with the *mis*use of *P*-values and with sloppy statistical inference generally.

3.1   The Significance Filter Exaggeration Machine

It is natural to assume that the effect size observed in an experiment is a good estimate of the true effect size, and in general that can be true. However, there are common circumstances where the observed effect size consistently overestimates the true, sometimes wildly so. The overestimation depends on the facts that experimental results exaggerating the true effect are more likely to be found statistically significant, and that we pay more attention to the significant results and are more likely to report them. The key to the effect is selective attention to a subset of results – the significant results – and so the process is appropriately called *the significance filter*.

If there is nothing untoward in the sampling mechanism,[11] sample means are unbiassed estimators of population means and sample-based standard deviations are nearly unbiassed estimators of population standard deviations.[12] Because of that we can assume that, on average, a sample mean provides a sensible 'guesstimate' for the population parameter and, to a lesser degree, so does the observed standard deviation. That is indeed the case for averages over all samples, but it cannot be relied upon for any particular sample. If attention has been drawn to a sample on the basis that it is 'statistically significant', then that sample is likely to offer an exaggerated picture of the true effect. The phenomenon is usually called *the significance filter*. The way it works is fairly easily described but, as usual, there are some complexities in its interpretation.

Say we are in the position to run an experiment 100 times with random samples of $n = 5$ from a single normally distributed population with mean $\mu = 1$ and standard deviation $\sigma = 1$. We would expect that, on average, the sample means, $\bar{x}$ would be scattered symmetrically around the true value of 1, and the sample-based standard deviations, s, would be scattered around the true value of 1, albeit slightly asymmetrically. A set of 100 simulations matching that scenario show exactly that result (see the left panel of Fig. 6), with the median of $\bar{x}$ being 0.97 and the median of s being 0.94, both of which are close to the expected values of exactly 1 and about 0.92, respectively. If we were to pay attention only to the results where the observed *P*-value was less than 0.05 (with the null hypothesis being that the population mean is 0), then we get a different picture because the values are very biassed (see the right panel of Fig. 6). Among the 'significant' results the median sample mean is 1.2 and the median standard deviation is 0.78.

The systematic bias of mean and standard deviation among 'significant' results in those simulations might not seem too bad, but it is conventional to scale the effect size as the standardised ratio $\bar{x}/s$,[13] and the median of that ratio among the 'significant' results is fully 50% larger than the correct value. What's more, the biasses get worse with smaller samples, with smaller true effect sizes, and with lower *P*-value thresholds for 'significance'.

It is notable that even the results with the most extreme exaggeration of effect size in Fig. 6 – 550% – would not be counted as an error within the Neyman–Pearsonian hypothesis testing framework! It would not lead to the false rejection of a true null or to an inappropriate failure to reject a false null and so it is neither a type I nor a type II

---

[11]That is not a safe assumption, in particular because a haphazard sample is not a random sample. When was the last time that you used something like a random number generator for allocation of treatments?

[12]The variance is unbiassed but the non-linear square root transformation into the standard deviation damages that unbiassed-ness. Standard deviations calculated from small samples are biassed toward underestimation of the true standard deviation. For example, if the true standard deviation is 1, the expected average observed standard deviation for samples of $n = 5$ is 0.94.

[13]That ratio is often called Cohen's *d*. Pharmacologists should pay no attention to Cohen's specifications of small, medium and large effect sizes (Cohen 1992) because they are much smaller than the effects commonly seen in basic pharmacological experiments.

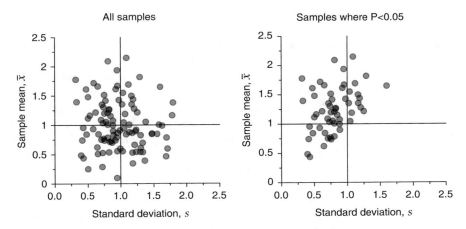

Fig. 6 The significance filter. The dots in the graphs are means and standard deviations of samples of $n = 5$ drawn from a normally distributed population with mean $\mu = 1$ and standard deviation $\sigma = 1$. The left panel shows all 100 samples and the right panel shows only the results where $P < 0.05$. The vertical and horizontal lines indicate the true parameter values. 'Significant' results tend to overestimate the population mean and underestimate the population standard deviation

error. But it is some type of error, a substantial error in estimation of the magnitude of the effect. The term *type M error* has been devised for exactly that kind of error (Gelman and Carlin 2014). A type M error might be underestimation as well as overestimation, but overestimation is more common in theory (Lu et al. 2018) and in practice (Camerer et al. 2018).

The effect size exaggeration coming from the significance filter is not a result of sampling, or of significance testing, or of *P*-values. It is a result of paying extra attention to a subset of all results – the 'significant' subset.

The significance filter presents a peculiar difficulty. It leads to exaggeration *on average*, but any particular result may well be close to the correct size whether it is 'significant' or not. A real-world sample mean of, say, $\bar{x} = 1.5$ might be an exaggeration of $\mu = 1$, it might be an underestimation of $\mu = 2$, or it might be pretty close to $\mu = 1.4$ and there would be no way to be certain without knowing μ, and if μ were known then the experiment would probably not have been necessary in the first place. That means that the possibility of a type M error looms over any experimental result that is interesting because of a small *P*-value, and that is particularly true when the sample size is small. The only way to gain more confidence that a particular significant result closely approximates the true state of the world is to repeat the experiment – the second result would not have been run through the significance filter and so its results would not have a greater than average risk of exaggeration and the overall inference can be informed by both results. Of course, experiments intended to repeat or replicate an interesting finding should take the possible exaggeration into account by being designed to have higher power than the original.

3.2   Multiple Comparisons

Multiple testing is the situation where the tension between global and local considerations is most stark. It is also the situation where the well-known jelly beans cartoon from XKCD.com is irresistible (Fig. 7). The cartoon scenario is that jelly beans were suspected of causing acne, but a test found "no link between jelly beans and acne ($P > 0.05$)", and so the possibility that only a certain colour of jelly bean causes acne is then entertained. All 20 colours of jelly bean are independently tested, with only the result from green jelly beans being significant, "($P < 0.05$)". The newspaper headline at the end of the cartoon mentions only the green jelly beans result, and it does that with exaggerated certainty. The usual interpretation of that cartoon is that the significant result with green jelly beans is likely to be a false positive because, after all, hypothesis testing with the threshold of $P < 0.05$ is expected to yield a false positive one time in 20, on average, when the null is true.

The more hypothesis tests there are, the higher the risk that one of them will yield a false positive result. The textbook response to multiple comparisons is to introduce 'corrections' that protect an overall maximum false positive error rate by adjusting the threshold according to the number of tests in the family to give protection from inflation of the family-wise false positive error rate. The Bonferroni adjustment is the best-known method, and while there are several alternative 'corrections' that perform a little better, none of those is nearly as simple. A Bonferroni adjustment for the family of experiments in the cartoon would preserve an overall false positive error rate of 5% by setting a threshold for significance of $0.05/20 = 0.0025$ in each of the 20 hypothesis tests.[14] It must be noted that such protection does not come for free, because adjustments for multiplicity invariably strip statistical power from the analysis.

We do not know whether the 'significant' link between green jelly beans and acne would survive a Bonferroni adjustment because the actual P-values were not supplied,[15] but as an example, a P-value of 0.003, low enough to be quite encouraging as the result of a significance test, would be 'not significant' according to the Bonferroni adjustment. Such a result that would present us with a serious dilemma because the inference supported by the local evidence would be apparently contradicted by global error rate considerations. However, that contradiction is not what it seems because the null hypothesis of the significance test P-value is a different null hypothesis from that tested by the Bonferroni-adjusted hypothesis test. The significance test null concerns only the green jelly beans whereas the null hypothesis of the Bonferroni is an omnibus null hypothesis that says that the link between green jelly beans on acne is zero *and* the link between purple jelly beans on

---

[14]You may notice that the first test of jelly beans without reference to colour has been ignored here. There is no set rule for saying exactly which experiments constitute a family for the purposes of correction of multiplicity.

[15]That serves to illustrate one facet of the inadequacy of reporting 'P less thans' in place of actual P-values.

Fig. 7 Multiple testing cartoon from XKCD, https://xkcd.com/882/

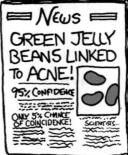

acne is zero *and* the link between brown jelly beans is zero, and so on. The *P*-value null hypothesis is local and the omnibus null is global. The global null hypothesis might be appropriate before the evidence is available (i.e. for power calculations and experimental planning), but after the data are in hand the local null hypothesis concerning just the green jelly beans gains importance.

It is important to avoid being blinded to the local evidence by a non-significant global. After all, the pattern of evidence in the cartoon is *exactly* what would be expected if the green colouring agent caused acne: green jelly beans are associated with acne but the other colours are not. (The failure to see an effect of the mixed jelly beans in the first test is easily explicable on the basis of the lower dose of green.) If the data from the trial of green jelly beans is independent of the data from the trials of other colours, then there is no way that the existence of those other data – or their analysis – can influence the nature of the green data. The green jelly bean data cannot logically have been affected by the fact that mauve and beige jelly beans were tested at a later point in time – the subsequent cannot affect the previous – and the experimental system would have to be bizarrely flawed for the testing of the purple or brown jelly beans to affect the subsequent experiment with green jelly beans. If the multiplicity of tests did not affect the data, then it is only reasonable to say that it did not affect the evidence.

The omnibus global result does not cancel the local evidence, or even alter it, and yet the elevated risk of a false positive error is real. That presents us with a dilemma and, unfortunately, statistics does not provide a way around it. Global error rates and local evidence operate in different logical spaces (Thompson 2007) and so there can be no strictly statistical way to weigh them together. All is not lost, though, because statistical limitations do not preclude thoughtful integration of local and the global issues when making inferences. We just have to be more than normally cautious when the local and global pull in different directions. For example, in the case of the cartoon, the evidence in the data favours the idea that green jelly beans are linked with acne (and if we had an exact *P*-value then we could specify the strength of favouring) but because the data were obtained by a method with a substantial false positive error rate we should be somewhat reluctant to take that evidence at face value. It would be up to the scientist in the cartoon (the one with safety glasses) to form a provisional scientific conclusion regarding the effect of green jelly beans, even if that inference is that any decision should be deferred until more evidence is available. Whatever the inference, the evidence, theory, the method, any other corroborating or rebutting information should all be considered and reported.

> A man or woman who sits and deals out a deck of cards repeatedly will eventually get a very unusual set of hands. A report of unusualness would be taken differently if we knew it was the only deal made, or one of a thousand deals, or one of a million deals, etc. – Tukey (1991, p. 133)

In isolation the cartoon experiments are probably only sufficient to suggest that the association between green jelly acne is worthy of further investigation (with the earnestness of that suggestion being inversely related to the size of the relevant *P*-

value). The only way to be in a position to report an inference concerning those jelly beans without having to hedge around the family-wise false positive error rate and the significance filter is to re-test the green jelly beans. New data from a separate experiment will be free from the taint of elevated family-wise error rates and untouched by the significance filter exaggeration machine. And, of course, *all* of the original experiments should be reported alongside the new, as well as reasoned argument incorporating corroborating or rebutting information and theory.

The fact that a fresh experiment is necessary to allow a straightforward conclusion about the effect of the green jelly beans means that the experimental series shown in the cartoon is a preliminary, exploratory study. Preliminary or exploratory research is essential to scientific progress and can merit publication as long as it is reported completely and openly as preliminary. Too often scientists fall into the pattern of misrepresenting the processes that lead to their experimental results, perhaps under the mistaken assumption that science has to be hypothesis driven (Medawar 1963; du Prel et al. 2009; Howitt and Wilson 2014). That misrepresentation may take the form of a suggestion, implied or stated, that the green jelly beans were the intended subject of the study, a behaviour described as *HARKing* for *h*ypothesising *a*fter the *r*esults are *k*nown, or *cherry picking* where only the significant results are presented. The reason that HARKing is problematical is that hypotheses cannot be tested using the data that suggested the hypothesis in the first place because those data *always* support that hypothesis (otherwise they would not be suggesting it!), and cherry picking introduces a false impression of the nature of the total evidence and allows the direct introduction of experimenter bias. Either way, focussing on just the unusual observations from a multitude is bad science. It takes little effort and few words to say that 20 colours were tested and only the green yielded a statistically significant effect, and a scientist can (should) then hypothesise that green jelly beans cause acne and test that hypothesis with new data.

3.3   P-hacking

P-hacking is where an experiment or its analysis is directed at obtaining a small enough *P*-value to claim significance instead of being directed at the clarification of a scientific issue or testing of a hypothesis. Deliberate P-hacking does happen, perhaps driven by the incentives built into the systems of academic reward and publication imperatives, but most P-hacking is accidental – honest researchers doing 'the wrong thing' through ignorance. P-hacking is not always as wrong as might be assumed, as the idea of P-hacking comes from paying attention exclusively to global consideration of error rates, and most particularly to false positive error rates. Those most stridently opposed to P-hacking will point to the increased risk of false positive errors, but rarely to the lowered risk of false negative errors. I will recklessly note that some categories of P-hacking look entirely unproblematical when viewed through the prism of local evidence. The local versus global distinction allows a more nuanced response to P-hacking.

Some P-hacking is outright fraud. Consider this example that has recently come to light:

> One sticking point is that although the stickers increase apple selection by 71%, for some reason this is a p value of .06. It seems to me it should be lower. Do you want to take a look at it and see what you think. If you can get the data, and it needs some tweeking, it would be good to get that one value below .05.
> – Email from Brian Wansink to David Just on Jan. 7, 2012. – Lee (2018)

I do not expect that any readers would find P-hacking of that kind to be acceptable. However, the line between fraudulent P-hacking and the more innocent P-hacking through ignorance is hard to define, particularly so given the fact that some behaviours derided as P-hacking can be perfectly legitimate as part of a scientific research program. Consider this cherry picked list[16] of responses to a *P*-value being greater than 0.05 that have been described as P-hacking (Motulsky 2014):

- Analyse only a subset of the data;
- Remove suspicious outliers;
- Adjust data (e.g. divide by body weight);
- Transform the data (i.e. logarithms);
- Repeat to increase sample size (n).

Before going any further I need to point out that Motulsky has a more realistic attitude to P-hacking than might be assumed from my treatment of his list. He writes: "If you use any form of P-hacking, label the conclusions as 'preliminary'." (Motulsky 2014, p. 1019).

Analysis of only a subset of the data is illicit if the unanalysed portion is omitted in order to manipulate the *P*-value, but unproblematical if it is omitted for being irrelevant to the scientific question at hand. Removal of suspicious outliers is similar in being only sometimes inappropriate: it depends on what is meant by the term "outlier". If it indicates that a datum is a mistake such as a typographical or transcriptional error, then of course it should be removed (or corrected). If an outlier is the result of a technical failure of a particular run of the experimental, then perhaps it should be removed, but the technical success or failure of an experimental run must not be judged by the influence of its data on the overall *P*-value. If with word outlier just denotes a datum that is further from the mean than the others in the dataset, then omit it at your peril! Omission of that type of outlier will reduce the variability in the data and give a lower *P*-value, but will markedly increase the risk of false positive results and it is, indeed, an illicit and damaging form of P-hacking.

Adjusting the data by standardisation is appropriate – desirable even – in some circumstances. For example, if a study concerns feeding or organ masses, then standardising to body weight is probably a good idea. Such manipulation of data

---

[16]There are nine specified in the original but I discuss only five: cherry picking!

should be considered P-hacking only if an analyst finds a too large P-value in unstandardised data and then tries out various re-expressions of the data in search of a low P-value, and then reports the results as if that expression of the data was intended all along. The P-hackingness of log-transformation is similarly situationally dependent. Consider pharmacological EC_{50}s or drug affinities: they are strictly bounded at zero and so their distributions are skewed. In fact the distributions are quite close to log-normal and so log-transformation before statistical analysis is appropriate and desirable. Log-transformation of EC_{50}s gives more power to parametric tests and so it is common that significance testing of $logEC_{50}$s gives lower P-values than significance testing of the un-transformed EC_{50}s. An experienced analyst will choose the log-transformation because it is known from empirical and theoretical considerations that the transformation makes the data better match the expectations of a parametric statistical analysis. It might sensibly be categorised as P-hacking only if the log-transformation was selected with no justification other than it giving a low P-value.

The last form of P-hacking in the list requires a good deal more consideration than the others because, well, statistics is complicated. That consideration is facilitated by a concrete scenario – a scenario that might seem surprisingly realistic to some readers. Say you run an experiment with $n = 5$ observations in each of two independent groups, one treated and one control, and obtain a P-value of 0.07 from Student's t-test. You might stop and integrate the very weak evidence against the null hypothesis into your inferential considerations, but you decide that more data will clarify the situation. Therefore you run some extra replicates of the experiment to obtain a total of $n = 10$ observations in each group (including the initial 5), and find that the P-value for the data in aggregate is 0.002. The risk of the 'significant' result being a false positive error is elevated because the data have had two chances to lead you to discard the null hypothesis. Conventional wisdom says that you have P-hacked. However, there is more to be considered before the experiment is discarded.

Conventional wisdom usually takes the global perspective. As mentioned above, it typically privileges false positive errors over any other consideration, and calls the procedure invalid. However, the extra data has added power to the experiment and lowered the expected P-value for any true effect size. From a local evidence point of view, increasing the sample increases the amount of evidence available for use in inference, which is a good thing. Is extending an experiment after the statistical analysis a good thing or a bad thing? The conventional answer is that it is a bad thing and so the conventional advice is don't do it! However, a better response might balance the bad effect of extending the experiment with the good. Consideration of the local and global aspects of statistical inference allows a much more nuanced answer. The procedure described would be perfectly acceptable for a preliminary experiment.

Technically the two-stage procedure in that scenario allows *optional stopping*. The scenario is not explicit, but it can be discerned that the stopping rule was, in effect, run $n = 5$ and inspect the P-value; if it is small enough, then stop and make inferences about the null hypothesis; if the P-value is not small enough for the stop but nonetheless small enough to represent some evidence against the null hypothesis, add an extra 5 observations to each group to give $n = 10$, stop, and analyse again. We do not know how low the interim P-value would have to be for the protocol to stop, and we do not know how high it could be and the extra data still be gathered, but no matter where those thresholds are set, such stopping rules yield false positive rates higher than the nominal critical value for stopping would suggest. Because of that, the conventional view (the global perspective, of course) is that the protocol is invalid, but it would be more accurate to say that such a protocol would be invalid unless the P-value or the threshold for a Neyman–Pearsonian dichotomous decision is adjusted as would be done with a formal *sequential test*. It is interesting to note that the elevation of false positive rate is not necessarily large. Simulations of the scenario as specified and with $P < 0.1$ as the threshold for continuing show that the overall false positive error rate would be about 0.008 when the critical value for stopping at the first stage is 0.005, and about 0.06 when that critical value is 0.05.

The increased rate of false positives (global error rate) is real, but that does not mean that the evidential meaning of the final P-value of 0.002 is changed. It is the same local evidence against the null as if it was obtained from a simpler one stage protocol with $n = 10$. After all, the data are *exactly the same* as if the experimenter had intended to obtain $n = 10$ from the beginning. The optional stopping has changed the global properties of the statistical procedure but not the local evidence which contained in the actualised data.

You might be wondering how it is possible that the local evidence be unaffected by a process that increases the global false positive error rate. The rationale is that the evidence is contained within the data but the error rate is a property of the procedure – evidence is local and error rates are global. Recall that false positive errors can only occur when the null hypothesis is true. If the null is true, then the procedure has increased the risk of the data leading us to a false positive decision, but if the null is false, then the procedure has *decreased* the risk of a false negative decision. Which of those has paid out in this case cannot be known because we do not know the truth of this local null hypothesis. It might be argued that an increase in the global risk of false positive decisions should outweigh the decreased risk of false negatives, but that is a value judgement that ought to take into account particulars of the experiment in question, the role of that experiment in the overall study, and other contextual factors that are unspecified in the scenario and that vary from circumstance to circumstance.

So, what can be said about the result of that scenario? The result of $P = 0.002$ provides moderately strong evidence against the null hypothesis, but it was obtained from a procedure with sub-optimal false positive error characteristics. That sub-optimality should be accounted for in the inferences that made from the evidence, but it is only confusing to say that it alters the evidence itself, because it is the data that contain the evidence and the sub-optimality did not change the data. Motulsky provides good advice on what to do when your experiment has the optional stopping:

- For each figure or table, clearly state whether or not the sample size was chosen in advance, and whether every step used to process and analyze the data was planned as part of the experimental protocol.
- If you used any form of P-hacking, label the conclusions as "preliminary."

Given that basic pharmacological experiments are often relatively inexpensive and quickly completed one can add to that list the option of also corroborating (or not) those results with a fresh experiment designed to have a larger sample size (remember the significance filter exaggeration machine) and performed according to the design. Once we move beyond the globalist mindset of one-and-done such an option will seem obvious.

3.4   What Is a Statistical Model?

I remind the reader that this chapter is written under the assumption that pharmacologists can be trusted to deal with the full complexity of statistics. That assumption gives me licence to discuss unfamiliar notions like the role of the statistical model in statistical analysis. All too often the statistical model is often invisible to ordinary users of statistics and that invisibility encourages thoughtless use of flawed and inappropriate models, thereby contributing to the misuse of inferential statistics like *P*-values.

A statistical model is what allows the formation of calibrated statistical inferences and non-trivial probabilistic statements in response to data. The model does that by assigning probabilities to potential arrangements of data. A statistical model can be thought of as a set of assumptions, although it might be more realistic to say that a chosen statistical model imposes a set of assumptions onto the experimenter.

> I have often been struck by the extent to which most textbooks, on the flimsiest of evidence, will dismiss the substitution of assumptions for real knowledge as unimportant if it happens to be mathematically convenient to do so. Very few books seem to be frank about, or perhaps even aware of, how little the experimenter actually *knows* about the distribution of errors in his observations, and about facts that are assumed to be known for the purposes of statistical calculations.
>
> – Colquhoun (1971, p. *v*)

Statistical models can take a variety of forms (McCullagh 2002), but the model for the familiar Student's *t*-test for independent samples is reasonably representative. That model consists of assumed distributions (normal) of two populations with parameters mean (μ_1 and μ_2) and standard deviation (σ_1 and σ_2),[17] and a rule for obtaining samples (e.g. a randomly selected sample of $n = 6$ observations from each

---

[17]The ordinary Student's *t*-test assumes that $\sigma_1 = \sigma_2$, but the Welch-Scatterthwaite variant relaxes that assumption.

population). A specified value of the difference between means serves as the null hypothesis, so $H_0 : \mu_1 - \mu_2 = \delta_{H_0}$. The test statistic is[18]

$$t = \frac{(\bar{x}_1 - \bar{x}_2) - \delta_{H_0}}{s_p\sqrt{1/n_1 + 1/n_2}}$$

where $\bar{x}$ is a sample mean and s_p is the pooled standard deviation. The explicit inclusion of a null hypothesis term in the equation for t is relatively rare, but it is useful because it shows that the null hypothesis is just a possible value of the difference between means. Most commonly the null hypothesis says that the difference between means is zero – it can be called a 'nill-null' – and in that case the omission of δ_{H_0} from the equation makes no numerical difference.

Values of t calculated by that equation have a known distribution when $\mu_1 - \mu_2 = \delta_{H_0}$, and that distribution is Student's t-distribution.[19] Because the distribution is known it is possible to define hypothesis test acceptance regions for any level of α for a hypothesis test, and any observed t-value can be converted into a P-value in a significance test.

An important problem that a pharmacologist is likely to face when using a statistical model is that it is just a model. Scientific inferences are usually intended to communicate something about the real world, not the mini world of a statistical model, and the connection between a model-based probability of obtaining a test statistic value and the state of the real world is always indirect and often inscrutable. Consider the meaning conveyed by an observed P-value of 0.002. It indicates that the data are strange or unusual compared to the expectations of the statistical model when the parameter of interest is set to the value specified by the null hypothesis. The statistical model expects a P-value of, say, 0.002 to occur only two times out of a thousand on average when the null is true. If such a P-value is observed, then one of these situations has arisen:

- a two in a thousand accident of random sampling has occurred;
- the null hypothesised parameter value is not close to the true value;
- the statistical model is flawed or inapplicable because one or more of the assumptions underlying its application are erroneous.

Typically only the first and second are considered, but the last is every bit as important because when the statistical model is flawed or inapplicable then the expectations of the model are not relevant to the real-world system that spawned the data. Figure 8 shows the issue diagrammatically. When we use that statistical inference to inform inferences about the real world we are implicitly assuming:

---

[18]Oh no! An equation! Don't worry, it's the only one, and, anyway, it is too late now to stop reading.

[19]Technically it is the central Student's t-distribution. When $\delta \neq \delta_{H_0}$ it is a non-central t-distribution (Cumming and Finch 2001).

Fig. 8 Diagram of inference using a statistical model

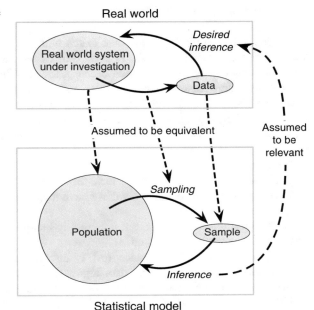

(1) that the real-world system that generated the data is an analogue to the population in the statistical model; (2) that the way the data were obtained is well described by the sampling rule of the statistical model; and (3) that the observed data is analogous to the random sample assumed in the statistical model. To the degree that those assumptions are erroneous there is degradation of the relevance of the model-based statistical inference to the real-world inference that is desired.

Considerations of model applicability are often limited to the population distribution (is my data normal enough to use a Student's *t*-test?) but it is much more important to consider whether there is a definable population that is relevant to the inferential objectives and whether the experimental units ("subjects") approximate a random sample. Cell culture experiments are notorious for having ill-defined populations, and while experiments with animal tissues may have a definable population, the animals are typically delivered from an animal breeding or holding facility and are unlikely to be a random sample. Issues like those mean that the calibration of uncertainty offered by statistical methods might be more or less uncalibrated. For good inferential performance in the real world, there has to be a flexible and well-considered linking of model-based statistical inferences and scientific inferences concerning the real world.

4   *P*-values and Inference

A *P*-value tells you how well the data match with the expectations of a statistical model when the null hypothesis is true. But, as we have seen, there are many considerations that have to be made before a low *P*-value can safely be taken to provide sufficient reason to say that the null hypothesis is false. What's more, inferences about the null hypothesis are not always useful. Royall argues that there are three fundamental inferential questions that should be considered when making scientific inferences (Royall 1997) (here paraphrased and re-ordered):

1. What do these data say?
2. What should I believe now that I have these data?
3. What should I do or decide now that I have these data?

Those questions are distinct, but not entirely independent and there is no single best way to answer to any of them.

A *P*-value from a significance test is an answer to the first question. It communicates how strongly the data argue against the null hypothesis, with a smaller *P*-value being a more insistent shout of "I disagree!". However, the answer provided by a *P*-value is at best incomplete, because it is tied to a particular null hypothesis within a particular statistical model and because it captures and communicates only some of the information that might be relevant to scientific inference. The limitations of a *P*-value can be thought of as analogous to a black and white photograph that captures the essence of a scene, but misses coloured detail that might be vital for a correct interpretation.

Likelihood functions provide more detail than *P*-values and so they can be superior to *P*-values as answers to the question of what the data say. However, they will be unfamiliar to most pharmacologists and they are not immune to problems relating to the relevance of the statistical model and the peculiarities of experimental protocol.[20] As this chapter is about *P*-values, we will not consider likelihoods any further, and those who, correctly, see that they might offer utility can read Royall's book (Royall 1997).

The second of Royall's questions, what should I believe now that I have these data?, requires integration of the evidence of the data with what was believed prior to the evidence being available. A formal statistical combination of the evidence with prior beliefs can be done using Bayesian methods, but they are rarely used for the analysis of basic pharmacological experiments and are outside the scope of this chapter about *P*-values. Considerations of belief can be assisted by *P*-values because when the data argue strongly against the null hypothesis one should be less inclined

---

[20]Royall (1997) and other proponents of likelihood-based inference (e.g. Berger and Wolpert 1988) make a contrary argument based on the likelihood principle and the (irrelevance of) sampling rule principle, but those arguments may fall down when viewed with the local versus global distinction in mind. Happily, those issues are beyond the scope of this chapter.

to believe it true, but it is important to realise that *P*-values do not in any way measure or communicate belief.

The Neyman–Pearsonian hypothesis test framework was devised specifically to answer the third question: it is a decision theoretic framework. Of course, it is a good decision procedure *only* when α is specified prior to the data being available, and when a loss function informs the experimental design. And it is only useful when there is a singular decision to be made regarding a null hypothesis, as can be the case in acceptance sampling and in some randomised clinical trials. A singular decision regarding a null hypothesis is rarely a sufficient inference from the collection of experiments and observations that typically make up a basic pharmacological studies and so hypothesis tests should not be a default analytical tool (and the hybrid NHST should not be used in any circumstance).

Readers might feel that this section has failed to provide a clear method for making inferences about any of the three questions, and they would be correct. Statistics is a set of tools to help with inferences and not a set of inferential recipes, scientific inferences concerning the real world have to be made by scientists, and my intention with this reckless guide to *P*-values is to encourage an approach to scientific inference that is more thoughtful than statistical significance. After all, those scientists invariably know much more than statistics does about the real world, and have a superior understanding of the system under study. Scientific inferences should be made after principled consideration of the available evidence, theory and, sometimes, informed opinion. A full evaluation of evidence will include both consideration of the strength of the local evidence and the global properties of the experimental system and statistical model from which that evidence was obtained. It is often difficult, just like statistics, and there is no recipe.

References

Baker M, Dolgin E (2017) Reproducibility project yields muddy results. Nature 541(7637): 269–270

Begley CG, Ellis LM (2012) Drug development: raise standards for preclinical cancer research. Nature 483(7391):531–533

Benjamin DJ, Berger JO, Johannesson M, Nosek BA, Wagenmakers EJ, Berk R, Bollen KA, Brembs B, Brown L, Camerer C, Cesarini D, Chambers CD, Clyde M, Cook TD, De Boeck P, Dienes Z, Dreber A, Easwaran K, Efferson C, Fehr E, Fidler F, Field AP, Forster M, George EI, Gonzalez R, Goodman S, Green E, Green DP, Greenwald AG, Hadfield JD, Hedges LV, Held L, Ho T-H, Hoijtink H, Hruschka DJ, Imai K, Imbens G, Ioannidis JPA, Jeon M, Jones JH, Kirchler M, Laibson D, List J, Little R, Lupia A, Machery E, Maxwell SE, McCarthy M, Moore DA, Morgan SL, Munafó M, Nakagawa S, Nyhan B, Parker TH, Pericchi L, Perugini M, Rouder J, Rousseau J, Savalei V, Schönbrodt FD, Sellke T, Sinclair B, Tingley D, Van Zandt T, Vazire S, Watts DJ, Winship C, Wolpert RL, Xie Y, Young C, Zinman J, Johnson VE (2018) Redefine statistical significance. Nat Hum Behav 2:6–10

Berger J, Sellke T (1987) Testing a point null hypothesis: the irreconcilability of P values and evidence. J Am Stat Assoc 82:112–122

Berger JO, Wolpert RL (1988) The likelihood principle. Lecture notes–Monograph Series. IMS, Hayward

Berglund L, Björling E, Oksvold P, Fagerberg L, Asplund A, Szigyarto CA-K, Persson A, Ottosson J, Wernérus H, Nilsson P, Lundberg E, Sivertsson A, Navani S, Wester K, Kampf C, Hober S, Pontén F, Uhlén M (2008) A genecentric Human Protein Atlas for expression profiles based on antibodies. Mol Cell Proteomics 7(10):2019–2027

Bhattacharya B, Habtzghi D (2002) Median of the p value under the alternative hypothesis. Am Stat 56(3):202–206

Birnbaum A (1977) The Neyman-Pearson theory as decision theory, and as inference theory; with a criticism of the Lindley-savage argument for Bayesian theory. Synthese 36(1):19–49

Bland JM, Bland DG (1994) Statistics notes: one and two sided tests of significance. Br Med J 309 (6949):248

Camerer CF, Dreber A, Holzmeister F, Ho T-H, Huber J, Johannesson M, Kirchler M, Nave G, Nosek BA, Pfeiffer T, Altmejd A, Buttrick N, Chan T, Chen Y, Forsell E, Gampa A, Heikensten E, Hummer L, Imai T, Isaksson S, Manfredi D, Rose J, Wagenmakers E-J, Wu H (2018) Evaluating the replicability of social science experiments in Nature and Science between 2010 and 2015. Nat Hum Behav 2:637–644

Cohen J (1992) A power primer. Psychol Bull 112(1):155–159

Colquhoun D (1971) Lectures on biostatistics. Oxford University Press, Oxford

Colquhoun D (2014) An investigation of the false discovery rate and the misinterpretation of p-values. R Soc Open Sci 1(3):140216

Cowles M (1989) Statistics in psychology: an historical perspective. Lawrence Erlbaum Associates, Inc., Mahwah

Cumming G (2008) Replication and p intervals: p values predict the future only vaguely, but confidence intervals do much better. Perspect Psychol Sci 3(4):286–300

Cumming G, Finch S (2001) A primer on the understanding, use, and calculation of confidence intervals that are based on central and noncentral distributions. Educ Psychol Meas 61(4): 532–574

Curtis M, Bond R, Spina D, Ahluwalia A, Alexander S, Giembycz M, Gilchrist A, Hoyer D, Insel P, Izzo A, Lawrence A, MacEwan D, Moon L, Wonnacott S, Weston A, McGrath J (2015) Experimental design and analysis and their reporting: new guidance for publication in BJP. Br J Pharmacol 172(2):3461–3471

Curtis MJ, Alexander S, Cirino G, Docherty JR, George CH, Giembycz MA, Hoyer D, Insel PA, Izzo AA, Ji Y, MacEwan DJ, Sobey CG, Stanford CC, Tiexeira MM, Wonnacott S, Ahluwalia A (2018) Experimental design and analysis and their reporting II: updated and simplified guidance for authors and peer reviewers. Br J Pharmacol 175(7):987–993. https://doi.org/10.1111/bph.14153

Drucker DJ (2016) Never waste a good crisis: confronting reproducibility in translational research. Cell Metab 24(3):348–360

du Prel J-B, Hommel G, Röhrig B, Blettner M (2009) Confidence interval or p-value?: Part 4 of a series on evaluation of scientific publications. Deutsches Ärzteblatt Int 106(19):335–339

Dubey SD (1991) Some thoughts on the one-sided and two-sided tests. J Biopharm Stat 1(1): 139–150

Fisher R (1925) Statistical methods for research workers. Oliver & Boyd, Edinburgh

Fisher R (1960) Design of experiments. Hafner, New York

Fraser H, Parker T, Nakagawa S, Barnett A, Fidler F (2018) Questionable research practices in ecology and evolution. PLoS ONE 13(7):e0200303

Freedman LS (2008) An analysis of the controversy over classical one-sided tests. Clin Trials 5(6): 635–640

García-Pérez MA (2016) Thou shalt not bear false witness against null hypothesis significance testing. Educ Psychol Meas 77(4):631–662

Gelman A, Carlin J (2014) Beyond power calculations. Perspect Psychol Sci 9(6):641–651

George CH, Stanford SC, Alexander S, Cirino G, Docherty JR, Giembycz MA, Hoyer D, Insel PA, Izzo AA, Ji Y, MacEwan DJ, Sobey CG, Wonnacott S, Ahluwalia A (2017) Updating the guidelines for data transparency in the British Journal of Pharmacology - data sharing and the use of scatter plots instead of bar charts. Br J Pharmacol 174(17):2801–2804

Gigerenzer G (1998) We need statistical thinking, not statistical rituals. Behav Brain Sci 21:199–200

Goodman SN (2001) Of P-values and Bayes: a modest proposal. Epidemiology 12(3):295–297

Goodman SN, Royall R (1988) Evidence and scientific research. Am J Public Health 78(12): 1568–1574

Halpin PF, Stam HJ (2006) Inductive inference or inductive behavior: Fisher and Neyman-Pearson approaches to statistical testing in psychological research (1940–1960). Am J Psychol 119(4): 625–653

Halsey L, Curran-Everett D, Vowler S, Drummond G (2015) The fickle p value generates irreproducible results. Nat Methods 12(3):179–185

Hoenig J, Heisey D (2001) The abuse of power: the pervasive fallacy of power calculations for data analysis. Am Stat 55:19–24

Howitt SM, Wilson AN (2014) Revisiting "Is the scientific paper a fraud?": the way textbooks and scientific research articles are being used to teach undergraduate students could convey a misleading image of scientific research. EMBO Rep 15(5):481–484

Hubbard R, Bayarri M, Berk K, Carlton M (2003) Confusion over measures of evidence (p's) versus errors (α's) in classical statistical testing. Am Stat 57(3):171–178

Huberty CJ (1993) Historical origins of statistical testing practices: the treatment of Fisher versus Neyman-Pearson views in textbooks. J Exp Educ 61:317–333

Hurlbert S, Lombardi C (2009) Final collapse of the Neyman-Pearson decision theoretic framework and rise of the neoFisherian. Ann Zool Fenn 46(5):311–349

Ioannidis JPA (2005) Why most published research findings are false. PLoS Med 2(8):e124

Johnson VE (2013) Revised standards for statistical evidence. Proc Natl Acad Sci 110(48): 19313–19317

Kobayashi K (1997) A comparison of one- and two-sided tests for judging significant differences in quantitative data obtained in toxicological bioassay of laboratory animals. J Occup Health 39(1): 29–35

Krueger JI, Heck PR (2017) The heuristic value of p in inductive statistical inference. Front Psychol 8:108–116

Laplace P (1812) Théorie analytique des probabilités

Lecoutre B, Lecoutre M-P, Poitevineau J (2001) Uses, abuses and misuses of significance tests in the scientific community: won't the Bayesian choice be unavoidable? Int Stat Rev/Rev Int Stat 69(3):399–417

Lee SM (2018) Buzzfeed news: here's how Cornell scientist Brian Wansink turned shoddy data into viral studies about how we eat, February 2018. https://www.buzzfeednews.com/article/stephaniemlee/brian-wansink-cornell-p-hacking.

Lehmann E (2011) Fisher, Neyman, and the creation of classical statistics. Springer, Berlin

Lenhard J (2006) Models and statistical inference: the controversy between Fisher and Neyman-Pearson. Br J Philos Sci 57(1):69–91. ISSN 0007-0882. https://doi.org/10.1093/bjps/axi152

Lew MJ (2012) Bad statistical practice in pharmacology (and other basic biomedical disciplines): you probably don't know P. Br J Pharmacol 166(5):1559–1567

Liu K, Meng X-L (2016) There is individualized treatment. Why not individualized inference? Annu Rev Stat Appl 3(1):79–111. https://doi.org/10.1146/annurev-statistics-010814-020310

Lombardi C, Hurlbert S (2009) Misprescription and misuse of one-tailed tests. Austral Ecol 34:447–468

Lu J, Qiu Y, Deng A (2018) A note on type s & m errors in hypothesis testing. Br J Math Stat Psychol. Online version of record before inclusion in an issue

McCullagh P (2002) What is a statistical model? Ann Stat 30(5):1125–1310

Medawar P (1963) Is the scientific paper a fraud? Listener 70:377–378

Motulsky HJ (2014) Common misconceptions about data analysis and statistics. Naunyn-Schmiedeberg's Arch Pharmacol 387(11):1017–1023

Neyman J, Pearson E (1933) On the problem of the most efficient tests of statistical hypotheses. Philos Trans R Soc Lond A 231:289–337

Nickerson RS (2000) Null hypothesis significance testing: a review of an old and continuing controversy. Psychol Methods 5(2):241–301

Nuzzo R (2014) Statistical errors: P values, the 'gold standard' of statistical validity, are not as reliable as many scientists assume. Nature 506:150–152

Royall R (1997) Statistical evidence: a likelihood paradigm. Monographs on statistics and applied probability, vol 71. Chapman & Hall, London

Ruxton GD, Neuhaeuser M (2010) When should we use one-tailed hypothesis testing? Methods Ecol Evol 1(2):114–117

Sackrowitz H, Samuel-Cahn E (1999) P values as random variables-expected P values. Am Stat 53:326–331

Senn S (2001) Two cheers for P-values? J Epidemiol Biostat 6(2):193–204

Shaw G, Nodder F (1789) The naturalist's miscellany: or coloured figures of natural objects; drawn and described immediately from nature

Strasak A, Zaman Q, Marinell G, Pfeiffer K (2007) The use of statistics in medical research: a comparison of the New England Journal of Medicine and Nature Medicine. Am Stat 61(1): 47–55

Student (1908) The probable error of a mean. Biometrika 6(1):1–25

Thompson B (2007) The nature of statistical evidence. Lecture notes in statistics, vol 189. Springer, Berlin

Trafimow D, Marks M (2015) Editorial. Basic Appl Soc Psychol 37(1):1–2. https://doi.org/10.1080/01973533.2015.1012991

Tukey JW (1991) The philosophy of multiple comparisons. Stat Sci 6(1):100–116

Voelkl B, Vogt L, Sena ES, Würbel H (2018) Reproducibility of preclinical animal research improves with heterogeneity of study samples. PLOS Biol 16(2):e2003693–13

Wagenmakers E-J (2007) A practical solution to the pervasive problems of p values. Psychonom Bull Rev 14(5):779–804

Wagenmakers E-J, Marsman M, Jamil T, Ly A, Verhagen J, Love J, Selker R, Gronau QF, Šmíra M, Epskamp S, Matzke D, Rouder JN, Morey RD (2018) Bayesian inference for psychology. Part I: theoretical advantages and practical ramifications. Psychon Bull Rev 25:35–57

Wasserstein RL, Lazar NA (2016) The ASA's statement on p-values: context, process, and purpose. Am Stat 70(2):129–133

Electronic Lab Notebooks and Experimental Design Assistants

Björn Gerlach, Christopher Untucht, and Alfred Stefan

Contents

Abstract

Documentation of experiments is essential for best research practice and ensures scientific transparency and data integrity. Traditionally, the paper lab notebook (pLN) has been employed for documentation of experimental procedures, but over the course of the last decades, the introduction of electronic tools has changed the research landscape and the way that work is performed. Nowadays, almost all data acquisition, analysis, presentation and archiving are done with electronic tools. The use of electronic tools provides many new possibilities, as well as challenges, particularly with respect to documentation and data quality. One of the biggest hurdles is the management of data on different devices with a

B. Gerlach (✉)
PAASP GmbH, Heidelberg, Germany
e-mail: bjoern.gerlach@paasp.net

C. Untucht
AbbVie Deutschland GmbH, Neuroscience Discovery, Ludwigshafen am Rhein, Germany
e-mail: christopher.untucht@abbvie.com

A. Stefan
AbbVie Deutschland GmbH, Information Research, Ludwigshafen am Rhein, Germany
e-mail: alfred.stefan@abbvie.com

© The Author(s) 2019
A. Bespalov et al. (eds.), *Good Research Practice in Non-Clinical Pharmacology and Biomedicine*, Handbook of Experimental Pharmacology 257,
https://doi.org/10.1007/164_2019_287

substantial amount of metadata. Transparency and integrity have to be ensured and must be reflected in documentation within LNs. With this in mind, electronic LNs (eLN) were introduced to make documentation of experiments more straightforward, with the development of enhanced functionality leading gradually to their more widespread use. This chapter gives a general overview of eLNs in the scientific environment with a focus on the advantages of supporting quality and transparency of the research. It provides guidance on adopting an eLN and gives an example on how to set up unique Study-IDs in labs in order to maintain and enhance best practices. Overall, the chapter highlights the central role of eLNs in supporting the documentation and reproducibility of experiments.

Keywords

Data management · Documenting experiments · Lab journal · Paper and electronic lab notebook · Study-ID

1   Paper vs. Electronic Lab Notebooks

Documentation of experiments is critical to ensure best practice in research and is essential to understand and judge data integrity. Detailed and structured documentation allows for researcher accountability as well as the traceability and reproducibility of data. Additionally, it can be used in the resolution of intellectual property issues. Historically, this had been performed by documentation in conventional paper laboratory notebooks (pLN). Lab notebooks (LN) are considered the primary recording space for research data and are used to document hypotheses, experiments, analyses and, finally, interpretation of the data. Originally, raw and primary data were recorded directly into the lab book and served as the basis for reporting and presenting the data to the scientific community (Fig. 1a), fulfilling the need for transparent and reproducible scientific work.

The situation has become much more complex with the entry of electronic tools into the lab environment. Most experimental data is now acquired digitally, and the overall amount and complexity of data have expanded significantly (Fig. 1b). Often, the data is acquired with a software application and processed with secondary tools, such as dedicated data analysis packages. This creates different databases, varying both in volume and the type of produced data. Additionally, there can be different levels of processed data, making it difficult to identify the unprocessed, or raw, data. The experimental data must also be archived in a way that it is structured, traceable and independent of the project or the data source. These critical points require new data management approaches as the conventional pLN is no longer an alternative option in today's digital environment.

Current changes in the type, speed of accumulation and volume of data can be suitably addressed by use of an electronic laboratory notebook (eLN). An eLN that can be used to document some aspects of the processes is the first step towards improvement of data management (Fig. 1b). Ultimately, eLNs will become a central tool for data storage and connection and will lead to improvements in transparency and communication between scientists, as it will be discussed in the last two sections of this chapter.

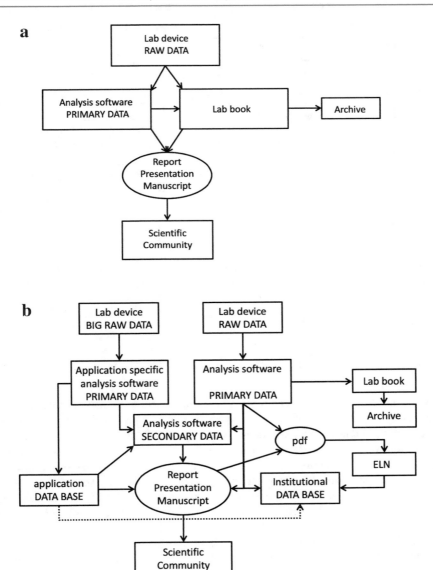

Fig. 1 (**a**) Traditional data flow: The lab book is an integral part of scientific data flow and serves as a collection of data and procedures. (**b**) An example of a more complex data flow might commonly occur today. Unfortunately, lab books and eLNs are often not used to capture all data flows in the modern laboratory

Although both paper and electronic LNs have advantages and disadvantages (Table 1), ultimately the overall switch of researchers from the use of pLNs to eLNs is inevitable. The widespread use of electronic tools for acquiring, analysing

Table 1 Overview of advantages and disadvantages of paper and electronic lab notebooks

Characteristic	Paper lab notebooks	Electronic lab notebooks
Data accessibility	Restricted access, needs copying, often inaccessible once a researcher leaves a lab	Advanced data sharing, direct links to all types of files
Data organization	Chronological order of experiments with page numbers	Flexible organization of experimental data, parallel documentation of different projects
Data presentation	All electronic data has to be printed or referenced	Direct access to large data files via links
Templates	None	Documentation facilitated by templates
Peer review and sharing information	Difficult	Viewing/editing permissions can be granted to individuals and groups of people
Search for keywords	Not possible	Possible
Ease of use	Easy to use, intuitive, flexible	Many different solutions, DIY
Audit trail	Witnessing with a signature	Witnessing with an electronic signature, not always straightforward with DYI eLN
IT support	Not required	Needs IT infrastructure, accessibility across devices is not always possible
Costs	Low	Can be expensive

and managing data renders traditional pLNs impractical in modern science. Nevertheless, the switch to eLN usage is gradual and will take time. Several groups documented and have published their experiences and discussed the biggest barriers for establishing the standardized use of eLNs. Kanza et al. carried out detailed analyses, based on surveys, focusing on potential obstacles to the application of eLNs in an academic laboratory setting (Kanza et al. 2017). According to their results, the major barriers are cost, ease of use and accessibility across devices. Certainly, these aspects seem to be valid at first glance and need to be considered carefully. A more detailed examination provides means to address these issues as progress has been made in the development of tailored solutions.

As mentioned, cost is often the most significant hurdle to overcome when adopting the use of an eLN. In this regard, pLNs have a clear advantage because they require close to no infrastructure. In contrast, there are multiple different eLN device options which vary widely, from freeware to complex solutions, which can have costs of up to thousands of euros per researcher. This is, however, evolving with the use of cloud-based software-as-a-service (SaaS) products which are available for a monthly subscription. One of the major advantages of SaaS is that the software version will always be up-to-date. Today, many of these SaaS solutions are available as online tools which can also be used within a firewall to have maximum

Table 2 Overview of different levels of complexity of laboratory notebooks

Type	Explanation
Paper lab notebook	Reporting of complete primary experimental data in paper format
Do-it-yourself electronic lab notebook	Most simple form of eLN. Many different possibilities exist that can be used as building blocks and combined to create a custom solution (e.g. OneNote, MS Office documents, Dropbox, OneDrive)
Dedicated electronic lab notebook	Sophisticated and dedicated software solutions for reporting and archiving complete primary experimental data
Systemic electronic lab notebook	As above including full lab inventory management system

protection of the data. Most digital solutions, however, require the appropriate hardware and infrastructure to be set up and updated.

In contrast to eLNs, pLNs do not require IT support and are, therefore, immediately usable, do not crash and are intuitive to use. However, in the majority of cases, the data cannot be used directly but must be printed, and a general search cannot be performed in the traditional lab book once the scientist who created it leaves the lab. Thus, the primary disadvantages of paper notebooks are the inability to share information, link electronic data, retrieve the information and search with keywords.

Different types of LNs are summarized in Table 2. A pLN can be seen as the primary place for reporting research. Simple electronic tools such as word processors, or do-it-yourself (DIY) solutions (Dirnagl and Przesdzing 2017), provide this in electronic format. A DIY eLN is a structured organization of electronic data without dedicated software, using instead simple software and data organization tools, such as Windows Explorer, OneNote, text editors or word processors, to document and organize research data (see the last two sections of this chapter for tips on how to create a DIY eLN). In contrast, dedicated and systemic versions of eLNs provide functionality for workflows, standard data entry as well as search and visualization tools. The systemic eLNs provide additional connections to laboratory information management systems (LIMS), which have the capability to organize information in different aspects of daily lab work, for example, in the organization of chemicals, antibodies, plasmids, clones, cell lines and animals and in calendars to schedule usage of devices. Following the correct acquisition of very large datasets (e.g. imaging, next-generation sequencing, high-throughput screens), eLN should support their storage and accessibility, becoming a central platform to store, connect, edit and share the information.

The traditional pLN is still widely used in academic settings and can serve as a primary resource for lab work documentation. Simple versions of eLNs improve the data documentation and management, for example, in terms of searching, accessibility, creation of templates and sharing of information. However, with an increasing amount of work taking place in a complex digital laboratory environment and the associated challenges with regard to data management, eLNs that can connect, control, edit and share different data sources will be a vital research tool.

2   Finding an eLN

Many different eLNs exist, and finding the appropriate solution can be tedious work. This is especially difficult when different research areas need to be covered. Several published articles and web pages provide guidance on this issue (are summarized in the following paragraphs with additional reading provided in Table 3). However, a dedicated one-size-fits-all product has, to date, not been developed. Several stand-

Table 3 Summary of published data describing selection and usage of eLNs

Author and year	Title	Description
Rubacha et al. (2011)	A review of electronic laboratory notebooks available in the market today	Very comprehensive analysis of eLNs on the market in 2011; distinguishes between five different categories
Nussbeck et al. (2014)	The laboratory notebook in the twenty-first century	A general advocacy for eLNs; good background information about the advantages
Oleksik and Milic-Frayling (2014)	Study of an electronic lab notebook design and practices that emerged in a collaborative scientific environment	Comprehensive user experience of establishing a DIY eLN (OneNote) in a physics research lab
Guerrero et al. (2016)	Analysis and implementation of an electronic laboratory notebook in a biomedical research institute	Quite comprehensive comparison of six eLNs with a more in-depth consideration of two devices (PerkinElmer and OneNote)
Kanza et al. (2017)	Electronic lab notebooks: can they replace paper?	Broad and informative background information on the use of eLNs
Riley et al. (2017)	Implementation and use of cloud-based electronic lab notebook in a bioprocess engineering teaching laboratory	Practical guidance on setting up an eLN (LabArchives) as an example in a teaching lab
Dirnagl and Przesdzing (2017)	A pocket guide to electronic laboratory notebooks in the academic life sciences	Experience report about setting up an eLN in an academic lab
May[a]	Companies in the cloud: Digitizing lab operations	Short description of several commercial tools connecting lab equipment with eLNs
Kwok[b]	How to pick an electronic laboratory notebook	Collection of several aspects to be considered when choosing an eLN
The Gurdon Institute[c]	Electronic lab notebooks – for prospective users	Comprehensive and up-to-date guidance on different eLNs
Atrium Research[d]	Electronic laboratory notebook products	Most comprehensive list of eLNs that are currently available on the market

[a]https://www.sciencemag.org/features/2017/02/companies-cloud-digitizing-lab-operations
[b]https://www.nature.com/articles/d41586-018-05895-3
[c]https://www.gurdon.cam.ac.uk/institute-life/computing/elnguidance
[d]http://atriumresearch.com/eln.html

alone solutions have been developed for dedicated purposes, and other eLNs provide the possibility to install "in programme applications". These additional applications can provide specific functionality for different research areas, e.g. for cloning.

Thus, the first course of action must be to define the needs of the research unit and search for an appropriate solution. This solution will most likely still not cover all specific requirements but should provide the functionality to serve as the central resource for multiple different types of data. In this way, the eLN is a centralized storage location for various forms of information and is embedded in the laboratory environment to allow transparency and easy data exchange. Ideally, an eLN is the place to store information, support the research process and streamline and optimize workflows.

The following points must be considered in order to identify the best solution:

- *Does the organization have the capability to introduce an eLN?* A research unit must judge the feasibility of eLN implementation based on different parameters:
  - *Financial resources*: Setting up an eLN can be cost intensive over the years, especially with the newly emerging model of "SaaS", in which the software is not bought but rented and paid for monthly.
  - *IT capability*: It is important to have the knowledge in the research unit to set up and provide training for users to adopt an eLN or at least to have a dedicated person who is trained and can answer daily questions and provide some basic training to new staff.
  - *Technical infrastructure*: The technical equipment to support the eLN has to be provided and maintained for the research unit, especially when hosted within a research unit and not on external servers.
- *At which level should the eLN be deployed and what is the research interest?* Clarification on these two considerations can dramatically reduce the number of potential eLNs which are suitable for application in the research lab. For example, a larger research unit covering more areas of research will need a general eLN, in contrast to a smaller group which can get an eLN ready tailored to their specific needs. An example of a field with very specific solutions is medicinal chemistry: An eLN for application in a lab of this type should have the functionality to draw molecules and chemical pathways, which will not be needed for an animal research or image analysis lab. Having a clear idea about the requirements of a research unit and necessary functionality will help in the choice of the correct eLN resource.
- *Which devices will be used to operate the eLN software?* Many researchers require the facility to create or update records "live" on the bench or other experimental areas (e.g. undertaking live microscopy imaging), as well as on other devices inside and outside of the lab (e.g. running a high-throughput robotic screening in a facility). Alternatively, researchers may want to use voice recognition tools or to prepare handwritten notes and then transcribe into a tidier, more organized record on their own computer. It should also be noted that some vendors may charge additional fees for applications to run their software on different device types.

- *Does the funding agreement require specific data security/compliance measures?* Some funding agencies require all data to be stored in a specific geographic location to ensure compliance with local data protection regulations (e.g. General Data Protection Regulation (GDPR)). Some eLN systems, however, are designed to store content and data only on their own servers (i.e. in "the cloud"), in which case a solution could be negotiated, or the provider cannot be used.
- *Consider exporting capabilities of the eLN.* This is a very critical point as the eLN market is constantly changing and companies might be volatile. The eLN market has existed for almost 20 years, during which time several solutions emerged and disappeared or were bought by other companies. With these changes, there are certain associated risks for the user, such as, for example, the fact that pricing plans and/or data formats might change or the eLN could be discontinued. Therefore, we consider it as an absolute requirement that all data can be easily exported to a widely readable format that can be used with other applications.

It is advisable to identify a dedicated person within the research unit to prepare an outline of the lab requirements and who will be educated about the available technical solutions and prepare a shortlist. Subsequently, it is necessary to test the systems. All eLN providers we have been in contact so far have been willing to provide free trial versions for testing. Additionally, they usually provide training sessions and online support for researchers to facilitate the adaptation of the system more quickly.

When choosing an eLN, especially in academia, an important question will always be the cost and how to avoid them. There is a debate about the usage of open-source software, that is, free versions with some restrictions, or building a DIY solution. A good resource in this context is the web page of Atrium Research & Consulting, a scientific market research and consulting practice which provides a comprehensive list of open-source solutions (http://atriumresearch.com/eLN.html). It is important to note that open source is not always a truly open source – often it is only free to use for non-commercial organizations. They can also come with some limitations, for example, that all modifications must be published on the "source website". However, the industry is ready to accept open source with many examples emerging, such as Tomcat or Apache, which are becoming in our experience de facto industry standards. Important factors to be considered for choosing open-source software are:

- The activity of the community maintaining the system (e.g. CyNote (https://sourceforge.net/projects/cynote/) and LabJ-ng (https://sourceforge.net/projects/labj/) was not updated for years, whereas eLabFTW (https://www.elabftw.net) or OpenEnventory (https://sourceforge.net/projects/enventory/) seems to be actively maintained.
- Maturity and structure of the documentation system.
- If the intended use is within the regulated environment, the local IT has to take ownership.

Open-source software, particularly in the light of best documentation practice, will generate risks, which need to be handled and mitigated. This includes ensuring the integrity of the quality system itself and guaranteeing data integrity and prevention of data loss. To our knowledge, most open-source solutions cannot guarantee this. This is one reason why a global open-source eLN doesn't exist: there is no system owner who could be held accountable and would take responsibility to ensure data integrity.

The SaaS approach often uses cloud services as an infrastructure, e.g. Amazon Web Services or Microsoft Azure. These services provide the foundation for the web-based eLNs and are very easy to set up, test and use. These systems have a significant benefit for smaller organizations, and it is worth investigating them. They provide truly professional data centre services at a more affordable price, attracting not only smaller research units but also big pharmaceutical companies to move into using SaaS solutions to save costs.

3   Levels of Quality for eLNs

In principle, the "quality features" of eLNs can be related to three categories which have to be considered when choosing or establishing an eLN: (1) System, (2) Data and (3) Support.

System-related features refer to functionality and capability of the software:

- *Different user rights*: The ability to create different user levels, such as "not share", read-only, comment, modify or create content within the eLN, is one of the big advantages. The sharing of information via direct access and finely tuned user levels allows for individual sharing of information and full transparency when needed.
- *Audit trail*: Here, a very prominent regulation for electronic records has to be mentioned, the CFR21 Part 11. CFR stands for Code of Federal Regulations, Title 21 refers to the US Food and Drug Administration (FDA) regulations, and Part 11 of Title 21 is related to electronic records and electronic signatures. It defines the criteria under which electronic records and electronic signatures are considered trustworthy, reliable and equivalent to paper records (https://www. accessdata.fda.gov/scripts/cdrh/cfdocs/cfcfr/CFRSearch.cfm?CFRPart=11).
  Hence, when choosing an eLN provider, it is important that the product adheres to these regulations since this will ensure proper record keeping. Proper record keeping in this case means that a full audit trail is implemented in the software application.
- *Archiving*: Scientists need to constantly access and review research experiments. Therefore, accessing and retrieving information relating to an experiment is an essential requirement. The challenge with archiving is the readability and accessibility of the data in the future. The following has to be considered when choosing an eLN to archive data:

- There is always uncertainty about the sustainability of data formats in the future, especially important for raw data when saved in proprietary formats (the raw data format dilemma resulted in several initiatives to standardize and harmonize data formats).
- The eLN can lead to dependency on the vendor and the system itself.
- Most current IT systems do not offer the capability to archive data and export them from the application. This results in eLNs becoming de facto data collection systems, slowing down in performance over time with the increased data volume.

Data-related features concern the ability to handle different scientific notations and nomenclatures, but most importantly, to support researchers in designing, executing and analysing experiments. For scientific notations and nomenclatures, the support of so-called namespaces and the ability to export metadata, associated with an experiment in an open and common data format, are a requirement. Several tools are being developed aiming to assist researchers in designing, executing and analysing experiments. These tools will boost and enhance quality of experimental data, creating a meaningful incentive to change from pLNs towards eLNs. The subsequent sections focus in detail on that point.

Support-related quality features affect the stability and reliability of the infrastructure to support the system and data. The most basic concern is the reliability and maintenance of the hardware. In line with that, fast ethernet and internet connections must be provided to ensure short waiting times when loading data. The security of the system must be ensured to prevent third-party intrusions, which can be addressed by including appropriate firewall protection and virus scans. Correspondingly, it is also very important to include procedures for systematic updates of the system. In contrast, this can be difficult to ensure with open-source software developed by a community. As long as the community is actively engaged, the software security will be normally maintained. However, without a liable vendor, it can change quickly and without notice. Only a systematic approach will ensure evaluation of risks and the impact of enhancements, upgrades and patches.

4   Assistance with Experimental Design

The design of experiments (DOE) was first introduced in 1920 by Ronald Aylmer Fisher and his team, who aimed to describe the different variabilities of experiments and their influence on the outcome. DOE also refers to a statistical planning procedure including comparison, statistical replication, randomization (chapter "Blinding and Randomization") and blocking. Designing an experiment is probably the most crucial part of the experimental phase since errors that are introduced here determine the development of the experiment and often cannot be corrected during the later phases. Therefore, consultation with experienced researchers, statisticians or IT support during the early phase can improve the experimental outcome. Many tools supporting researchers during the DOE planning phase already exist. Although most of these tools have to be purchased, there are some freely available, such as the scripting language R, which is widely used among statisticians and offers a wide range of additional scripts (Groemping 2018).

An example of an interactive tool for the design and analysis of preclinical in vivo studies is an online tool based on the R application (R-vivo) MANILA (MAtched ANImaL Analysis). R-vivo is a browser-based interface for an R-package which is mainly intended for refining and improving experimental design and statistical analysis of preclinical intervention studies. The analysis functions are divided into two main subcategories: pre-intervention and post-intervention analysis. Both sections require a specific data format, and the application automatically detects the most suitable file format which will be used for uploading into the system. A matching-based modelling approach for allocating an optimal intervention group is implemented, and randomization and power calculations take full account of the complex animal characteristics at the baseline prior to interventions. The modelling approach provided in this tool and its open-source and web-based software implementations enable researchers to conduct adequately powered and fully blind preclinical intervention studies (Laajala et al. 2016).

Another notable solution for in vivo research is the freely available Experimental Design Assistant (EDA) from the National Centre for the Replacement, Refinement and Reduction of Animals in Research (NC3R). NC3R is a UK-based scientific organization aiming to find solutions to replace, refine and reduce the use of animals in research (https://www.nc3rs.org.uk/experimental-design-assistant-eda). As one of the goals, NC3R has developed the EDA, an online web application allowing the planning, saving and sharing of individual information about experiments. This tool allows an experimental plan to be set up, and the engine provides a critique and makes recommendations on how to improve the experiment. The associated website contains a wealth of information and advice on experimental designs to avoid many of the pitfalls that have been previously identified, including such aspects as the failure to avoid subjective bias, using the wrong number of animals and issues with randomization. This support tool and the associated resources allow for better experimental design and increase the quality in research, presenting an excellent example for electronic study design tools. A combined usage of these online tools and eLN would improve the quality and reproducibility of scientific research.

5   Data-Related Quality Aspects of eLNs

An additional data-related quality consideration, besides support with experimental design, concerns the metadata. Metadata is the information important for understanding experiments by other researchers. Often, the metadata is generated by the programme itself and needs a manual curation by the researcher to reach completeness. In principle, the more metadata is entered into the system, the higher the value of the data since it makes it possible to be analysed and understood in greater detail, thus being reproducible in subsequent experiments.

Guidance on how to handle metadata is provided in detail in chapter "Data Storage" in the context of the FAIR principles and ALCOA. In brief, ALCOA, and lately its expanded version ALCOAplus, is the industry standard used by the FDA, WHO, PIC/S and GAMP to give guidance in ensuring data integrity. The guidelines point to several important aspects, which also explain the acronym:

- *Attributable*: Who acquired the data or performed an action and when?
- *Legible, Traceable and Permanent*: Can you read the data and any entries?
- *Contemporaneous*: Was it recorded as it happened? Is it time-stamped?
- *Original*: Is it the first place that the data is recorded? Is the raw data also saved?
- *Accurate*: Are all the details correct?

The additional "plus" includes the following aspects:

- *Complete*: Is all data included (was there any repeat or reanalysis performed on the sample)?
- *Consistent*: Are all elements documented in chronological order? Are all elements of the analysis dated or time-stamped in the expected sequence?
- *Enduring*: Are all recordings and notes preserved over an extended period?
- *Available*: Can the data be accessed for review over the lifetime of the record?

These guiding principles can be used both for paper and electronic LNs, and the questions can be asked by a researcher when documenting experiments. However, eLNs can provide support to follow these regulations by using templates adhering to these guiding principles, increasing transparency and trust in the data. The electronic system can help to manage all of these requirements as part of the "metadata", which is harder to ensure with written lab notebooks or at least requires adaptation of certain habits.

Having these principles in place is the first important step. Then, it needs to be considered that the data will be transferred between different systems. Transferring data from one specialized system to another is often prone to errors, e.g. copy-and-paste error, loss of metadata, different formats and so on. Thus it must always be considered when the information is copied, especially when data are processed within one programme and then stored in another. It has also to be ensured that data processing can be reproduced by other researchers. This step should be documented in LNs, and eLNs can provide a valuable support by linking to different versions of a file. There are myriad of applications to process raw data which come with their own file formats. Ideally, these files are saved in a database (eLN or application database), and the eLN can make a connection with a dedicated interface (Application programming interfaces, API) between the analysis software and the database. This step ensures that the metadata is maintained. By selecting the data at the place of storage yet still using a specialized software, seamless connection and usage of the data can be ensured between different applications. For some applications, this issue is solved with dedicated data transfer protocols to provide a format for exporting and importing data. The challenge of having a special API is recognized by several organizations and led to the formation of several initiatives, including the Pistoia Alliance (https://www.pistoiaalliance.org) and the Allotrope Foundation (https://www.allotrope.org). These initiatives aim to define open data formats in cooperation with vendors, software manufactures and pharmaceutical companies. In particular, Allotrope has released data formats for some technologies which will be a requirement for effective and sustainable integration of eLNs into the lab environment. Therefore, it is worth investing time to find solutions for supporting storage, maintenance and transfer of research data.

6 The LN as the Central Element of Data Management

Today's researcher faces a plethora of raw data files that in many cases tend to stay within separated data generation and analysis systems. In addition, the amount of data a scientist is able to generate with one experiment has increased exponentially. Large data sets coming from omics and imaging approaches generate new data flows in scientific work that are not captured by eLNs at all, primarily due to a lack of connectivity. Thus, the eLN needs a structured approach to connect all experimental data with the raw data (Fig. 2). This can most likely still be achieved with a pLN, but support from digital tools seems to be obviously advantageous, which can be in the form of a dedicated eLN or even a DIY approach. In any case, the approach will have to evolve from a scientist's criteria of an integrated data management system meeting several requirements: Documentation of intellectual property generation, integrated raw data storage and linking solutions and enhanced connectivity with lab equipment, colleagues and the scientific community.

Several eLNs provide a user interface for document generation: Independent of the kind of document created, the eLN should be the starting point by creating a unique Study-ID and will add directly, or at least link, the credentials of the scientist to the document. Some eLNs require the input of some basic information such as the project, methods used or co-workers involved and will compile a master project file with some essential metadata. Existing files like protocols or data packages can then be linked in the master file. The master project file within the eLN acts as a map for the landscape of scientific experiments belonging to one project. In a DIY approach, these steps can also be created manually within a folder system (e.g. OneNote, Windows Explorer, etc.) and a master file created in the form of a word processor file (e.g. MS Office Word) or a page within OneNote. Of course, this can also be achieved within a pLN, but, again, it is more effort to structure the information and keep an overview. In all cases, each step of the scientific experiment process can then be automatically tracked either by the eLN or manually to allow for fast data location and high reproducibility. Even if copied elsewhere, recovery and identification of ownership is easy with the help of the unique Study-ID. At the end of a series of experiments or projects, the researcher or an artificial intelligence within the eLN decides which files to pool in a PDF file, and after proofreading, somebody can mark the project as completed. Thereby, the final document, best in the form of a PDF file, will not only contain some major files but also the master file with the Study-IDs of

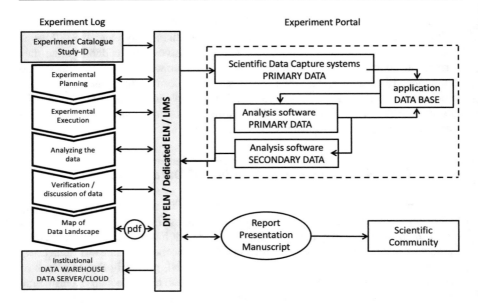

Fig. 2 Workflow for the lab environment with the eLN/LIMS being the central element. Each experimental setup will start with a unique Study-ID entry into the eLN in the experimental catalogue which will be used throughout the whole experiment and allow for tagging during all steps. The eLN will be the hub between the experimental procedure (left) and the data collection and reporting (right). The eLN should collect all different types of data or at least provide the links to the respective storage locations. One of the last steps is the summary of the experiment in the "Map of the Data Landscape" in the form of a PDF file. Next-generation dedicated eLNs could themselves be used to create such a document, thereby providing a document for reporting with the scientific community and storage in the data warehouse

all data sets that are searchable and identifiable across the server of an organization. This will also allow smart search terms making the organization of experimental data very straightforward.

Unique Study-IDs for projects across an organization are key prerequisites to cross-link various sets of documents and data files into a project (see the next section on how to set up). Using an eLN as the managing software connecting loose ends during the data generation period of a project and even right from the beginning will free the researcher from the hurry to do so at the end of the project. This type of software can easily be combined with an existing laboratory inventory management system (LIMS) not only fulfilling the documentation duties but also adding the value of project mapping.

In any case, enhanced connectivity is key for successful transparency. Enhanced connectivity can be understood as connectivity across different locations and between different scientists. In terms of different locations, a researcher can access data, such as the experimental protocol, within the eLNs from the office and lab areas. A tablet can be placed easily next to a bench showing protocols that could be amended by written or spoken notes. Timer and calculation functions are nothing new but form an essential part of daily lab work and need to be easily accessible. In very advanced applications, enhanced connectivity is implemented using camera assistance in lab goggles capturing essential steps of the experiment. If coupled to optical character recognition (OCR) and artificial intelligence (AI) for identifying the devices and materials used and notes taken, a final photograph protocol might form part of the documentation. Information on reagent vessels would be searchable and could easily be reordered. Connectivity to compound databases would provide information about molecular weight with just one click and can be used to help a researcher check compound concentrations during an experiment. A tablet with the facility to connect to lab devices can be used to show that the centrifuge next door is ready or that there is available space at a sterile working bench or a microscope. These benefits allow seamless, transparent and detailed documentation without any extra effort.

Connectivity between researchers within an organization can be optimized in a different way by using electronic tools. At some point, results may need to be presented on slides to others in a meeting. Assuming best meeting practice is applied, these slides will be uploaded to a folder or a SharePoint. After a certain period, it may become necessary to access again the data presented in the files uploaded to the SharePoint. SharePoint knowledge usually lacks connectivity to raw data as well as to researchers that have not been part of a particular meeting. However, a unique identifier to files of a project will allow for searchability and even could render SharePoint data collections redundant. Feedback from a presentation and results of data discussion might be added directly to a set of data rather than to a presentation of interpreted raw data. The whole process of data generation, its interpretation and discussion under various aspects followed by idea generation, re-analyses and new experiments can become more transparent with the use of electronic tools. An eLN may also potentially be used to enhance clarity with respect to the ownership of ideas and experimental work.

---

7   Organizing and Documenting Experiments

Unique Study-IDs are important components for appropriate and transparent documentation and data management. Creating a structured approach for the accessibility of experimental data is reasonably straightforward, although it requires discipline. Unique Study-IDs should be used in all lab environments independent of the LN used. The DIY eLN is discussed here as an example by which to structure the data, which can be achieved with a generic software such as OneNote, as described by Oleksik and Milic-Frayling (2014), or can be implemented within Windows Explorer and a text editing programme, such as Word. Dedicated eLNs will provide the Study-ID automatically, but in the case of a DIY eLN (or LN), this must be set up and organized by the researcher. Several possibilities exist, and one can be creative, yet consistent, to avoid any confusion. The Study-ID will not only help to identify experiments but will also assist a scientist in structuring their data in general. Therefore, it should be kept simple and readable.

Table 4 Implementing a unique Study-ID system, an example

Number of experimental categories	Description of the experimental category	Number of the experiment within the category	Example file name	Unique identifier (further differentiation possible to label material)
01	IHC – brain	01	01.01 effect of treatment X	01.01
		02	01.02 confirmatory experiment for 01.01	01.02
		…	…	…
02	SH-SY5Y	01	02.01 titration of rapamycin to test fibre length	02.01
		02	02.02 analysis of gene expression after treatment Y	02.02
		…	…	…
03	Primary cortical cells	01	03.01 Western blot analysis to determine expression of proteins after treatment G	03.01
		02	03.02 pull-down assay of X to test interaction with Y	03.02
		…	…	…
04	…	…	…	…

A possible labelling strategy could be to include the date, with several formats possible, depending on preference (e.g. DDMMYY, YY-MM-DD, YYMMDD). It is, however, advisable to agree on one format which is then used throughout the research unit. The latter one (YYMMDD), for example, has the advantage that files are directly ordered according to the day if used as the prefix of a file name or, even better, if used as a suffix. When used as a suffix, files with the same starting name are always ordered according to the date. If several experiments are planned on the same day, a short additional suffix in the form of a letter can distinguish the experiments (YYMMDD.A and YYMMDD.B). Such a unique identifier could be created on the day when the experiment was planned and should always be used throughout the experiment to identify experimental materials (such as in labelling cell culture plates, tubes, immunohistochemistry slides, Western blots and so on). It is also advisable to add some more information to the name of the folder to get a better overview and use only the Study-ID and short numbers for temporary items or items with little labelling space (e.g. for cell culture plates and tubes).

Another possibility to categorize experiments is by using "structured" numbers and categories created by the researcher and then using running numbers to create unique IDs. Table 4 provides examples of potential categories. The researcher can be creative here and set up a system to fit their requirements. The main purpose is to create a transparent system to be able to exactly trace back each sample in the lab.

This was found to be especially useful for samples that might need to be analysed again, e.g. cell lysates which were immunoblotted a second time half a year later for staining with a different antibody. To achieve this level of organization, the tubes only need to be labelled with one number consisting of a few digits. To take an example from Table 4, six tubes were labelled with the Study-ID (03.01) plus an additional number differentiating the tube (e.g. 03.01.1 to 03.01.6). This approach clearly identified the lysates from experiment 03.01 which is the lysates of primary cortical cells treated with substance "G" under six different sets of conditions. If such a labelling system is used for all experimental materials in the lab, it will ensure that each item can be unambiguously identified.

This unique identifier should be used for all files created on every device during experiments. In case this type of system is adopted by several researchers in a research unit, another prefix, e.g. initials, or dedicated folders for each researcher have to be created to avoid confusion. The system can be adopted for the organization of the folders housing all the different experiments. To more easily understand the entire content of an experiment, there should be a master file in each folder providing the essential information on the experiment. This master file can be a text file which is always built from the same template and contains the most important information about an experiment. Templates of this type are easy to create, and they can simplify lab work, providing an advantage for eLNs over the pLN. The template for the master file should contain at least the following sections or links to locations for retrieving the information:

(a) Header with unique ID
(b) Name of the researcher
(c) Date
(d) Project
(e) Aim of the experiment
(f) Reagents and materials
(g) Experimental procedure
(h) Name or pathway to the storage folder of raw data (if different to the parent folder)
(i) Results
(j) Analysis
(k) Conclusions
(l) References

The master file should be saved in PDF format and should be time-stamped to properly document its time of creation. The usage of a master file requires agreement by the organization, perseverance from the researcher and control mechanisms. Depending on the organization and the methods applied regularly, written standard operating procedures (SOPs) can ensure a common level of experimental quality. These files can be easily copied into the master file, and only potential deviations need to be documented. It is recommended to back up or archive the complete experimental folder. How this can be achieved depends on the infrastructure of the facility. One possibility would be that the files are archived for every researcher and year or only if researchers leave the lab. This creates a lot of flexibility and makes searching for certain experiments much more convenient than in a pLN. This type of simple electronic organization combines the advantage of a pLN and an eLN: All electronic files can be stored in one place or directly linked, and it is electronically searchable, thus increasing transparency without the association of any additional costs.

Well-structured scientific data can be more easily transferred to an eLN, and even changes in the eLN system will not have a devastating effect on original file organization. Based on the requirements for data documentation, we suggest the following steps to a better data documentation policy that ultimately will improve data reproducibility:

1. Identify the best system for your needs:
   There are different means by which data can be documented within a lab environment. Identifying the best approach for the specific requirements in a lab saves time and resources by optimizing workflows.
2. Structure your data:
   Well-structured data increases the possibility to find and retrieve the information. Using a unique Study-ID is a good step towards achieving this.
3. Structure the data sources and storage locations:
   Organizing data-storage locations, connecting them to experimental documentation and considering backup solutions are important for transparency.
4. Agree to and follow your rules:
   Agree on minimal operational standards in the research unit that fulfil the long-term needs, e.g. adherence to ALCOA, design and documentation of experiments or IP ownership regulations.
5. Revise your data strategy and search for improvements:
   Search for tools that allow for better connectivity and simplify documentation.

In summary, documentation is the centrepiece for best research practices and has to be properly performed to create transparency and ensure data integrity. The adoption of eLNs along with the establishment of routinely applied habits will facilitate this best practice. The researchers themselves have to invest time and resources to identify the appropriate tool for their research unit by testing different vendors. Once the right tool is identified, only regular training and permanent encouragement will ensure a sustainable documentation practice.

References

Dirnagl U, Przesdzing I (2017) A pocket guide to electronic laboratory notebooks in the academic life sciences [Version 1; Referees: 4 Approved]. Referee Status 1–11. https://doi.org/10.12688/f1000research.7628.1

Groemping U (2018) CRAN task view: design of experiments (DoE) & analysis of experimental data. https://CRAN.R-project.org/view=ExperimentalDesign

Guerrero S, Dujardin G, Cabrera-Andrade A, Paz-y-Miño C, Indacochea A, Inglés-Ferrándiz M, Nadimpalli HP et al (2016) Analysis and Implementation of an electronic laboratory notebook in a Biomedical Research Institute. PLoS One 11(8):e0160428. https://doi.org/10.1371/journal. pone.0160428

Kanza S, Willoughby C, Gibbins N, Whitby R, Frey JG, Erjavec J, Zupančič K, Hren M, Kovač K (2017) Electronic lab notebooks: can they replace paper? J Cheminf 9(1):1–15. https://doi.org/ 10.1186/s13321-017-0221-3.

Laajala TD, Jumppanen M, Huhtaniemi R, Fey V, Kaur A, Knuuttila M, Aho E et al (2016) Optimized design and analysis of preclinical intervention studies in vivo. Sci Rep. https://doi. org/10.1038/srep30723

Nussbeck SY, Weil P, Menzel J, Marzec B, Lorberg K, Schwappach B (2014) The laboratory notebook in the twenty-first century. EMBO Rep 6(9):1–4. https://doi.org/10.15252/embr. 201338358.

Oleksik G, Milic-Frayling N (2014) Study of an electronic lab notebook design and practices that 592 emerged in a collaborative scientific environment. Proc ACM Conf Comput Support Coop Work CSCW:120–133. https://doi.org/10.1145/2531602.2531709

Riley EM, Hattaway HZ, Felse PA (2017) Implementation and use of cloud-based electronic lab notebook in a bioprocess engineering teaching laboratory. J Biol Eng:1–9. https://doi.org/10. 1186/s13036-017-0083-2

Rubacha M, Rattan AK, Hosselet SC (2011) A review of electronic laboratory notebooks available in the market today. J Lab Autom 16(1):90–98. https://doi.org/10.1016/j.jala.2009.01.002

Data Storage

Christopher Frederick Isambard Blumzon
and Adrian-Tudor Pănescu

Contents

Abstract

While research data has become integral to the scholarly endeavour, a number of challenges hinder its development, management and dissemination. This chapter follows the life cycle of research data, by considering aspects ranging from

C. F. I. Blumzon (✉)
Figshare, London, UK
e-mail: c.george@digital-science.com; chris@figshare.com

A.-T. Pănescu
Figshare, London, UK

"Gheorghe Asachi" Technical University of Iaşi, Iaşi, Romania
e-mail: tudor@figshare.com

© The Author(s) 2019
A. Bespalov et al. (eds.), *Good Research Practice in Non-Clinical Pharmacology and Biomedicine*, Handbook of Experimental Pharmacology 257,
https://doi.org/10.1007/164_2019_288

storage and preservation to sharing and legal factors. While it provides a wide overview of the current ecosystem, it also pinpoints the elements comprising the modern research sharing practices such as metadata creation, the FAIR principles, identifiers, Creative Commons licencing and the various repository options. Furthermore, the chapter discusses the mandates and regulations that influence data sharing and the possible technological means of overcoming their complexity, such as blockchain systems.

Keywords

Blockchain · FAIR · Identifier · Licence · Metadata · Preservation · Repository · Reproducibility · Storage

1 Introduction

The evolution of scientific research has been recently shaped by the so-called reproducibility crisis, a phenomenon brought to light by a number of studies that failed to replicate previous results (see, e.g. Eklund et al. 2016; Phillips 2017). This highlighted the necessity of making available the research data underlying studies published in more traditional mediums, such as journals, articles and conference papers, practice which was promptly mandated by both funding bodies (European Commission 2017; National Institutes of Health 2018) and the publishing industry (Elsevier 2018; Springer Nature 2018).

This has left the main actors of scholarly communication, researchers, in an interesting but also possibly difficult position. While the necessity of making data available, especially when it is generated by publicly funded research or it presents high-impact consequences, is uncontested, a number of challenges remain as neither the technical, legal or societal environments were fully prepared for this prospect.

In a 2017 study across Springer Nature and Wiley authors, less than half of the respondents reported sharing research data frequently, with more than 30% rarely or never sharing (Digital Science 2017). Various reasons for the lack of sharing have been identified, ranging from the lack of experience, technical knowledge or time, to fear of shaming (in case errors are discovered in the published dataset), or competition to published results (Federer et al. 2015; Youngseek and Zhang 2015). While the latter are more difficult to overcome, requiring profound changes across the whole scholarly communication spectrum, the more practical aspects can be solved by either technical means or specialised guidance.

This chapter attempts to provide a high level overview of all components encompassing research data sharing while detailing some of the more important aspects, such as storage, metadata, or data anonymisation.

2 Data Storage Systems

While data storage has a history closely linked to that of computing, research data presents a handful of new and unique challenges, especially when it comes to persistence and privacy. Usually most technical details will be handled by specialised professionals, but gathering a basic understanding of the inner workings, advantages and limitations of the various options can help devise data management plans customised to the needs of each research project, community or subject area.

2.1 Types of Storage

The first aspects of storage that need to be considered are the actual medium and employed technology; currently, the most prevalent options are:

- Magnetic (hard disk drives (HDD), magnetic tapes): data is stored using the magnetization patterns of a special surface.
- Optical (compact disks, Blu-ray): data is stored in deformities on a circular surface which can be read when being illuminated by a laser diode.
- Semiconductor: data is stored using semiconductor-based integrated circuits. While traditionally this type of technology was used for *volatile* storage (data is lost if electric power is not supplied, as opposed to magnetic or optical storage), so-called solid-state drives (SSD) are now included in consumer computers, offering a non-volatile option with superior access speeds to their magnetic counterparts.

When considering these options, various aspects need to be accounted for, such as convenience, costs and reliability. For example, while tape drives tend to be cheaper than hard disks (a 2016 analysis determined that 1 gigabyte of tape storage costed $0.02 opposed to $0.033 for HDD (Coughlin 2016)), they also exhibit slow data retrieval rates and require specialised hardware.

Reliability is one of the most important aspects when considering scientific data, as loss of information can lead to delays or even experiment failures. While in the early days of solid-state drives these encountered higher failure rates than HDD counterparts, a 2016 study puts them at comparable or even lower rates; under 2% of SSDs fail in their first usage year (Schroeder et al. 2016). Reliability can also be determined by brand and models; Klein (2017) determined an average 1.94% annual failure rate, but with a maximum at over 14% for a certain model. As no technology can offer absolute guarantees regarding reliability, other protection mechanisms, such as backups, need to be considered, these being discussed in the next section.

2.2   Features of Storage Systems

Most often the underlying technology for storing data can be of less relevance for the scholarly and scientific pursuit, but other characteristics can play an important role when choosing a solution.

The location of the data storage is the first considered aspect. Storing data on a local machine is advantageous as it allows the researcher to quickly access it, but might place obstacles when attempting to share it with a larger team, and also requires that the owner of the machine is fully responsible for its preservation.

Storage facilities managed at the institutional level, such as storage area network (SAN) systems, move the burden of managing data storage from the individual researcher to specialised personnel, providing higher reliability and enhanced possibilities for sharing data among peers.

Finally, data can be stored off-site in specialised facilities; this model became prominent with the advent of *cloud* systems, such as Amazon Web Services, Microsoft Azure or Google Cloud Platform, and has benefits in terms of reliability, scalability and accessibility. This might be preferred when the individual researcher or institution does not possess the required resources for managing a storage solution, when large quantities of data need to be stored, or when data needs to be shared across a large network of collaborators. At the same time, the privacy and legal implications need to be considered, given that a third party usually handles the storage; for a more in-depth discussion on this, see Sect. 4. It is worth noting that *cloud* deployments can also be managed by governmental bodies or similar official entities, this alleviating some of the legal issues (for example, the Australian National Research Data Storage Services provides such facilities to researchers in Australia, including storage of sensitive data, such as clinical trial sets (Australian Research Data Commons 2018)).

From a technical point of view, the choice of a storage solution needs to account for the following:

- Redundancy: as noted previously, no storage system can be guaranteed to function without faults and thus it is important that data is copied and stored on different systems simultaneously. The higher the number of copies and the broader their distribution, the higher the guarantee for their persistence is.
- Persistence and preservation: simply having data stored on a medium does not provide guarantees that, over time, it would not become inaccessible. For example, both tape drives and hard disks can become demagnetised, hence corrupting the stored data. This phenomenon is frequently described as *bit rot*. Hence, data needs to be periodically tested and, if problems arise, fixed. A common method for detecting issues employs so-called checksums, fingerprints of data files which change when even 1 byte switches value. If a file is detected to have changed, it is usually replaced with a redundant copy.

- Transformation: as technology evolves, so do the methods for storing data, this also leading to deprecation; for example, floppy disks are rarely used nowadays, despite being ubiquitous just a few years back. Storage and archival systems need to account for this and migrate data to current technological requirements, while ensuring that its contents are not semantically modified.

Of course, the emphasis on each of these requirements depends on the characteristics of the underlying data; for example, for raw data the transformation aspect might be less relevant, as that is not the final considered form of the data, but redundancy could play a more important role due to its sole existence as a research artefact.

2.3   Data File Formats

The file formats and even the structure and organisation of research data will most often be enforced by various laboratory instruments and software used for producing it. Nevertheless, it might be beneficial to apply transformations to the raw outputs in order to ensure their persistence over time, their ease of use, and the possibility for others to work with them.

While no silver bullet exists for file formats, a number of considerations can help choosing the right option. A first question regards the choice between proprietary and open-source file formats. Proprietary formats place a barrier for other collaborators that need to access the data file, as they might require special software or even hardware licences in order to read and modify these; for example, the Microsoft Office formats for storing tabular data (Excel files such as XLS and XLSX) can be opened only by certain applications, while comma-separated value (CSV) files can be manipulated by any text editor.

Another point to consider regards the standardisation of formats; file formats which are backed up by an established standard provide higher guarantees in terms of accessibility and preservation over time, as clear rules on how data is encapsulated and structured are defined. For example, the Digital Imaging and Communications in Medicine (DICOM) format is the de facto method for storing and transmitting medical information; using it guarantees that any systems that implements the standard can fully read the data files. Similarly, the Clinical Data Interchange Standards Consortium (CDISC) (2018) has developed a number of standards encompassing the whole research workflow, such as the Clinical Data Acquisition Standards Harmonization (CDASH), which establishes data collection procedures, or the Study Data Tabulation Model (STDM), a standard for clinical data organisation and formatting. As a counterexample, the CSV format does not have complete standards behind it, and thus, particularities can arise at both the structural and semantic levels.

Finally, the effects on the research life cycle need to be considered. Producing the data files is most often only the first step in the research workflow. Data files will need to be processed, shared with peers and even published alongside more traditional outputs, such as journal articles. While it is highly improbable that data will maintain the same format along the whole cycle (published data rarely includes the whole initial raw dataset), informed choices can aid the process.

2.4   Dataset Structure and Organisation

Another important aspect regards the organisation and splitting of datasets. While a single file might seem a simpler method for storing a dataset, issues will arise when it grows in size, and it needs to be processed or transferred to other systems. Similarly, a large number of files can pose issues with navigating and understanding the structure; making a choice needs again to be an informed process.

The first point to consider is the dimension of the dataset. Large instances, especially those exceeding 1 terabyte, might prove difficult to handle. Consider for example the need to transfer such a large file over the network[1] and the possibility that the connection will drop during the operation; most often the transfer will need to be restarted, wasting the effort. In such cases, splitting the dataset might prove to be a better solution, as each smaller file can be transferred individually and independently of the others, network issues requiring only the retransfer of unsent files. A number of storage systems include an option for so-called chunked file transfers, where the system automatically splits larger files in smaller blocks, allowing these to be transferred independently and at any point in time.

In cases where a large number of files constitute a dataset, it is important to describe the overall structure such that other applications or human users can understand it. Traditionally, *folders* are used for categorising and structuring content, but these can prove ineffective in describing the full organisation, and certain systems might not even implement this facility. A common solution to this issue is including separate file(s) describing the structure, usually called *manifests*, along with the datasets. Preferably these would follow a standard structure and semantics, and for this purpose standards such as BagIt[2] and Metadata Encoding and Transmission Standard (METS)[3] have been established. Along with the structural description, these files can also contain technical information (e.g. checksums) that might ease other processes along the scholarly workflow, such as preservation.

3   Metadata: Data Describing Data

Storing research data can have many purposes, from facilitating study replication to allowing further hypotheses to be tested. Nevertheless, archiving only the data points, with no information regarding their purpose, provenance or collection

---

[1]Even over an optical fibre network connection 1 terabyte of data will require over 1 h to transfer.

[2]https://tools.ietf.org/html/draft-kunze-bagit-16.

[3]https://www.loc.gov/standards/mets/.

method, will exponentially decrease their value over time, as both other researchers and the authors of the data will find impossible to reuse them without further information about the syntax and semantics.

Metadata, *data about data*, is the information created, stored and shared in order to describe objects (either physical or digital), facilitating the interaction with said objects for obtaining knowledge (Riley 2017). Metadata can describe various aspects of the underlying dataset, and it is often useful to split the various attributes based on their purpose.

Descriptive metadata, such as the title, creators of the dataset or description, is useful for allowing others to find and achieve a basic understanding of the dataset. Often linked to this is the *licencing and rights* metadata that describes the legal ways in which the data can be shared and reused; it includes the copyright statement, rights holder and reuse terms.

Technical metadata, which most often includes information on the data files, such as their formats and size, is useful for transferring and processing data across systems and its general management. *Preservation* metadata will often enhance the technical attributes by including information useful for ensuring that data remains accessible and usable, such as the checksum or replica replacement events (see Sect. 2.2). Finally, *structural* metadata describes the way in which data files are organised and their formats.

Given its complexity, producing metadata can become a significant undertaking, its complexity exceeding even that of the underlying data in certain cases. This is one of the reasons for which the standardisation of metadata has become mandatory, this happening at three main levels.

At the structural level, standardisation ensures that, on one hand, a minimum set of attributes is always attached to datasets and, on the other, that enough attributes are present for ensuring proper description of any possible artefact, no matter its origin, subject area or geographical location. Multiple such standards have been developed, from more succinct ones, such as the DataCite Schema[4] or the Dublin Core Metadata Element Set,[5] to more extensive, such as MARC21[6] or the Common European Research Information Format (CERIF).[7] The usage of these standards might vary across subject areas (e.g. the *Document, Discover and Interoperate (DDI)* standard is targeted at survey data in social, behavioural and health sciences (DDI Alliance 2018)) or the main purpose of the metadata (e.g. the METS standard emphasises technical, preservation and structural metadata more than the DataCite schema).

At the semantic level, the focus is on ensuring that the language used for describing data observes a controlled variability both inside a research area and across domains. For example, the CRediT vocabulary (CASRAI 2012) defines

---

[4]https://schema.datacite.org/meta/kernel-4.1/.

[5]http://dublincore.org/documents/dces/.

[6]https://www.loc.gov/marc/bibliographic/.

[7]See https://www.eurocris.org/cerif/main-features-cerif.

various roles involved in research activities, Friend of a Friend (FOAF) establishes the terminology for describing and linking persons, institutions and other entities (Brickley and Miller 2014), and the Multipurpose Internet Mail Extensions (MIME) standard defines the various file formats (Freed Innosoft and Borenstein 1996).

The third point considered from a standardisation point of view involves the formats used for storing and transferring metadata. The Extensible Markup Language (XML)[8] is one of the most prevalent formats, almost all standards providing a schema and guidance on employing it. The JavaScript Object Notation (JSON)[9] format is also starting to gain traction, both due to its pervasiveness in web services nowadays and also due to certain initiatives, such as schema.org which use it as the de facto output format.

3.1   Unique and Persistent Identifiers

One important aspect of metadata, considered by most standards, vocabularies and formats relates to the usage of identifiers. Similar to social security numbers for humans or serial numbers for devices, when it comes to research data, the aim of identifiers is to uniquely and persistently describe it. This has become a stringent necessity in the age of Internet, both due to the requirement to maintain resources accessible for periods of times of the order of years or even decades, no matter the status or location of the system preserving them at any discrete moment,[10] and also due to the necessity of linking various resources across systems. Thus, various elements of research data started receiving identifiers, various initiatives and standards becoming available.

Even before the prevalence of research data sharing, bibliographic records received identifiers, such as International Standard Book Numbers (ISBN) for books and International Standard Serial Numbers (ISSN) for periodicals. For research data, Archival Resource Keys (ARK) and Handles[11] are more prevalent, as these mechanisms facilitate issuing new identifiers and, thus, are more suited for the larger volume of produced records.

The Digital Object Identifier (DOI) system[12] is quickly emerging as the industry standard; it builds upon the Handle infrastructure, but adds an additional dimension over it, namely, *persistence* (DOI Foundation 2017). At a practical level, this is implemented using a number of processes that ensure that an identified object will remain available online (possibly only at the metadata level) even when the original

---

[8]https://www.w3.org/XML/.

[9]https://www.json.org/.

[10]Similar to bit rot, link rot describes the phenomenon of web addresses becoming unavailable over time, for example, due to servers going offline. This can pose significant issues for research artefacts, which need to remain available for longer periods of time due to their societal importance; nevertheless, link rot was proven to be pervasive across scholarly resources (Sanderson et al. 2011).

[11]http://handle.net/.

[12]https://doi.org.

holding server becomes unavailable and the resource needs to be transferred else-where. A DOI is linked to the metadata of the object and is usually assigned when the object becomes public. The metadata of the object can be updated at any time and, for example, the online address where the object resides, could be updated when the object's location changes; so-called resolver applications are in charge of redirecting accesses of the DOI to the actual address of the underlying object.

A second important dimension of research outputs relates to persons and institutions. ORCiD is currently the most widespread identifier for researchers, with over 5 million registrations (ORCID 2018), while the International Standard Name Identifier (ISNI)[13] and the Global Research Identifier Database (GRID)[14] provide identifiers for research institutions, groups and funding bodies.

Identifiers have been developed for other entities of significant importance in terms of sharing and interoperability. For example, the Protein Data Bank provides identifiers for the proteins, nucleic acids and other complex assemblies (RCSB PDB 2018), while GenBank indexes genetic sequences using so-called accession numbers (National Center for Biotechnology Information 2017).

Research Resource Identifiers (RRID) (FORCE11 2018) aim to cover a wider area, providing identifiers for any type of asset used in the scientific pursuit; the current registry includes entities ranging from organisms, cells and antibodies to software applications, databases and even institutions. Research Resource Identifiers have been adopted by a considerable number of publishing institutions and are quickly converging towards a community standard.

The main takeaway here is that, it is in general better to use a standard unique and possibly persistent identifier for describing and citing a research-related entity, as this will ensure both its common understanding and accessibility over time.

3.2   The FAIR Principles

As outlined, producing quality metadata for research data can prove to be an overwhelming effort, due to the wide array of choices in terms of standards and formats, the broad target audience or the high number of requirements. To overcome this, the community has devised the FAIR principles (Wilkinson et al. 2016), a concise set of recommendations for scientific data management and stewardship which focuses on the *aims* of metadata.

The FAIR principles are one of the first attempts to systematically address the issues around data management and stewardship; they were formulated by a large consortium of research individuals and organisations and are intended for both data producers and data publishers, targeting the promotion of maximum use and reuse of data. The acronym *FAIR* stands for the four properties research data should present.

---

[13]http://www.isni.org/.

[14]https://grid.ac.

Findability relates to the possibility of coming across the resource using one of the many Internet facilities. This requires that the attached metadata is rich enough (e.g. description and keywords are crucial for this), that a persistent identifier is associated and included in the metadata and that all this information is made publicly available on the Internet.

Accessibility mostly considers the methods through which data can be retrieved. As such, a standard and open protocol, like the ones used over the Internet, should be employed. Moreover, metadata should always remain available, even when the object ceases to be, in order to provide the continuity of the record.

Interoperability considers the ways in which data can be used, processed and analysed across systems, both by human operators and machines. For this, metadata should both "use a formal, accessible, shared, and broadly applicable language for knowledge representation" (FORCE11 2015) and standardised vocabularies.[15]

Moreover, the interoperability guideline requires that metadata contains *qualified* references to other metadata. This links both the persistent and unique identifiers described earlier, but also to the relations between them, the foundation of *Linked Data*, a concept introduced by the inventor of the World Wide Web, Tim Berners-Lee. This concept relies heavily on the Resource Description Framework (RDF) specification which allows describing a *graph* linking pieces of information (W3C RDF Working Group 2014). The linked data concept is of utmost importance to the scholarly workflow, as it can provide a wider image over scientific research, as proven by projects such as SciGraph, which defines over one billion relationships between entities such as journals, institutions, funders or clinical trials (Springer Nature 2017), or SCHOLIX[16] which links research data to the outputs that reference it.

Finally, the FAIR principles mandate that research data should be *reusable*, thus allowing for study replicability and reproducibility. For this, it requires that metadata contains accurate and relevant attributes (e.g. it describes the columns of tabular data) and information about its provenance. Moreover, it touches on certain legal aspects, such as the need for clear and accessible licencing and adherence to "*domain-relevant community standards*", such as, for example, the requirements on patient data protection.

4   Legal Aspects of Data Storage

There are a number of legal aspects to consider regarding the storage and sharing of research data; certain elements will differ depending on the geographic location. This section outlines the main points to consider.

---

[15]Here the principles become *recursive*, mandating that vocabularies describing FAIR datasets should themselves follow the same principles, see FORCE11 (2015).

[16]https://scholix.org.

4.1   Anonymisation of Research Data

Broadly, anonymisation allows data to be shared while preserving privacy. Anonymity is not be confused with confidentiality, although the two are linked. Anonymity is the process of not disclosing the identity of a research participant or the author of a particular view or opinion. Confidentiality is the process of not disclosing to other parties opinions or information gathered in the research process (Clark 2006).

The process of anonymising research data requires that key identifiers are changed or masked. An individual's identity can be disclosed from *direct identifiers* such as names or geographic information or *indirect identifiers* which, when linked with other available data, could identify someone, like occupation or age. Anonymisation should be planned in the early stages of research, or costs can become burdensome later. Anonymisation considerations should be built in when gaining consent for data sharing.

One of the challenges of anonymisation is balance. Going too far could result in important information being missed or incorrect conclusions being drawn, all the while balancing the potential of reidentification. If the research data is for public release, the probability of potential reidentification needs to be low. It may be acceptable for this probability to be higher for private or semi-public as other controls can be put in place (El Emam et al. 2015).

For example, in the USA the Health Insurance Portability and Accountability Act of 1996 (HIPAA) directly addresses anonymisation concerns (U.S. Department of Health and Human Services 2017); it requires that systems and repositories that handle such information need to ensure physical, technical and administrative safeguards that meet the obligations laid out in the Act.

4.2   Legal Frameworks to Consider

As mentioned earlier, the legal frameworks that need to be considered will vary dependent on geography. Three important frameworks to consider are the General Data Protection Regulation (GDPR)[17] in the EU, the UK Data Protection Act 1998/ 2018[18] and the Patriot ACT in the USA.[19]

The EU General Data Protection Regulation (GDPR) along with the new UK Data Protection Act came into force on May 25, 2018, and governs the processing (holding or using) of personal data in the UK. Although not specifically aimed at research, some changes will still need to be considered. GDPR has a clearer definition of personal data which is that personal data is about living people from which

---

[17]https://www.eugdpr.org/.

[18]https://www.gov.uk/government/collections/data-protection-act-2018.

[19]https://www.justice.gov/archive/ll/highlights.htm.

they can be identified. As well as data containing obvious *identifiers*, such as name and date of birth, this includes some genetic, biometric and online data if unique to an individual. Data that has been pseudonymised (with identifiers separated), where the dataset and identifiers are held by the same organisation, is still personal data.

The UK Data Protection Act 1998 and its update in 2018 applies in Scotland, England, Wales and Northern Ireland. The Act gives individuals rights of access to request copies of their personal data collected by a researcher. It requires that any processors of personal data must comply with eight principles, which make sure that personal data are:

1. Fairly and lawfully processed
2. Processed for limited purposes
3. Adequate, relevant and not excessive
4. Accurate and up to date
5. Not kept for longer than is necessary
6. Processed in line with your rights
7. Secure
8. Not transferred to other countries without adequate protection

There are exceptions for personal data collected as part of research. It can be retained indefinitely if needed and can be used for other purposes in some circumstances. People should still be informed if the above applies.

Sensitive data also falls under UK Data Protection rules. Sensitive data includes but is not limited to race or ethnic origin, political opinion, religious beliefs or physical or mental health. Sensitive data can only be processed for research purposes if explicit consent (ideally in writing) has been obtained, the data is in substantial public interest and not causing substantial damage and distress, or if the analysis of racial/ethnic origins is for purpose of equal opportunities monitoring.

The legal definition of personal data is complex and is affected by the act's subsequent update in 2018 and GDPR, but the simplest and safest definition is of any information about a living, identifiable individual. This is relevant to anonymisation, as if research data is appropriately anonymised, then the UK Data Protection act will no longer apply. Institutions generally have a Data Protection Officer which should be utilised to address any specific concerns about research outputs.

The PATRIOT Act was signed into law in the USA in 2001. This legislation, again, not specifically aimed at research, has impact when it comes to data storage and grants the potential for the US government to have access to data stored by US cloud servers providers. A common misconception is that avoidance of US-located servers solves the problem, which is only partially accurate. This act would, in theory, allow US judicial authorities and intelligence agencies to request data stored in cloud services outside of the USA. The police, the judiciary and intelligence agencies are able in one way or another to request information from higher education and research institutions and any other parties concerned (van Hoboken et al. 2012).

From a legal perspective, access to cloud data cannot be denied and "cloud service providers can give no guarantees in this respect" (van Hoboken et al. 2012). In practice, access can take place in one of two ways:

- If the cloud service provider is subject to US jurisdiction, a request for release of the data can be made directly to the service provider company in the USA.
- If the cloud service provider is not subject to US jurisdiction, data may be retrieved from the service provider or the institution or with the assistance of relevant local judicial authorities or intelligence agencies.

4.3   Licencing

As is the case with any type of scientific output, research data requires a framework upon which sharing, along with proper attribution, can be achieved. What sets data apart from say, journal papers, is that it can be reused in more ways and such copyright protocols suited for citing research can prove insufficient when considering, for example, the extraction of new hypothesis from existing datasets. This is why new means of licencing and enforcing copyright have either been devised or borrowed from other domains where reuse is common. When data is shared, the original copyright owner usually retains the copyright (UK Data Service 2012), but a licence can be applied in order to describe how the data can be reused. It is important to note that when no proper licencing terms are applied, content cannot be redistributed or reused (Brock 2018).

The Creative Commons (CC) suite of licencing options[20] is one of the most popular for research data; the model consists of a modular set of clauses which can be aggregated for obtaining licences with varying degrees of freedom in terms of reuse. As such, the CC BY licence allows unrestricted reuse as long as attribution is given to the original authors, while more restrictive options such as CC BY-NC-SA or CC BY-NC-ND disallow either using a different licence for derivative work (SA, *share-alike*) or no derivatives (ND) at all, respectively, with a supplementary clause forbidding the usage of the data for any commercial interest (NC, *no commercial*).

Apart from these, other licencing options have been devised for more specific use cases. For example, research software can use one of the deeds commonly employed across the software development ecosystem, such as the MIT licence[21] or the GNU General Public License (GPL).[22] Another example relates to licences developed by national bodies, in order to ensure better compliance with the regional laws and regulations; such instances include the European Union Public License (EUPL)[23] or the Open Government License (OGL).[24] Finally, data can be placed in the public domain, forgoing any copyright or reuse terms; such content can be associated with a notice such as the one provided by Creative Commons as CC0.[25]

---

[20]https://creativecommons.org/.

[21]https://opensource.org/licenses/MIT.

[22]https://www.gnu.org/licenses/gpl-3.0.en.html.

[23]https://joinup.ec.europa.eu/collection/eupl.

[24]http://www.nationalarchives.gov.uk/doc/open-government-licence/version/3/.

[25]https://creativecommons.org/share-your-work/public-domain/cc0/.

While in general researchers should aim for allowing unrestricted use of their data, as also stipulated by the FAIR principles, this is of course not always possible or desirable due to various ethical, regulatory or even technical reasons. In such cases, consulting with personnel specialised in licencing and copyright issues, such as a librarian or lawyer, might be desirable, in order to avoid issues with certain deeds that might place undesirable obstacles on reuse. For example, the no commercial (NC) clause in the CC suite can disallow the use of data not only by commercial corporations but also by research institutions generating minimal amounts of revenue (Klimpel 2013).

4.4   Blockchain: A Technical Solution for Legal Requirements

As the legal requirements around the research area become more complex, technical solutions for alleviating them are currently being researched. The aim of these is to simplify the workflows and maintain a low entry point for users lacking legal expertise while also ensuring the required level of compliance.

In recent years, the blockchain technology has been examined as a potential mean of solving some of these issues, the movement in this direction being fuelled by the increase in interest due to usage in the financial domain (e.g. Bitcoin). At a very high level, the blockchain allows recording data and verifying its authenticity without the need for a central authority. In the most popular implementations, each participant in a blockchain holds a copy of the whole record set, and each record is linked to a previous one by a cryptographic value, distilled using the records' content; thus, even the most trivial change would propagate across the whole chain, signalling the modification.

The basic idea behind blockchains can prove useful in areas where authenticity, provenance and anonymization are important, research being one of them. In Digital Science and van Rossum (2017), the authors have identified a number of ways in which blockchains could be implemented across the scholarly workflow, such as:

- Hypothesis registration: allow researchers to signal a discovery, proof or new dataset while providing evidence of ownership in any future instance.
- Study preregistration: while committing research plans before executing them is already a practice (see, e.g. the "Preregistration Challenge"[26]), it can be difficult to ensure that these plans are not modified while the experiments are ongoing, in order to mask potential discrepancies with the actual results; a blockchain system could easily detect such a change.
- Digital rights management: a blockchain system can easily record ownership, and a related technology, *smart contracts*, a system for defining, actioning and enforcing a set of clauses, can be used to ensure that future usage of data respects and attributes ownership along with any associated stipulations (Panescu and

---

[26]https://cos.io/prereg/.

Manta 2018). This could prove useful also in terms of counting citations and understanding how data is further reused.

- Data anonymisation and provenance: this can prove to be of utmost importance for medical and pharmacological research, where, on one hand, stringent requirements on patient privacy and data anonymisation are in place, and, on the other hand, the origin of the data should be verifiable. The use of cryptographic controls and the distributed architecture of blockchain systems can help with these challenges.

The application of blockchain technology to various real-world problems is still in its infancy, with many challenges still to overcome. Nevertheless this space is one to closely follow, as it can provide glimpses onto the future of research data sharing.

5   Overview of Research Data Repositories and Tools

What data sharing means is rarely explicitly defined. Questions arise such as:

- How *raw* should the data be?
- Where and how should data be shared?
- How should data be standardised and structured to make it useful?

To complicate matters, these answers will differ dependent on the particular discipline of the person asking the question. Successes in this area like the sharing of DNA sequences via Genbank (National Center for Biotechnology Information 2017) and functional magnetic resonance imaging (fMRI) scans via the Human Connectome Project[27] are research outputs that lend themselves to standardisation. These successes may be difficult to replicate in other disciplines.

Pharmacology presents some particular challenges to data sharing, with common research outputs like raw traces of electrophysiological measurements and large imaging files potentially too unwieldy for existing solutions. For this very reason, The British Journal of Pharmacology (BJP) has a recommended but not mandated policy of data sharing (George et al. 2017).

Look at the wider world of repositories for research data, there are many available options. This section will look at some of the reasons to share research data and how this may influence the discovery path to choosing a repository for publication. It will also look at some of the key features to consider which will have different levels of importance dependent on usage requirements.

There are a number of other considerations to take into account when looking at where to share research data outside of a feature analysis only. There are a variety of certifications to look at, starting with the Data Seal of Approval/CoreTrustSeal. CoreTrustSeal is a non-profit whose stated aim is to promote sustainable and

---

[27]http://www.humanconnectomeproject.org/.

trustworthy data infrastructures. The 16 questions asked on assessment are an excellent set of questions to think about when deciding where and how to store data (Data Seal of Approval, ICSU World Data System 2016).

Additional certifications to look out for are the Nestor seal[28] and ISO16363[29]; their aim is to provide stricter guidelines concerning the processes data repositories should follow in order to ensure higher guarantees in terms of data and metadata preservation and accessibility.

5.1   Repositories

As alluded to earlier, there are more general repositories for data sharing and numerous very specific repositories. Dependent on the reasons for data sharing, there are numerous methods to choosing where and how to share data. A good starting point are tools such as FAIRsharing[30] from the University of Oxford e-Research centre which has additional resources outside of repositories including data policies and standards, and r3data,[31] an extensive index which can be browsed by subject and country.

Publisher mandates are one of the most common cases for data sharing when publishing an article. Some journals require and others may only recommend that data behind the paper is shared alongside publication. In these instances, the journal will often have a list of approved repositories that they either directly work and integrate with or simply recommend. Examples can be found from Nature (2018, Recommended Repositories) and Elsevier (2018, Link with data repositories).

Funder mandates are an ever-increasing force (Digital Science 2017) in open research data sharing, with some funders opting for a policy that all research data that can be shared must be. In certain cases, funders will have even more stringent requirements, demanding so-called data management plans (Netherlands Organisation for Scientific Research 2015), which detail the whole life cycle of research data, from collection to publication.

Research data sharing in life sciences is commonly scrutinised in terms of adherence to the ALCOA principles, a series of guidelines adopted by a number of large funding and regulatory institutions such as the US Food and Drug Administration, National Institutes of Health or the World Health Organisation. These principles establish a number of properties quality data and, by transitivity, the systems generating, processing and holding it should exhibit:

- Attributable: the generator of the data should be clearly identified.
- Legible: records should remain readable, especially by human users, over time.

---

[28]http://www.dnb.de/Subsites/nestor/EN/Siegel/siegel.html.

[29]https://www.iso.org/standard/56510.html.

[30]https://fairsharing.org/.

[31]https://doi.org/10.17616/R3D.

- Contemporaneous: data points should be recorded as near as possible to the event that generated them, thus avoiding any loss of information.
- Original: analyses should be performed using the initially collected data, or a strictly controlled copy, to avoid, for example, errors generated from the transcription of handwritten data to an electronic record.
- Accurate: data should be complete, correct and truthful of the event being recorded.
- Complete[32]: data should not be deleted at any point.
- Consistent: research data should be recorded in a chronological manner.
- Enduring: research data should be preserved for an extended period of time.
- Available: research data should remain accessible for the lifetime of either the final research artefact (e.g. research paper) or physical product (e.g. a drug).

Of course, data may be shared outside of any mandate to do so. This may be to contribute to the increasing success and importance of Open Science movement (Ali-Khan et al. 2018), increase the visibility and reach of the work or to safely ensure the long-term availability of the data or myriad other reasons. In this situation, the wider world of repositories is available. One of the first decisions would be to choose whether to opt for a subject-specific repository or one of the more general repositories available.

Subject-specific repositories have the advantage of potentially having functionality and metadata schemas directly related to your field as well as the advantage of having a targeted audience of potential viewers. Examples of these include:

- Clinical Trials[33] is a database of privately and publicly funded clinical studies conducted around the world.
- The SICAS Medical Image Repository[34] host medical research data and images.
- The TCIA[35] is a service which de-identifies and hosts a large archive of medical images of cancer accessible for public download.

While there are numerous comparisons of generalist data repositories available (see Amorim et al. 2015; Stockholm University Library 2018; Dataverse Project 2017), the rapid pace of development seen on these platforms can mean these are difficult to maintain, and thus repositories should be evaluated at the time of use.

Figshare[36] was launched in January 2011 and hosts millions of outputs from across the research spectrum. It is free to use, upload and access. It has a number of different methods of framing data from single items to themed collections. Figshare

---

[32]The last four points correspond to a later addition to the ALCOA principles, ALCOA Plus.

[33]https://clinicaltrials.gov.

[34]https://www.smir.ch.

[35]http://www.cancerimagingarchive.net.

[36]https://figshare.com/.

features a range of file visualisation options which may be attractive dependent on the particular files used.

Zenodo[37] was launched in March 2013 is non-profit general repository from OpenAIRE and CERN. It is free to use and access. It is a mature, well-featured product and of particular interest may be the Communities functionality. These curated groups allow for elements of subject-specific repositories to be catered for.

DataDryad[38] was launched in January 2008 and is a non-profit repository which is more tied to the traditional publication process. While all types of data are accepted, they must be linked to a published or to-be-published paper. Outputs on DataDryad are free to download and reuse; uploads incur a submission fee as opposed to the examples above. There are significant integrations with journals, and this option is worth considering if this is of significant importance.

Other options to consider are institutional or national data repositories. Data repositories at institutions are becoming increasingly prevalent and should be investigated as either the sole method of publication or as reference record to the destination of choosing. Examples of these include:

- Cambridge University Data Repository[39]
- Sheffield University Data Repository[40]
- Monash University Data Repository[41]

National and even international repositories are again an area that is still in its infancy but are under active development in many countries around the world. Examples of these include:

- The Norwegian Centre for Research Data[42]
- The UK Data Service[43]
- Swedish National Data Service[44]
- European Open Science Cloud (EOSC)[45]

The repository chosen should be one that works well with FAIR principles outlined previously. While there are many factors to consider that will have different weightings dependent on use, some good general areas to consider include:

---

[37]https://zenodo.org/.

[38]https://datadryad.org.

[39]https://www.repository.cam.ac.uk.

[40]https://www.sheffield.ac.uk/library/rdm/orda.

[41]https://monash.figshare.com/.

[42]http://www.nsd.uib.no/nsd/english/index.html.

[43]https://www.ukdataservice.ac.uk.

[44]https://snd.gu.se/en.

[45]https://www.eoscpilot.eu/.

- Embargo and access options: does the repository allow the ability to grant different levels of access conditions to the files and/or metadata? Can data be shared privately? Can the files, metadata or both be embargoed?
- Licences: does the repository provide access to the necessary licence options needed for publication of the data? Can you add your own licence?
- Metrics and citations: what type of metrics does the repository track? Do they report the metrics to any tracking bodies? Do they track citations of the data? Do they track nontraditional online attention, e.g. altmetrics?
- Availability: what is the sustainability model of the repository? What guarantees do they provide about the continued availability of the data? Is the data easily accessible programmatically to allow for ease of export?

References

Ali-Khan SE, Jean A, MacDonald E, Gold ER (2018) Defining success in open science. MNI Open Res 2:2. Gates Found Author Manuscript. https://doi.org/10.12688/mniopenres.12780.1

Amorim RC, Castro JA, da Silva JR, Ribeiro C (2015) A comparative study of platforms for research data management: interoperability, metadata capabilities and integration potential. In: Rocha A, Correia A, Costanzo S, Reis L (eds) New contributions in information systems and technologies, Advances in intelligent systems and computing, vol 353. Springer, Cham

Australian Research Data Commons (2018) Australian National Data Service – what we do. https://web.archive.org/web/20180319112156/https://www.ands.org.au/about-us/what-we-do. Accessed 11 Sept 2018

Brickley D, Miller L (2014) FOAF vocabulary specification 0.99. http://xmlns.com/foaf/spec/ https://web.archive.org/web/20180906035208/http://xmlns.com/foaf/spec/. Accessed 6 Sept 2018

Brock J (2018) 'Bronze' open access supersedes green and gold. Nature Index. https://web.archive.org/web/20180313152023/https://www.natureindex.com/news-blog/bronze-open-access-supersedes-green-and-gold. Accessed 13 Sept 2018

CASRAI (2012) CRediT. https://casrai.org/credit/. Accessed 6 Sept 2018

Clark A (2006) Anonymising research data (NCRM working paper). ESRC National Centre for Research Methods. Available via NCRM. http://eprints.ncrm.ac.uk/480/

Clinical Data Interchange Standards Consortium (2018) CDISC standards in the clinical research process. https://web.archive.org/web/20180503030958/https://www.cdisc.org/standards. Accessed 21 Oct 2018

Coughlin T (2016) The costs of storage. Forbes. https://web.archive.org/web/20170126160617/https://www.forbes.com/sites/tomcoughlin/2016/07/24/the-costs-of-storage/. Accessed 19 Aug 2018

Data Seal of Approval, ICSU World Data System (2016) Core trustworthy data repositories requirements. https://web.archive.org/web/20180906181720/https://www.coretrustseal.org/wp-content/uploads/2017/01/Core_Trustworthy_Data_Repositories_Requirements_01_00.pdf. Accessed 6 Sept 2018

Dataverse Project (2017) A comparative review of various data repositories. https://web.archive.org/web/20180828161907/https://dataverse.org/blog/comparative-review-various-data-repositories. Accessed 6 Sept 2018

DDI Alliance (2018) Document, discover and interoperate. https://web.archive.org/web/20180828151047/https://www.ddialliance.org/. Accessed 6 Sept 2018

Digital Science (2017) The state of open data 2017 – a selection of analyses and articles about open data, curated by Figshare. Digital Science, London

Digital Science, van Rossum J (2017) Blockchain for research. Digital Science, London

DOI Foundation (2017) DOI system and the handle system. https://web.archive.org/web/20180112115303/https://www.doi.org/factsheets/DOIHandle.html. Accessed 6 Sept 2018

Eklund A, Thomas EN, Knutsson H (2016) Cluster failure: why fMRI inferences for spatial extent have inflated false-positive rates. PNAS 113:7900. https://doi.org/10.1073/pnas.1602413113

El Emam K, Rodgers S, Malin B (2015) Anonymising and sharing individual patient data. BMJ 350:h1139. https://doi.org/10.1136/bmj.h1139

Elsevier (2018) Sharing research data. https://web.archive.org/web/20180528101029/https://www.elsevier.com/authors/author-services/research-data/. Accessed 6 Sept 2018

European Commission, Directorate-General for Research & Innovation (2017) 2020 programme – guidelines to the rules on open access to scientific publications and open access to research data in horizon 2020, version 3.2. https://web.archive.org/web/20180826235248/http://ec.europa.eu/research/participants/data/ref/h2020/grants_manual/hi/oa_pilot/h2020-hi-oa-pilot-guide_en.pdf. Accessed 15 July 2018

Federer L, Lu Y-L, Joubert DJ, Welsh J, Brandy B (2015) Biomedical data sharing and reuse: attitudes and practices of clinical and scientific research staff. PLoS One 10:e0129506. https://doi.org/10.1371/journal.pone.0129506

FORCE11 (2015) The FAIR data principles. https://web.archive.org/web/20180831010426/https://www.force11.org/group/fairgroup/fairprinciples. Accessed 6 Sept 2018

FORCE11 (2018) Resource identification initiative. https://web.archive.org/web/20181005133039/https://www.force11.org/group/resource-identification-initiative. Accessed 21 Oct 2018

Freed Innosoft N, Borenstein N (1996) Multipurpose internet mail extensions (MIME) part two: media types. https://web.archive.org/web/20180819062818/https://tools.ietf.org/html/rfc2046. Accessed 19 Aug 2018

George C et al (2017) Updating the guidelines for data transparency in the British Journal of Pharmacology – data sharing and the use of scatter plots instead of bar charts. Br J Pharm 174:2801. https://doi.org/10.1111/bph.13925

Klein A (2017) Backblaze hard drive stats for 2016. https://web.archive.org/web/20180611041208/https://www.backblaze.com/blog/hard-drive-benchmark-stats-2016/. Accessed 19 Aug 2018

Klimpel P (2013) Consequences, risks and side-effects of the license modules "non-commercial use only – NC". https://web.archive.org/web/20180402122740/https://openglam.org/files/2013/01/iRights_CC-NC_Guide_English.pdf. Accessed 6 Sept 2018

National Center for Biotechnology Information (2017) GenBank. https://web.archive.org/web/20180906065340/https://www.ncbi.nlm.nih.gov/genbank/. Accessed 6 Sept 2018

National Institutes of Health (2018) NIH public access policy details. https://web.archive.org/web/20180421191423/https://publicaccess.nih.gov/policy.htm. Accessed 6 Sept 2018

Netherlands Organisation for Scientific Research (2015) Data management protocol. https://web.archive.org/web/20170410095202/http://www.nwo.nl/en/policies/open+science/data+management. Accessed 21 Oct 2018

ORCID (2018) ORCID. https://web.archive.org/web/20180904023120/https://orcid.org/. Accessed 6 Sept 2018

Panescu A-T, Manta V (2018) Smart contracts for research data rights management over the Ethereum blockchain network. Sci Technol Libr 37:235. https://doi.org/10.1080/0194262X.2018.1474838

Phillips N (2017) Online software spots genetic errors in cancer papers. Nature 551:422. https://doi.org/10.1038/nature.2017.23003

RCSB PDB (2018) Protein data bank. https://www.rcsb.org/. Accessed 15 July 2018

Riley J (2017) Understanding metadata – what is metadata, and what is it for? NISO, Baltimore

Sanderson R, Phillips M, Van de Sompel H (2011) Analyzing the persistence of referenced web resources with memento. https://arxiv.org/abs/1105.3459

Schroeder B, Lagisetty R, Merchant A (2016) Flash reliability in production: the expected and the unexpected. In: Proc of 14th USENIX Conference on File and Storage Technology (FAST 16)

Springer Nature (2017) SN SciGraph. https://web.archive.org/web/20180823083626/https://www.springernature.com/gp/researchers/scigraph. Accessed 6 Sept 2018

Springer Nature (2018) Research data support. https://web.archive.org/web/20180309193834/https://www.springernature.com/gp/authors/research-data-policy. Accessed 6 Sept 2018

Stockholm University Library (2018) Data repositories. https://web.archive.org/web/20180828161808/https://www.su.se/english/library/publish/research-data/data-repositories. Accessed 6 Sept 2018

U.S. Department of Health and Human Services (2017) Research. https://web.archive.org/web/20180828205916/https://www.hhs.gov/hipaa/for-professionals/special-topics/research/index.html. Accessed 6 Sept 2018

UK Data Service (2012) Copyright for data sharing and fair dealing. https://web.archive.org/web/20180828160153/https://www.ukdataservice.ac.uk/manage-data/rights/sharing. Accessed 28 Aug 2018

van Hoboken J, Arnback A, van Ejik NANM (2012) Cloud computing in higher education and research institutions and the USA Patriot Act. SSRN. https://doi.org/10.2139/ssrn.2181534

W3C RDF Working Group (2014) Resource description framework (RDF). https://www.w3.org/RDF/. Accessed 6 Sept 2018

Wilkinson MD et al (2016) The FAIR guiding principles for scientific data management and stewardship. Sci Data 3:160018. https://doi.org/10.1038/sdata.2016.18

Youngseek K, Zhang P (2015) Understanding data sharing behaviors of STEM researchers: the roles of attitudes, norms, and data repositories. Libr Inf Sci Res 37:189. https://doi.org/10.1016/j.lisr.2015.04.006

Design of Meta-Analysis Studies

Malcolm R. Macleod, Ezgi Tanriver-Ayder, Kaitlyn Hair, and Emily Sena

Contents

Abstract

Any given research claim can be made with a degree of confidence that a phenomenon is present, with an estimate of the precision of the observed effects and a prediction of the extent to which the findings might hold true under different experimental or real-world conditions. In some situations, the certainty and precision obtained from a single study are sufficient reliably to inform future research decisions. However, in other situations greater certainty is required. This might be the case where a substantial research investment is planned, a pivotal claim is to be made or the launch of a clinical trial programme is being considered. Under these circumstances, some form of summary of findings across studies may be helpful.

Summary estimates can describe findings from exploratory (observational) or hypothesis testing experiments, but importantly, the creation of such summaries is, in itself, observational rather than experimental research. The process is therefore particularly at risk from selective identification of literature to be included, and this can be addressed using systematic search strategies and pre-specified criteria for inclusion and exclusion against which possible contributing data will be assessed. This characterises a systematic review (in contrast to nonsystematic or narrative reviews). In meta-analysis, there is an attempt to provide a quantitative summary of such research findings.

M. R. Macleod (✉) · E. Tanriver-Ayder · K. Hair · E. Sena
Centre for Clinical Brain Sciences, University of Edinburgh, Edinburgh, UK
e-mail: malcolm.macleod@ed.ac.uk

© The Author(s) 2019
A. Bespalov et al. (eds.), *Good Research Practice in Non-Clinical Pharmacology and Biomedicine*, Handbook of Experimental Pharmacology 257,
https://doi.org/10.1007/164_2019_289

299

Keywords

Reproducibility · Sytematic review · Translation

1   Principles of Systematic Review

1. *Search strategy*: the objective is to identify all possible sources of relevant information, so that they can contribute to the research summary. Informal searches have a number of weaknesses:

   (a) There is a risk of the preferential identification of work in high-impact journals. We know that the quality of work published in such journals is no higher than that in the rest of the literature and that a premium on novelty means that the findings in such journals tend to be more extreme than in other journals. This has been shown, for instance, in gene association studies in psychiatry (Munafo et al. 2009).

   (b) While English is still, largely, the language of science, searches which are limited to the English language literature will miss those studies published in other languages. For research conducted in countries where English is not the first language, there is likely to be a difference in the "newsworthiness" of work published in the English literature compared with the domestic literature, with work published in English being unrepresentative of the whole.

   (c) Where there is not a clear articulation of inclusion and exclusion criteria, de facto judgements may be made about eligibility based on convenience or information source, and eligibility criteria may drift with emerging understanding of the literature. This is essentially a data-led approach, and while it is sometimes appropriate, it needs to be apparent.

   (d) There should be articulation in advance of the research types to be included. Should conference abstracts be considered? In spite of their brevity, they do sometimes include sufficient information to contribute outcome data to meta-analysis. There is an increasing pre-peer-reviewed literature, most notably bioRxiv, often described in as much detail as a formal journal paper. Reviewers should decide this in advance, and in general in fast-moving fields, it is preferable to consider both of these sources if possible.

   (e) This decision also has implications for the number of databases to be searched. PubMed is easy to use, is widely accessible and provides good coverage of much of the published literature. However, conference abstracts and preprints are not reliably retrieved, and if these are important, then the use of, for instance, EMBASE and Google Scholar, or perhaps direct searching within bioRxiv, is preferred. Importantly, the Google Scholar algorithm is based in part on that user's search history and will differ between individuals. Therefore, while potentially useful, it does not provide a reproducible search strategy and should not be used as the main or only search engine. As registries of animal experiments become more widely used, searching of these may provide useful information about the proportion of studies which have been initiated but not (at least not yet) published.

The SRA-polyglot tool (http://crebp-sra.com/#/polyglot) developed by the Bond University Centre for Research in Evidence-Based Practice allows the syntax of search strings to be converted between the requirements of several different databases.

2. *Deduplication*

   (a) Where more than one database is searched, it is inevitable that some articles will appear in more than one search result, and it is important to identify such duplication. The earliest this can be done, the more work is saved; and in some large multi-author reviews, duplicate publications may persist to very late stages of the review. Bibliographic software such as EndNote has deduplication facilities which require manual curation, as does the SRA-dedupe tool developed by the Bond University Centre for Research in Evidence-Based Practice. However, emerging experience in our group suggests that completely automated deduplication may be achieved with a high degree of precision using the RecordLinkage (https://cran.r-project.org/web/packages/RecordLinkage/index.html) R package with additional filters built into the code to maximise the number of duplicates detected without removing false duplicate records.

3. *Protocol registration (PROSPERO) and publication*

   (a) Systematic review is observational research. There is opportunity therefore for hypothesising after results are known ("HARKing") – that is, for the intention of the study to be changed in the light of observed data, with a claim made, the data supported what the investigators had been looking for all along, and for flexibility in data analysis (choosing the analysis technique that delivers $p < 0.05$), and for shifts in the entire purpose of the study. Say, for example, we were interested in the effect of maternal deprivation in the first trimester on blood pressure in adult offspring, but found many more studies using maternal deprivation in the third trimester and switched to studying that. These flexibilities increase the risk of identifying spurious associations and devalue the findings of systematic review. Researchers should articulate, in advance, the population to be studied, the hypothesis, the intervention of interest, the statistical analysis plan and the primary outcome measure. These should be recorded in a registry such as PROS-PERO, which has a dedicated platform for reviews of animal studies (https://www.crd.york.ac.uk/prospero/#guidancenotes_animals).

   (b) For more complex reviews, it may be worth considering publication of a protocol manuscript, giving the opportunity to articulate in greater detail the background to the study and the approach to be used; and some journals have adopted the Registered Reports format, where the protocol is reviewed for methodological quality, with an undertaking to accept the final manuscript regardless of results, as long as the methodology described in the protocol has been followed (see https://cos.io/rr for further discussion).

4. *Ensuring reviews are up to date*

   (a) Depending on the resources available, systematic reviews may take as much as 2 years to complete. Given the pace of scientific publication, this means that the findings may be out of date before the review is even published. One approach is to update the search once data extraction from the originally identified studies is complete, but this should be performed before any data analysis, and the intention to update the search, perhaps conditional on the original search being above a certain age, should be articulated in a study protocol.

(b) An alternative approach is to conduct a living systematic review (Elliott et al. 2017). In this the intention is that the review is continually updated as new information becomes available. Automation of many of the key steps means that much of this can be done in the background, with little human intervention required (Thomas et al. 2017). At present the critical stage which resists automation is the extraction of outcome data, but even here the use of machine assistance may have much to offer; a pilot study suggests time saving of over 50% in data extraction, with gains in accuracy (Cramond et al. 2018). It is now possible to imagine completely automated living reviews, right through to a continually updated web-based dissemination of review findings.

(c) Such reviews raise important issues about statistical analysis and versions of record. For the former, the concern is that sequential statistical analysis of an enlarging dataset raises the false discovery rate. The problem is similar to those encountered in interim analyses in clinical trials, but because data might continue to accumulate indefinitely, approaches such as alpha spending used in clinical trials would not be appropriate. Possible approaches include either adopting a Bayesian approach, with priors informed by the first formal meta-analysis, or a rationing of formal statistical testing at milestones of data accumulation, for instance, with each doubling of the amount of data available (Simmonds et al. 2017).

(d) For a version of record, there needs to be a persisting digital identifier, with the possibility to recreate the data which contributed to that analysis. One approach would be to allow research users to create a snapshot of the analysis, with its own DOI and linked public domain data repository, with the snapshot labelled to indicate findings from the last formal statistical analysis and with the informal updated analysis. This would provide transparency to the provenance of the claims made.

5. *Machine learning for citation screening*

(a) Any bibliographic search represents a compromise between sensitivity and specificity – a highly sensitive search will identify all relevant studies and many more which are irrelevant; and attempts to increase specificity reduce sensitivity. For most systematic reviews, the proportion of relevant search results is around 10–20%. For some reviews, particularly "broad and shallow" reviews or living reviews, the work required in screening citations can be substantial. For instance, our search for a review of the animal modelling of depression returned more than 70,000 "hits", and one for the modelling of Alzheimer's disease returned over 260,000 "hits". In such cases the burden of human screening is prohibitive.

(b) The task of identifying citations is well suited to machine learning. Essentially an automated tool extracts features from the text such as word frequency and topics described, determines the representation of these features in a learning set of included versus excluded citations and makes a prediction of the probability that any given citation should be included. This can then be tested in a validation set and the sensitivity and specificity of various cut-off scores determined. By varying the cut-off score, the user can choose the levels of sensitivity and specificity which best meets their needs. Our practice is to choose the cut-off which provides sensitivity of 95% (roughly equivalent to human screening) and to observe the sensitivity achieved. If this is not sufficient, we increase the size of the training set in an attempt to secure better performance.

(c) There is a further elaboration to improve performance. Although the training sets have usually been defined through dual screening (i.e. two humans have independently adjudicated the citation, and disagreements have been reconciled by a third screener), errors still occur. Such errors pollute the training sets and reduce machine performance. Citations in the training set where there is greatest mismatch between human decision and machine prediction are those most likely to represent human errors, and so identifying these for further human screening to identify errors leads to improved performance – in the depression example (Bannach-Brown et al. 2019), increasing sensitivity from 86.7% to 88.3% while achieving sensitivity of 98.7%, resulting in a reduction in the burden of screening of over 1,000 citations.

6. *Text mining to partition and annotate the literature*

(a) Particularly in a broad and shallow review, there is often a need to categorise studies according to the disease model, the experimental intervention or the outcome measure reported. In all reviews, it may be helpful to annotate studies according to the reporting of measures – such as blinding or randomisation – which might reduce the risk of bias. This can be done either on title and abstract only or can consider the full text if this is available. The basic approach is to use a dictionary-based approach, determining the frequency with which a specific word or phrase appears. In our experience, this is usually sufficient for disease model, experimental intervention and the outcome measure reported – probably because there is a very limited number of ways in which such details are reported. Annotation for risks of bias is more challenging, because there are more ways in which such details can be described. More sophisticated textual analysis using regular expressions – where a word or phrase is detected in proximity to (or without proximity to) other words or phrases – can be used to detect the reporting of, for instance, blinding and randomisation, with a reasonable degree of precision (Bahor et al. 2017). However, performance at the level of the individual publication is not perfect, and access to full text is required. In the clinical trial literature, tools using more sophisticated machine learning approaches have been described (Marshall et al. 2016), and we and others are currently exploring the performance of similar approaches to the in vivo literature.

7. *Wide and shallow reviews and narrow and deep reviews*

   (a) Reviews can serve diverse purposes, from the very focussed investigation of the effect of a specific drug on a specific outcome in a specific disease model to more broad ranging reviews of a field of research. It is usually too burdensome for a review to be both wide and deep, but wide and shallow reviews can serve important purposes in describing a field of research; reporting the range of outcomes reported, drugs tested and models employed; and reporting of risks of bias, without a detailed meta-analysis. These can be critically important in designing future research questions, for instance, in determining priorities for future narrow and deep reviews. Indeed, by making available datasets from wide and shallow reviews with living searches, machine learning for citation screening and text mining to identify drugs and models of interest in "Curated current contents" (see below), these reviews can be a launch pad for those wishing to conduct narrow and deep reviews in particular areas, with much of the burden of searching and citation screening already performed.

2   Principles of Meta-Analysis

1. *Measures of effect size*

   (a) Usually we are interested in measuring differences in outcomes between two or more experimental cohorts. This might be a difference in, for instance, infarct volume in an animal model of stroke, or of cognitive performance in animal models of dementia, or of ejection fraction in animal models of myocardial ischaemia. It is very unusual for the outcome measure used to function as a ratio scale across the different experimental designs presented (a 5 mm^3 reduction in infarct volume has very different meaning in a mouse compared with a rat or a cat), and so simply taking the raw outcome measure is seldom appropriate.

   (b) Another approach is to calculate a "standardised mean difference" (SMD), where the difference is expressed as a proportion of the pooled standard deviation (Cohen's D), sometimes with a correction factor to account for small group sizes (Hedges G). If groups are large enough, the measured pooled standard deviation reflects the underlying biological variability in the phenomenon under study and is independent of the scale used; it can therefore be used to convert between scales. For example, if the variation in daily temperature recordings is 3.6°F and is also 2.0°C, then we can establish that 1.8°F = 1.0°C.

   (c) However, when group size is smaller, the measured pooled standard deviation reflects both underlying variability and a measurement error. In a simple simulation of 100 control groups with 10 animals each, the observed standard deviation ranged from 51% to 172% of the modelled value, giving substantial imprecision if this was used as the yardstick to scale the effect size (unpublished simulation).

(d) An alternative approach is to calculate a "normalised mean difference" (NMD) by mapping the observed outcomes onto a ratio scale where 0 is the outcome expected from an unlesioned, normal animal and 1 is the outcome observed in a lesioned, untreated animal (usually the control group). The effect size can then be expressed as the proportional or percentage improvement in the treatment group, with a pooled standard deviation on the same scale derived from that observed in the treatment and control groups. So a drug that reduced infarct volume in a rat from 300 to 240 mm^3 would be considered to have the same magnitude of effect as one that reduced infarct volume in a mouse from 25 to 20 mm^3.

(e) This NMD approach also has shortcomings. Firstly, although some outcome measures such as infarct volume appear as a ratio scale, the range of possible infarct volumes in a rat has a minimum at zero and a maximum at the volume of the intracranial cavity, so we expect floor and ceiling effects. Secondly, many behavioural outcomes are measured on scales which are ordinal rather than interval or ratio scales, where parametric approaches are considered less appropriate. Finally, this approach can only be used where outcome in non-lesioned ("normal") animals is either presented or can be inferred – for some outcomes (e.g. spontaneous motor activity), these data may not be available. Also, if the purpose is to summarise the impact of disease modelling rather than of the effect of an intervention in a disease model, the NMD approach is not possible.

(f) Nonetheless, where an NMD approach is possible, it is preferred. It has fewer relevant weaknesses than the alternative approaches, and it is a more powerful approach when you are interested in identifying differences between groups of studies (see Sect. 3).

2. *Giving different studies different weights*
(a) The calculation of a summary estimate of effect could be as simple as presenting the median observed effect or a mean value from the observed effects. However, this approach would give the same weight to small and large studies, to precise and imprecise studies.

(b) To address this, meta-analysis adjusts the weight which each study is given. In the simplest approach, studies are weighted according to the inverse of their observed variance. More precise studies – and this will generally also be the larger studies – are accorded greater importance than imprecise (usually smaller) studies. This is termed "fixed effects meta-analysis" and is appropriate where all studies are essentially asking the same question – we expect the differences between studies to be due simply to sampling error and that the true underlying results of these studies are the same.

(c) In reviews of animal studies, it is unusual for this to be the case; drugs are tested in different species, at different doses, in models of different severities and at different times in relation to when disease modelling was initiated.

We are therefore not so much interested in an "average" effect, but rather in how the observed effect varies under different circumstances. The true underlying results of included studies are likely to be different.

(d) To account for this, we can use random effects meta-analysis. Here the principle is that we make a statistical observation of the differences between studies (the heterogeneity) and compare this to the differences expected if the studies were all drawn from the same population (i.e. if all the observed variation was within studies). The difference between these estimates is the between-study variability, expressed as τ^2 ("tau squared"). Studies are then weighted by the inverse of the variance within that study and the between-study variance τ^2. Because τ^2 is constant across studies, if there is large between-study variation, this contributes a major, fixed component of study weights; and so the meta-analysis becomes more like a simple average. Where τ^2 is measured as zero, the meta-analysis behaves as a fixed effects meta-analysis.

(e) Importantly, the choice between fixed and random effects approaches should be made in advance, on the basis of investigator expectations of whether they expect there to be differences in true effect sizes between studies, rather than being decided once the data have been collected.

3. *Establishing differences between studies*

(a) As discussed above, the primary purpose of meta-analyses of in vivo data is not to come to some overall estimate of effect, but rather to gain a better understanding of differences in effect size between different types of studies. There are a number of approaches to this. Firstly we will outline these different approaches and then consider the strengths and weaknesses of each.

(b) Partitioning heterogeneity: In this approach, the overall heterogeneity between studies is calculated as the weighted sum of the squared deviations from the fixed effects estimate. The studies are then divided (partitioned) according to the variable of interest, and meta-analysis is performed within each group. From this we calculate the within-group heterogeneity as the weighted sum of the squared deviations from the fixed effects estimate within that group. We can then add together all of these "within-group heterogeneities" and subtract this from the overall heterogeneity. What remains, the between-group heterogeneity, is interpreted as the differences which are "explained" by our partitioning, and the significance of such differences can be tested using the χ^2 ("chi squared") statistic with $n-1$ degrees of freedom, where n is the number of partitions.

(c) Univariate meta-regression: Here we seek to model observed outcome (the dependent variable) in a simple regression equation. Firstly, we label each study for its status for the category of interest. Where this is a binary variable (present or absent), studies are labelled 0 or 1. For continuous variables such as weight, dose or time, it may be appropriate to offer these directly to the model, if you consider the response will be linear (or could be transformed to a linear response) or you could divide the studies into categories, for instance, in tertiles or quartiles of the distribution of values. For these and other categorical variables, we then create a series of dummy variables where each value of the category is either present or absent. With this approach we have redundant information – if there are three categories, and a study does not belong to A or B, it must belong to category C. It is our practice to censor the category which is the largest and to consider these as a reference category included in the baseline (and accounted for in the constant term (β_0) of the regression equation).

(d) Univariate meta-regression is essentially a linear regression, except that the best fitting model is chosen based on the minimisation of the weighted deviations from the model, with weights calculated as described above – so more precise studies are given greater weight. The constant (β_0) is an estimate of the treatment effect in the base case (usually describing the most commonly observed value for the category in question), and the other β-coefficients give an estimate of the different efficacies observed for other values of the category being studied). These coefficients are reported with their standard errors, from which it is possible to determine whether the coefficient is significantly different from zero. Most software packages (such as R metafor and STATA metareg) are also able to provide 95% confidence intervals for efficacy according to each of the modelled values within the category.

(e) Multiple meta-regression: In this extension, instead of one variable being offered, multiple variables can be offered simultaneously. As with other regression approaches, this can be done with unselected variables or with variables selected following univariate meta-regression, and it is possible to include interaction terms if this is desired. There is much discussion about both the number of instances of a variable within a category required for valid analysis and the number of categories which might be included. Having a small number of variables within a category will lead to imprecision in the estimate of the β-coefficient, but otherwise is unlikely to have deleterious consequences. For the number of variables which might be modelled, there is a general consensus that this should be no more than 10% of the number of observations, although the provenance of this recommendation is unknown to us.

(f) Tools: Most software packages have packages developed to support these approaches. The flexibility of R, and in particular the ability to embed R code within shinyapps, makes this a particularly attractive approach, and our shinyapp, developed to support the SyRF platform, is available at https://camarades.shinyapps.io/meta-analysis-app-syrf/.

(g) Choosing the best approach: Meta-analysis is a well-established technique, and many books and guides (e.g. the *Cochrane Handbook*, https://training.cochrane.org/handbook) are available. However, there are important differences between datasets derived from human clinical trials and those from animal studies. Broadly, human reviews include a relatively small number of studies each including a large number of subjects, addressing a reasonably well-focussed question. There may be substantial heterogeneity of subjects (different ages, sex, disease severity, geographical location, treatment centre) within a single study. In contrast, in animal reviews, there are usually a large number of individually small studies, and there may be much less focus (different drug doses because investigators have studied dose-response relationship, different stages or severity of disease, different species, different disease models, different outcome measures). Within each study, however, there is less heterogeneity, often sung animals of the same sex, age and weight of identical genetic background kept in the same cages on the same diet and undergoing identical study-related procedures.

It turns out these differences affect the performance of the statistical approaches used. Firstly, SMD estimates of effect size are less precise, as discussed in 1(c) above. In estimating the overall effect, NMD estimation of effect size has substantially greater power.

As well as having an impact on the effect size, this also has an impact on the attributed weight; studies which (through sampling error) have underestimated variance are given too much weight and (because calculated heterogeneity is the weighted squared deviation from the fixed effects estimate) contribute disproportionately to the observed heterogeneity. Following partitioning, the fixed effects estimate within that partition will move substantially towards overweighted studies (because they carry so much weight), and the within-group heterogeneity will fall substantially as a result.

(h) This gives a large artefactual increase in the between-study heterogeneity, which results in false-positive test of significance. In simulation studies we have recently shown that this false-positive rate, for NMD estimates of effect size, is over 50% (Wang et al. 2018). SMD is not affected to quite the same extent, but the power of that approach is limited. In contrast, in those simulations, both univariate and multivariable meta-analyses have acceptable false-positive rates (at around the expected level of 5%); and here the power of the NMD approach is again higher than SMD approaches (Wang et al. 2018).

(i) However, for reasons given above, it may not always be possible to calculate NMD effect sizes, and insistence on this approach would lead to exclusion of some studies. The best approach here depends on the number and the proportion of studies which would have to be excluded; if this number is less than around 30% of the total, and the studies to be excluded are in other respects typical of the included studies, then exclusion with NMD analysis provides greater statistical power. If however more than 30% of studies would be excluded, or these studies have specific features of interest not represented elsewhere in the dataset, it may be better to accept some diminution of power.

4. *Approaches to identifying publication bias*

(a) The Soviet Union had two key newspapers, *Izvestia* and *Pravda*. An old Russian joke held that *Izvestia* meant "News" and *Pravda* meant "Truth", and that meant there was no truth in *Izvestia* and no news in *Pravda*. The scientific literature is similarly afflicted by a focus on things which are newsworthy, but not necessarily true. Because our experiments (biases in the underlying study designs notwithstanding) sample underlying truth, our experiments are approximations to that truth. The results of our sampling are likely to follow a normal distribution, with some overstating and some understating the observed effect and most being about right. If our publication model only communicates a subset of our experimental results – selected, for instance, on the basis of statistical "significance" – then the literature will mislead. Rosenthal described this as the file drawer problem, where the 5% of studies which were falsely positive were in the journals and the 95% which were truly negative were in the file drawers of the investigators. His statement contains a latent suggestion that the problem may be due as much to investigators not seeking publication, rather than journals rejecting neutral or negative findings. Certainly, there is evidence from human clinical trials that this may be the case (Chan et al. 2014).

(b) In meta-analysis, we have the advantage of seeing a collection of publications rather than a single publication. If there is an underlying effect, we would expect to see a distribution of estimates around that true underlying effect, with more precise studies giving estimates closer to the true effect and less precise studies giving more variable estimates. A funnel plot is a graphical representation of effect size plotted against a measure of precision, and asymmetry is suggestive of "small study" effects, which include but are not limited to publication bias. As well as visual inspection, it is possible to analyse this mathematically using either Egger regression or the iterative "trim and fill" approach.

(c) Each of these approaches requires using a measure of precision, and because SMD effect size estimates are based in part on a consideration of precision, this leads to constraints in the possible values represented in a funnel plot determined in part by the number of subjects in each study. In clinical research this "n" is highly variable, and so few studies have the exact same n. In contrast, most animal studies are small, and many studies will have the same number of subjects. This leads to funnel plots showing studies with the same "n" describing curves easily seen in visual inspections. Analysis of several existing datasets using both SMD and NMD approaches and simulation studies modelling the presence of publication bias approaches have shown significant publication bias is more frequently found with SMD (rather than NMD) estimates of effect size. The simulation studies suggested that this was due to increased false-positive results with SMD analysis (Zwetsloot et al. 2017), and the authors suggested that, if it is not possible to use NMD effect size estimates, alternative measure of precision such as the square root of the number of experimental subjects should be used instead.

(d) Selective outcome reporting bias: Unlike human clinical trials publications, most publications describing in vivo research report findings from more than one experimental cohort, and – like human studies – they often describe more than one outcome from each cohort or the same outcome measured at different times. This gives substantial opportunities for selective reporting of outcome data and is a main factor in recommendations that ex ante study protocols should specify which outcome measure will be considered the primary yardstick of success or failure and a listing of all the outcomes which will be measured.

The extent of such selective outcome reporting can be estimated by seeking evidence of an excess of significant studies. Essentially the approach is to establish an overall measure of effect, to then estimate, based on the characteristics of identified studies, the number of positive studies one would expect to observe and then to compare this to the number of positive studies actually observed. Any excess significance might be due to data coercion in individual experiments (asymmetric exclusion of outliers, flexibility in statistical tests applied) or to the selective non-reporting of outcomes which do not reach statistical significance. Tsilidis et al. have applied this approach to the in vivo neuroscience literature (Tsilidis et al. 2013) and suggest that in some fields, up to 75% of experimental outcomes may be unreported.

5. *How complete are the data?*

Useful biomedical research informs either further research or policy decisions. Further research may involve seeking to apply the findings in a different research domain, for instance, where findings from in vivo research provide motivation for a human clinical trial. This is conventionally called "translational research". Alternatively, if there are not yet sufficient data to support such translation, there may be motivation to conduct further research in the same domain, which one might term "cis-lational research", or to decide that further research is likely to be fruitless. Getting these decisions right is critical and depends not only on the findings of individual experiments but also on an assessment of the "maturity", the completeness of the data portfolio being assessed. There are of course no precise boundaries, but in principle at least it should be possible to predict the chances of successful translation or of appropriate discontinuation. This might allow more rational research investment decisions to be made. The thresholds of evidence required for translation or discontinuation will of course differ according to circumstance; a lower threshold for translation would be appropriate, for instance, for the development of a treatment for Ebola virus infection than for the common cold.

To date, we have not had the tools to allow a quantitative assessment of a data portfolio against these thresholds, and such decisions have been largely qualitative, based instead on habit, experience and expert opinion. However, systematic review and meta-analysis are beginning to offer novel approaches. The optimal approach is not yet clear, but in our reviews both of tissue plasminogen activator (tPA) and of hypothermia in focal cerebral ischaemia, we have mature datasets, developed because investigators have been interested either in the effectiveness of co-treatments or where these interventions have been used as a positive control.

(a) Assessing the impact of known variables of interest and their beta coefficients: In some fields there is reasonably clear consensus around a range of circumstance under which efficacy should be observed in animal studies to justify attempted translation to human clinical trials. For instance, the stroke community, in the Stroke Therapy Academic Industry Roundtable (STAIR), suggested that efficacy be observed in more than one species, for both structural (infarct volume) and functional (neurobehavioral) outcomes, in animals with comorbidities (STAIR 1999). Using meta-regression it is possible to go beyond the basic requirement that efficacy be observed in such circumstances, to consider also the precision of the estimate of efficacy in each species, for both structural and functional outcomes, for animals with comorbidities and so on. This can be established using meta-regression, the factor of interest being the precision of the estimates of the beta coefficients for each of these features. For instance, it might be considered desirable that the impact of co-morbidity be estimated to within a 5% difference in infarct volume.

(b) The precision of the estimate of the impact of species on the efficacy of tPA in reducing infarct volume changed over time, increasing as more experiments were available for analysis. The figure shows the precision of the estimate of effect in different species and how this changed as the literature grew. For simplicity we show the estimates when 25, 50, 75% and all of the data (by date of publication) were included. If it were considered important to have a precise estimate of efficacy in primates, then further experiments are required. If knowing the difference between rats and mice is all that is important, then the data for species can be considered mature (Fig. 1).

(c) Total variability and marginal change in τ^2: We know that, even when offered a large number of potential independent variables, meta-regression is able to explain only a modest proportion of the observed variability. We consider that this is due to the impact of other variables which might either be

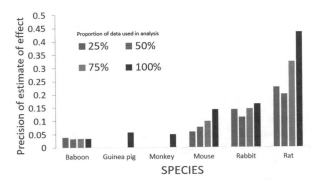

Fig. 1 Increasing precision in estimating the impact of species: because tPA is often tested in combination with other drugs, the literature is particularly mature. This allows us to observe changes in the precision of the estimation of the impact of the species of the experimental animal (the inverse of the standard error of the beta coefficient derived from meta-regression) as the amount of available data grows. In the figure we show precision after 75, 150 and 225; and finally the complete set of 301 experiments were available for analysis

latent (unreported or unknown to the investigators) or too sparsely represented to be sampled (for instance, between-lab effects). Under these circumstances, it would be interesting to know whether an additional experiment would add valuable new information or whether the data can be considered complete.

(d) As discussed above, the between-study differences are measured using a statistic called τ^2, for which different measures are available. The computationally simplest approach is the DerSimonian and Laird estimator. This is derived from the observed Cochrane's Q and gives an estimate of the heterogeneity adjusted for the number of observations. However, it tends to give biased estimates when sample size is small, and alternative approaches such as restricted maximum likelihood (REML) are now widely available and are probably more appropriate in the context of meta-analysis of animal data.

(e) When the evidence in support of a hypothesis is immature, we expect that additional experiments – through deliberate or accidental differences in the circumstance of testing – will add to the value of τ^2. Conversely, when a field is mature, additional experiments will add little useful additional information and will not increase the observed τ^2. We can therefore track – again using the tPA and hypothermia datasets described above – how τ^2 changes as new studies are added.

(f) When we do this, an interesting biphasic pattern emerges. At first there is a rapid increase in observed τ^2, followed by a decline, followed by another increase (although not to the same peak as the first rise), after which the value is relatively constant. We think that the first rise in heterogeneity reflects differences in experimental design (for instance, using different drug doses to characterise dose-response relationships) and heterogeneity as different research teams seek to replicate the originator finding. The fall occurs, we propose, as the community unites or coalesces around designs where efficacy is reliably observed. The second rise, we propose, occurs as investigators seek to extend the range of circumstances under which efficacy is seen, to identify the limits to efficacy. Finally, the plateau occurs when investigators have tested all relevant circumstances and represents the maturation of the evidence. Under this schema, evidence for efficacy cannot be considered mature until τ^2 has plateaued.

(g) Using datasets from systematic reviews of NXY059 (Macleod et al. 2008), FK506 (Macleod et al. 2005), nicotinamide (Macleod et al. 2004), tirilazad (Sena et al. 2007), IL1RA (McCann et al. 2016), hypothermia (van der Worp et al. 2007) and tPA (Sena et al. 2010), which include varying numbers of experiments, we have performed cumulative random effects meta-analysis and investigated the changes in the heterogeneity as more studies are added. As the number of included studies increases, all the datasets show the expected increase in Cochrane's Q. However, for both I^2 (the percentage of the variability in effect sizes that is due to variability between studies rather than just random sampling error) and when Q is adjusted for the number of included studies, there is first an increase with small number of studies, followed by a slow decline and stabilisation as more studies are included. Using cumulative meta-regression with inclusion of explanatory variables in the analysis shows an increasing precision in the estimates of beta coefficients with inclusion of more studies. Similarly, the cumulative between-study variability (measured using the REML estimate of τ^2 explained by an explanatory variable shows an initial peak with a later decreasing trend, where it gradually stabilises, suggesting that saturation of evidence has been reached. These preliminary findings using seven preclinical datasets suggest that the systematic characterisation of heterogeneity within stroke datasets relating to important community-identified requirements for the circumstances in which efficacy is observed, when considered alongside the size of effects observed, might form the basis of a useful guide to inform decisions to proceed with further clinical testing.

(h) It is inconceivable that a drug will show efficacy under all conceivable circumstances of testing; and (for decisions to embark on human clinical trials at least) it is important that the limits to efficacy are established. Therefore, where a cohort of animal studies shows evidence for efficacy but little or no heterogeneity, this should raise concern – it is scarcely credible that a drug always works and much more likely that the range of circumstances under which efficacy has been tested has been too narrow reliably to define the characteristics of drug response.

(i) This is important; the GRADE approach to evidence synthesis considers that heterogeneity in a body of evidence is a bad thing and that the strength of evidence-based recommendations should be downgraded in the presence of heterogeneity. While this may be true for very tightly defined clinical questions, it is in our view certainly not the case when summarising a group of animal studies.

6. *Examples*

(a) *Disease models*: Systematic review and meta-analysis can be used to summarise published work using a particular disease model. For instance, Currie and colleagues examined the literature on bone cancer-induced pain (Currie et al. 2013). Across 112 studies they found substantial increases in pain-related behaviours, most commonly measured using mechanical allodynia, along with reduced body weight and reduced locomotion, but no change in reported food intake. There was also evidence of changes in the spinal cord, each reported by more than five publications, of astrocytosis, and increased c-Fos, substance P (NK1) receptor internalisation, dynorphin, IL-1b and TNF-a.

(b) *Drugs*: Rooke et al. reported (Rooke et al. 2011) the effect of dopamine agonists in animal models of Parkinson's disease. For drugs tested in more than one publication, all drugs in common clinical use showed evidence of substantial efficacy, with ropinirole, rotigotine, apomorphine, lisuride and pramipexole having more efficacy (in the point estimate) than the 95% confidence limits of the overall estimate for all drugs combined. However, as discussed above these estimates have limited value, and random allocation to group was reported by 16% of publications (16%), blinded assessment of outcome by 15% and a sample size calculation by <1%. Across all neurobehavioural outcomes, there was an inverse relationship between study quality and effect size, and reporting of blinded assessment of outcome was associated with significantly smaller effect sizes.

(c) *Outcome measures*: Egan et al. conducted a systematic review of publications reporting the efficacy of drugs tested in animal models of Alzheimer's disease (Egan et al. 2016). As well as describing the variety of neurobehavioural and histological outcomes which had been reported, they gave particular focus to the use of the Morris water maze. Reporting of experimental details was generally incomplete; 16% of studies did not report the size of the water maze used, and in those that did, this ranged from 85 cm to 200 cm. 35% of studies did not report water temperature, and in those that did, this ranged from 16°C to 28°C. The number of acquisition trials per day ranged from 2 to 12 and was unreported in 23%, and the number of days training ranged from 1 to 15 and was unreported in 13. Remarkably, in 57 publications describing the probe phase, there were 59 different approaches used to demonstrate efficacy, suggesting a degree of flexibility in analysis and reporting. Only 36% of experiments reported randomisation to intervention or control, and only 24% of experiments reported the blinded assessment of outcome. Overall, reported efficacy was significantly higher in non-randomised and in non-blinded studies.

(d) *Risks of bias*: Following the publication of the neutral SAINT II trial (Shuaib et al. 2007), we conducted a systematic review of published in vivo data on the efficacy of NYY-059 (Macleod et al. 2008). Reporting of measures to reduce the risk of bias was again low, with lower estimates of improvement in infarct volume in those studies which reported randomisation, in those which reported the blinded conduct of the experiment and in those which reported the blinded assessment of outcome. These findings were supported by a later individual animal meta-analysis which also had access to unpublished industry data (Bath et al. 2009).

In later work we examined reporting of risks of bias in work published in leading journals and, separately, in work from leading UK institutions (Macleod et al. 2015). We found journal impact factor to be no guarantee of study quality, and in fact randomisation was less frequently reported in high-impact journals. At an institutional level, only 1 of 1,173 publications from leading UK institutions reported 4 aspects of study design (randomisation, blinding, reporting of inclusions and exclusions and sample size calculations) identified by Landis et al. as being critical to allowing readers to judge the provenance of the findings presented, and 68% of publications reported not one of these.

(e) *Power calculations*: Appropriate design of animal experiments includes consideration of how many subjects should be included. Formal power calculations require assertion of the minimum effect size of interest which the investigator would like to be able to detect, their tolerance of the risk of missing a true result and the variability of the outcome measure used. As well as giving some indication of the possible statistical variance which might be observed when a lab uses a model or outcome measure for the first time, knowledge of the performance of different outcome measures testing broadly similar behavioural substrates can inform refinement of experimental designs to reduce animal pain and suffering. For instance, as part of a systematic review of animal studies modelling chemotherapy-induced peripheral neuropathy, Currie et al. compared the statistical performance of different approaches to measuring mechanical allodynia, showing superiority of electronic over mechanical von Frey testing (Currie et al. 2019).

(f) *Curated current contents*: Borrowing from the concept of "Living" systematic reviews (Elliott et al. 2017), real-time information in a given field can be summarised on an online platform which presents the up-to-date results visually. Ideally, such a platform should be interactive, allowing any research user (a biomedical researcher, a funder, an institution) not only to gain a quick overview of the field but also to filter the results in a way which is most relevant to them, e.g. by specific models or treatments of interest, by reporting quality or by year of publication. Two recent examples are our RShiny applications which summarise the literature on animal models of depression (https://abannachbrown.shinyapps.io/preclinical-models-of-depression/) and animal models of chemotherapy-induced peripheral neuropathy (https://khair.shinyapps.io/CIPN/).

3   Summary

The amount of relevant in vivo data is substantial, and nonsystematic attempts to summarise what is already known may draw misleading conclusions. Because the selection of included information is an objective process, it is not possible critically to appraise the conclusions drawn, other than by reference to the reputation of the authors (as indeed is the case with the current work). Systematic review offers a transparent approach to identifying relevant information such that it would be possible for others to replicate the approach. Such reviews also allow ascertainment of the features of a body of work, which might lead to suggestions for how a field might seek improvement. Meta-analysis allows a quantitative summary of overall effects, any association between various study design factors and observed outcome, an assessment of the likelihood of publication bias and recommendations for sample size calculations for future experiments.

While the process is burdensome, the value of the information obtained is substantial, and emerging automation tools are likely substantially to reduce the costs, and the time taken, for systematic review and meta-analysis.

References

Bahor Z, Liao J, Macleod MR, Bannach-Brown A, McCann SK, Wever KE et al (2017) Risk of bias reporting in the recent animal focal cerebral ischaemia literature. Clin Sci (Lond) 131 (20):2525–2532. https://doi.org/10.1042/CS20160722

Bannach-Brown A, Przybyla P, Thomas J, Rice ASC, Ananiadou S, Liao J et al (2019) Machine learning algorithms for systematic review: reducing workload in a preclinical review of animal studies and reducing human screening error. Syst Rev 8(1):23. https://doi.org/10.1186/s13643-019-0942-7

Bath PM, Gray LJ, Bath AJ, Buchan A, Miyata T, Green AR (2009) Effects of NXY-059 in experimental stroke: an individual animal meta-analysis. Br J Pharmacol 157(7):1157–1171

Chan AW, Song F, Vickers A, Jefferson T, Dickersin K, Gotzsche PC et al (2014) Increasing value and reducing waste: addressing inaccessible research. Lancet 383(9913):257–266. https://doi.org/10.1016/S0140-6736(13)62296-5

Cramond F, O'Mara-Eves A, Doran-Constant L, Rice AS, Macleod M, Thomas J (2018) The development and evaluation of an online application to assist in the extraction of data from graphs for use in systematic reviews. Wellcome Open Res 3:157. https://doi.org/10.12688/wellcomeopenres.14738.3

Currie GL, Delaney A, Bennett MI, Dickenson AH, Egan KJ, Vesterinen HM et al (2013) Animal models of bone cancer pain: systematic review and meta-analyses. Pain 154(6):917–926

Currie GL, Angel-Scott HN, Colvin L, Cramond F, Hair K, Khandoker L et al (2019) Animal models of chemotherapy-induced peripheral neuropathy: a machine-assisted systematic review and meta-analysis. PLoS Biol 17(5):e3000243. https://doi.org/10.1371/journal.pbio.3000243

Egan KJ, Vesterinen HM, Beglopoulos V, Sena ES, Macleod MR (2016) From a mouse: systematic analysis reveals limitations of experiments testing interventions in Alzheimer's disease mouse models. Evid Based Preclin Med 3:e00015

Elliott JH, Synnot A, Turner T, Simmonds M, Akl EA, McDonald S et al (2017) Living systematic review: 1. Introduction-the why, what, when, and how. J Clin Epidemiol 91:23–30. https://doi.org/10.1016/j.jclinepi.2017.08.010.

Macleod MR, O'Collins T, Howells DW, Donnan GA (2004) Pooling of animal experimental data reveals influence of study design and publication bias. Stroke 35(5):1203–1208

Macleod MR, O'Collins T, Horky LL, Howells DW, Donnan GA (2005) Systematic review and metaanalysis of the efficacy of FK506 in experimental stroke. J Cereb Blood Flow Metab 25 (6):713–721

Macleod MR, van der Worp HB, Sena ES, Howells DW, Dirnagl U, Donnan GA (2008) Evidence for the efficacy of NXY-059 in experimental focal cerebral ischaemia is confounded by study quality. Stroke 39(10):2824–2829

Macleod MR, Lawson MA, Kyriakopoulou A, Serghiou S, de WA, Sherratt N et al (2015) Risk of bias in reports of in vivo research: a focus for improvement. PLoS Biol 13(10):e1002273

Marshall IJ, Kuiper J, Wallace BC (2016) RobotReviewer: evaluation of a system for automatically assessing bias in clinical trials. J Am Med Inform Assoc 23(1):193–201. https://doi.org/10.1093/jamia/ocv044

McCann SK, Cramond F, Macleod MR, Sena ES (2016) Systematic review and Meta-analysis of the efficacy of Interleukin-1 receptor antagonist in animal models of stroke: an update. Transl Stroke Res 7(5):395–406. https://doi.org/10.1007/s12975-016-0489-z

Munafo MR, Stothart G, Flint J (2009) Bias in genetic association studies and impact factor. Mol Psychiatry 14(2):119–120

Rooke ED, Vesterinen HM, Sena ES, Egan KJ, Macleod MR (2011) Dopamine agonists in animal models of Parkinson's disease: a systematic review and meta-analysis. Parkinsonism Relat Disord 17(5):313–320

Sena E, Wheble P, Sandercock P, Macleod M (2007) Systematic review and meta-analysis of the efficacy of tirilazad in experimental stroke. Stroke 38(384):391

Sena ES, Briscoe CL, Howells DW, Donnan GA, Sandercock PA, Macleod MR (2010) Factors affecting the apparent efficacy and safety of tissue plasminogen activator in thrombotic occlusion models of stroke: systematic review and meta-analysis. J Cereb Blood Flow Metab 30 (12):1905–1913

Shuaib A, Lees KR, Lyden P, Grotta J, Davalos A, Davis SM et al (2007) NXY-059 for the treatment of acute ischemic stroke. N Engl J Med 357(6):562–571

Simmonds M, Salanti G, McKenzie J, Elliott J (2017) Living systematic review N. living systematic reviews: 3. Statistical methods for updating meta-analyses. J Clin Epidemiol 91:38–46. https://doi.org/10.1016/j.jclinepi.2017.08.008

STAIR (1999) Recommendations for standards regarding preclinical neuroprotective and restorative drug development. Stroke 30(12):2752–2758

Thomas J, Noel-Storr A, Marshall I, Wallace B, McDonald S, Mavergames C et al (2017) Living systematic reviews: 2. Combining human and machine effort. J Clin Epidemiol 91:31–37. https://doi.org/10.1016/j.jclinepi.2017.08.011

Tsilidis KK, Panagiotou OA, Sena ES, Aretouli E, Evangelou E, Howells DW et al (2013) Evaluation of excess significance bias in animal studies of neurological diseases. PLoS Biol 11(7):e1001609

van der Worp HB, Sena ES, Donnan GA, Howells DW, Macleod MR (2007) Hypothermia in animal models of acute ischaemic stroke: a systematic review and meta-analysis. Brain 130 (Pt 12):3063–3074

Wang Q, Liao J, Hair K, Bannach-Brown A, Bahor Z, Currie GL et al (2018) Estimating the statistical performance of different approaches to meta-analysis of data from animal studies in identifying the impact of aspects of study design. bioRxiv:256776. https://doi.org/10.1101/256776

Zwetsloot PP, van der Naald M, Sena ES, Howells DW, IntHout J, de Groot JA et al (2017) Standardized mean differences cause funnel plot distortion in publication bias assessments. elife 6:10. https://doi.org/10.7554/eLife.24260

Publishers' Responsibilities in Promoting Data Quality and Reproducibility

Iain Hrynaszkiewicz (ID)

Contents

I. Hrynaszkiewicz (✉)
Public Library of Science (PLOS), Cambridge, UK
e-mail: ihrynaszkiewicz@plos.org

© The Author(s) 2019
A. Bespalov et al. (eds.), *Good Research Practice in Non-Clinical Pharmacology and Biomedicine*, Handbook of Experimental Pharmacology 257,
https://doi.org/10.1007/164_2019_290

319

Abstract

Scholarly publishers can help to increase data quality and reproducible research by promoting transparency and openness. Increasing transparency can be achieved by publishers in six key areas: (1) understanding researchers' problems and motivations, by conducting and responding to the findings of surveys; (2) raising awareness of issues and encouraging behavioural and cultural change, by introducing consistent journal policies on sharing research data, code and materials; (3) improving the quality and objectivity of the peer-review process by implementing reporting guidelines and checklists and using technology to identify misconduct; (4) improving scholarly communication infrastructure with journals that publish all scientifically sound research, promoting study registration, partnering with data repositories and providing services that improve data sharing and data curation; (5) increasing incentives for practising open research with data journals and software journals and implementing data citation and badges for transparency; and (6) making research communication more open and accessible, with open-access publishing options, permitting text and data mining and sharing publisher data and metadata and through industry and community collaboration. This chapter describes practical approaches being taken by publishers, in these six areas, their progress and effectiveness and the implications for researchers publishing their work.

Keywords

Data sharing · Open access · Open science · Peer review · Publishing · Reporting guidelines · Reproducible research · Research data · Scholarly communication

1   Introduction

Scholarly publishers have a duty to maintain the integrity of the published scholarly record. Science is often described as self-correcting, and when errors are identified in the published record, it is the responsibility of publishers to correct them. This is carried out by publishing corrections, expressions of concern or, sometimes, retracting published articles. Errors in published research can be honest, such as typographical errors in data tables or broken links to source material, but errors also result from research misconduct, including fraudulent or unethical research, and plagiarism. Only a small fraction – less than 0.1% (Grieneisen and Zhang 2012) – of published research is retracted, and papers are more likely to be retracted due to misconduct, than honest error (Fang et al. 2012).

However, the numbers of reported corrections and retractions do not account for the more pressing issue: that a large proportion of published – assumed accurate – research results are not reproducible, when reproducibility and replicability are tenets of science. Pharmaceutical companies have reported that fewer than 25% of the results reported in peer-reviewed publications could be reproduced in their labs (Prinz et al. 2011). A survey of 1,500 researchers found that more than half of respondents could not reproduce their own results and more than 70% could not reproduce the results of others (Baker 2016). An economic analysis in 2015

Table 1 Causes of poor reproducibility and poor data quality in preclinical research

	Relevant chapters elsewhere in this textbook
Conduct of research	
Experimental design	Chapters "Guidelines and Initiatives for Good Research Practice", "Learning from Principles of Evidence-Based Medicine to Optimize Nonclinical Research Practices", "General Principles of Preclinical Study Design", "Blinding and Randomization", "Out of Control? Managing Baseline Variability in Experimental Studies with Control Groups", "Building Robustness Intro Translational Research", and "Design of Meta-Analysis Studies"
Quality control	Chapters "Quality of Research Tools", "Quality Governance in Biomedical Research", and "Costs of Implementing Quality in Research Practice"
Lab supervision and training	
Adherence to ethical standards	Chapter "Good Research Practice: Lessons from Animal Care and Use"
Culture of publishing some results, and not others	Chapter "Resolving the Tension Between Exploration and Confirmation in Preclinical Biomedical Research"
Reporting of research	
Completeness of methods descriptions	Chapters "Minimum Information and Quality Standards for Conducting, Reporting, and Organizing In Vitro Research", and "Minimum Information in In Vivo Research"
Accuracy of images, figures and graphs	
Availability of research data, protocols, computer code	Chapters "Quality of Research Tools", "Electronic Lab Notebooks and Experimental Design Assistants", and "Data Storage"
Statistical reporting	Chapter "A Reckless Guide to P-Values: Local Evidence, Global Errors"
Publication of all scientifically sound results, regardless of their outcome	

estimated that irreproducible preclinical research costs US $28 billion per year (Freedman et al. 2015).

There are numerous causes of irreproducibility and suboptimal data quality (Table 1). Some of these causes relate to how research is conducted and supervised, and others relate to how well or completely research is reported. Data quality and reproducibility cannot be assessed without complete, transparent reporting of research and the availability of research outputs which can be reused. Scholarly publishers have a responsibility to promote reproducible research (Hrynaszkiewicz et al. 2014) but are more able to influence the reporting of research than the conduct of research. Transparency is a precursor to reproducibility and can be supported by journals and publishers (Nature 2018).

Implementation of greater transparency in, reporting and reuse potential of, research by publishers can be achieved in several ways:

1. Understanding researchers' problems and motivations
2. Raising awareness and changing behaviours
3. Improving the quality, transparency and objectivity of the peer-review process
4. Better scholarly communication infrastructure and innovation
5. Enhancing incentives
6. Making research publishing more open and accessible

This chapter describes practical approaches being taken by publishers, in these six areas, to achieve greater transparency and discusses their progress and effectiveness and the implications for researchers.

2 Understanding Researchers' Problems and Motivations

Publishers wishing to increase transparency and reproducibility need to understand the problems (or "challenges") researchers have in practising reproducible and transparent research. Sharing of research data is essential for reproducible research, and between 2016 and 2018, several large surveys of researchers were conducted by publishing and publishing technology companies, providing insights into researchers' reported data sharing practices and behaviours, as well as insight into what motivates researchers to share, or not share, research data.

A survey conducted in 2017, and published in 2018, by the publisher Springer Nature explored the "practical challenges" researchers have in data sharing, which received 7,719 responses, one of the largest of its kind. Seventy-six percent of respondents reported that the discoverability of their research data is important to them, and 63% had previously shared data associated with a peer-reviewed article. However, researchers also reported common problems in sharing their data, including difficulties in "organising data in a presentable and useful way" (46% of respondents), being unsure about licencing and copyright of data (37%) and not knowing which data repository to use (33%). A lack of time (26%) and being unable to cover costs of data sharing (19%) were also commonly cited (Stuart et al. 2018) (Fig. 1).

Disciplinary differences were also identified in the survey. Biological science researchers reported the highest levels of data sharing (75%), and medical science researchers reported that copyright and licencing (data ownership) issues were their biggest challenge. Medical science researchers were also most likely to report concerns about data sensitivity and misuse, and concerns about protecting research participants, consistent with other surveys (Rathi et al. 2012) of clinical researchers.

Surveys from the publishers Wiley (Wiley Open Science Researcher Survey 2016) and Elsevier (Berghmans et al. 2017) and the publishing technology company Digital Science (Science et al. 2017) have found similar results regarding the proportion of researchers who report they share data and the ways in which

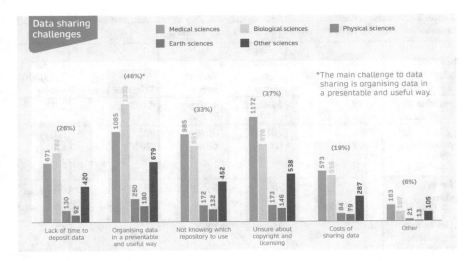

Fig. 1 Organising data in a presentable and useful way was the most common problem for researchers in data sharing, in a large survey ($n = 7,719$). Figure adapted from Stuart et al. (2018)

researchers share data (Table 2). The most common ways of sharing data that were reported tend to be suboptimal, with email being the most common method for private data sharing (Allin 2018) and journal supplementary materials being most common for public data sharing (email is not secure enough for private data sharing; data repositories are preferred over supplementary materials for public data sharing) (Michener 2015).

2.1 Understanding Motivations to Share Data

Sharing research data has been associated with an increase in the number of citations that researchers' papers receive (Piwowar et al. 2007; Piwowar and Vision 2013; Colavizza et al. 2019) and an increase in the number of papers that research projects produce (Pienta and Alter 2010). Some researchers report that increased academic credit (Science et al. 2017), and increased visibility of their research (Wiley Open Science Researcher Survey 2016; Schmidt et al. 2016), motivates them to share research data. Publishers and other service providers to researchers can help to both solve problems and increase motivations, in particular those relating to academic credit, impact and visibility (see Sect. 6).

Table 2 Seventy percent of researchers report that they share data but only 26% use data repositories when the results of five large surveys are combined

Survey responses and findings	Springer Nature global survey[a]	Springer Nature Japan survey[b]	Elsevier survey	Wiley survey	Digital Science survey
Number of respondents	7,719	1,966	1,200	4,668	2,352
Year published (year conducted)	2018 (2017)	2018 (2018)	2017 (2016)	2017 (2016)	2017 (2017)
Level of data sharing reported %	63	95[b]	64	69	60
Use of data repositories reported %	21	25	13	41	30
Most common data sharing problem	Organising data in a presentable and useful way	Concerns about misuse of data	Data ownership	Intellectual property or confidentiality issues	Data sensitivity

[a]Focused on sharing data associated with peer-reviewed publications rather than data sharing in general
[b]Explicitly included private (peer-to-peer) data sharing and public data sharing

3 Raising Awareness and Changing Behaviours

Scholarly publishers and journals can help to raise awareness of issues through their wide or community-focused readership – with editorials, opinion pieces and conference and news coverage. Behavioural change can be created by changing journal and publisher policies, as researchers are motivated to comply with them when submitting papers (Schmidt et al. 2016).

3.1 Journal Policies

Journal policies and guides to authors include large amounts of information covering topics from manuscript formatting, research ethics and conflicts of interest. Many journals and publishers have, since 2015, endorsed – and are beginning to implement – the Transparency and Openness Promotion (TOP) guidelines. The TOP guidelines are a comprehensive but aspirational set of journal policies and include eight modular standards, each with three levels of increasing stringency including transparency in data, code and protocols (Nosek et al. 2014). A summary table of the requirements is available in the public domain from the Center for Open Science (Table 3). Full compliance with the TOP guidelines is

Table 3 TOP guidelines summary table[a]

	Not implemented	Level I	Level II	Level III
Citation standards	Journal encourages citation of data, code and materials or says nothing	Journal describes citation of data in guidelines to authors with clear rules and examples	Article provides appropriate citation for data and materials used consistent with journal's author guidelines	Article is not published until providing appropriate citation for data and materials following journal's author guidelines
Data transparency	Journal encourages data sharing or says nothing	Article states whether data are available and, if so, where to access them	Data must be posted to a trusted repository. Exceptions must be identified at article submission	Data must be posted to a trusted repository, and reported analyses will be reproduced independently prior to publication
Analytic methods (code) transparency	Journal encourages code sharing or says nothing	Article states whether code is available and, if so, where to access it	Code must be posted to a trusted repository. Exceptions must be identified at article submission	Code must be posted to a trusted repository, and reported analyses will be reproduced independently prior to publication
Research materials transparency	Journal encourages materials sharing or says nothing	Article states whether materials are available and, if so, where to access them	Materials must be posted to a trusted repository. Exceptions must be identified at article submission	Materials must be posted to a trusted repository, and reported analyses will be reproduced independently prior to publication
Design and analysis transparency	Journal encourages design and analysis transparency or says nothing	Journal articulates design transparency standards	Journal requires adherence to design transparency standards for review and publication	Journal requires and enforces adherence to design transparency standards for review and publication
Study preregistration	Journal says nothing	Article states whether preregistration of study exists and,	Article states whether preregistration of study exists and, if so, allows journal	Journal requires preregistration of studies and provides link and badge in article to

(continued)

Table 3 (continued)

	Not implemented	Level I	Level II	Level III
		if so, where to access it	access during peer review for verification	meeting requirements
Analysis plan preregistration	Journal says nothing	Article states whether preregistration of study exists and, if so, where to access it	Article states whether preregistration with analysis plan exists and, if so, allows journal access during peer review for verification	Journal requires preregistration of studies with analysis plans and provides link and badge in article to meeting requirements
Replication	Journal discourages submission of replication studies or says nothing	Journal encourages submission of replication studies	Journal encourages submission of replication studies and conducts results blind review	Journal uses registered reports as a submission option for replication studies with peer review prior to observing the study outcomes

[a]Reproduced and available from the Center for Open Science, under a Creative Commons public domain CC0 waiver

typically a long-term goal for journals and publishers, and implementation of the requirements is tending to happen in progressive steps, with most progress being made initially in policies for sharing of research data.

3.1.1 Standardising and Harmonising Journal Research Data Policies

While availability of research data alone does not enable reproducible research, unavailability of data (Ioannidis et al. 2009) and suboptimal data curation (Hardwicke et al. 2018) have been shown to lead to failures to reproduce results. Historically, relatively few journals have had research data policies, and, where policies have existed, they have lacked standards and consistency, which can be confusing for researchers (authors) and research support staff (Naughton and Kernohan 2016; Barbui 2016). In 2016 Springer Nature, which publishes more than 2,500 journals, begun introducing standard, harmonised research data policies to its journals (Hrynaszkiewicz et al. 2017a). Similar initiatives were introduced by some of the other largest journal publishers Elsevier, Wiley and Taylor and Francis in 2017, greatly increasing the prevalence of journal data sharing policies. These large publishers have offered journals a controlled number (usually four or five) of data policy types, including a basic policy with fewer requirements compared to the more stringent policies (Table 4).

Providing several options for journal data policy is necessary because, across multiple research disciplines, some research communities and their journals are more able to introduce strong data sharing requirements than others. In parallel to these

Table 4 Summary of Springer Nature journal data policy types and examples of journals with those policy types

Policy type/ level	Policy summary	Example journal	Weblink
1	Data sharing is encouraged	*Cardiovascular Drugs and Therapy*	https://www.springer.com/ medicine/cardiology/journal/ 10557
2	Data sharing and evidence of data sharing and data availability statements are encouraged	*Clinical Drug Investigation*	https://www.springer.com/ adis/journal/40261
3	Data sharing encouraged and data statements are required	*Nature*	http://www.nature.com/ authors/policies/data/data-availability-statements-data-citations.pdf
4	Data sharing, evidence of data sharing, data availability statements and peer review of data required	*Scientific Data*	https://www.nature.com/sdata/ policies/data-policies

individual publisher's data policy initiatives, a global collaboration of publishers, and other stakeholders in research, have created a master research data policy framework that supports all journal and publisher requirements (Hrynaszkiewicz et al. 2017b, 2019).

There have also been research data policy initiatives from communities of journals and journal editors. In 2010 journals in ecology and evolutionary biology joined in supporting a Joint Data Archiving Policy (JDAP) (Whitlock et al. 2010), Public Library of Science (PLOS) introduced a strong data sharing policy to all its journals in 2014, and in 2017 the International Committee of Medical Journal Editors (ICMJE) introduced a standardised data sharing policy (Taichman et al. 2017) for its member journals, which include *BMJ*, *Lancet*, *JAMA* and the *New England Journal of Medicine*. The main requirement of the ICMJE policy was not to mandate data sharing but for reports of clinical trials to include a data sharing statement.

Data sharing statements (also known as data availability statements) are a common feature of journal and publisher data policies. They provide a statement about where data supporting the results reported in a published article can be found – including, where applicable, hyperlinks to publicly archived datasets analysed or generated during the study. Many journals and publishers provide guidance on preparing data availability statements (e.g. https://www. springernature.com/gp/authors/research-data-policy/data-availability-statements/ 12330880). All Public Library of Science (PLOS), Nature and BMC journals require data availability statements (Colavizza et al. 2019). Some research funding agencies – including the seven UK research councils (UK Research and Innovation 2011) – also require the provision of data availability statements in published articles.

Experimental pharmacology researchers publishing their work in 2019 and beyond, regardless of their target journal(s), should be prepared at minimum to provide a statement regarding the availability and accessibility of the research data that support the results of their papers.

Code and Materials Sharing Policies

To assess data quality and enable reproducibility, transparency and sharing of computer code and software (and supporting documentation) are also important – as is, where applicable, the sharing of research materials. Materials include samples, cell lines and antibodies. Journal and publisher policies on sharing code, software and materials are becoming more common but are generally less well evolved and less widely established compared to research data policies.

In 2015 the Nature journals introduced a policy across all its research titles that encourages all authors to share their code and provide a "code availability" statement in their papers (Nature 2015). *Nature Neuroscience* has taken this policy further, by piloting peer review of code associated with research articles in the journal (Nature 2017). Software-focused journals such as the *Journal of Open Research Software* and *Source Code for Biology and Medicine* tend to have the most stringent requirements for availability and usability of code.

3.2   Effectiveness of Journal Research Data Policies

Journal submission guidelines can increase transparent research practices by authors (Giofrè et al. 2017; Nuijten et al. 2017). Higher journal impact factors have been associated with stronger data sharing policies (Vasilevsky et al. 2017). Stronger data policies that mandate and verify data sharing by authors, and require data availability statements, are more effective at ensuring data are available long term (Vasilevsky et al. 2017) compared to policies that passively encourage data sharing (Vines et al. 2013). Many journal policies ask authors to make supporting data available "on reasonable request", as a minimum requirement. This approach to data sharing may be a necessity in medical research, to protect participant privacy, but contacting authors of papers to obtain copies of datasets is an unreliable method of sharing data (Vanpaemel et al. 2015; Wicherts et al. 2006; Savage and Vickers 2009; Rowhani-Farid and Barnett 2016). Using more formal, data sharing (data use) agreements can improve authors' willingness to share data on request (Polanin and Terzian 2018), and guidelines on depositing clinical data in controlled-access repositories have been defined by editors and publishers, as a practical alternative to public data sharing (Hrynaszkiewicz et al. 2016). Publishers are also supporting editors to improve policy effectiveness and consistency of implementation (Graf 2018).

4 Improving the Quality, Transparency and Objectivity of the Peer-Review Process

The reporting of research methods, interventions, statistics and data on harms of drugs, in healthcare research, and the presentation of results in journal articles, has repeatedly been found to be inadequate (Simera et al. 2010). Increasing the consistency and detail of reporting key information in research papers, with reporting guidelines and checklists, supports more objective assessment of papers in the peer-review process.

4.1 Implementation of Reporting Guidelines

The prevalence and endorsement of reporting guidelines, catalogued by the EQUA-TOR Network (http://www.equator-network.org), in journals has increased substantially in the last decade. Reporting guidelines usually comprise a checklist of key information that should be included in manuscripts, to enable the research to be understood and the quality of the research to be assessed. Reporting guidelines are available for a wide array of study designs, such as randomised trials (the CON-SORT guideline), systematic reviews (the PRISMA guidelines) and animal preclinical studies (the ARRIVE guidelines; discussed in detail in another chapter in this volume).

The positive impact of endorsement of reporting guidelines by journals has however been limited (Percie du Sert et al. 2018), in part due to reporting guidelines often being implemented by passive endorsement on journal websites (including them in information for authors). Some journals, such as the medical journals *BMJ* and *PLOS Medicine*, have mandated the provision of certain completed reporting guidelines, such as CONSORT, as a condition of submitting manuscripts. Endorsement and implementation of reporting guidelines has been more prevalent in journals with higher impact factors (Shamseer et al. 2016). More active interventions to enforce policy in the editorial process are generally more effective, as demonstrated with data sharing policies (Vines et al. 2013), but these interventions are also more costly as they increase demands on authors' and editors' time. For larger, multidisciplinary journals publishing many types of research, identifying and enforcing the growing number of relevant reporting guidelines, which can vary from paper to paper that is submitted, is inherently more complex and time-consuming. These processes of checking manuscripts for adherence to guidelines can however be supported with artificial intelligence tools such as https://www.penelope.ai/.

An alternative approach to this problem taken by the multidisciplinary science journal *Nature* was to introduce a standardised editorial checklist to promote transparent reporting that could be applied to many different study designs and research disciplines. The checklist was developed by the journal in collaboration with researchers and funding agencies (Anon 2013) and is implemented by professional editors, who require that all authors complete it. The checklist elements focus on experimental and analytical design elements that are crucial for the interpretation of research results. This includes description of methodological parameters that can introduce bias or influence robustness and characterisation of reagents that may be subject to biological variability, such as cell lines and antibodies. The checklist has led to improved reporting of risks of bias in in vivo research and improved reporting of randomisation, blinding, exclusions and sample size calculations; however in vitro data compliance was not improved, in an independent assessment of the checklist's effectiveness (Macleod and The NPQIP Collaborative Group 2017; The NPQIP Collaborative Group 2019).

In 2017 the *Nature* checklist evolved into two documents, a "reporting summary" that focuses on experimental design, reagents and analysis and an "editorial policy checklist" that covers issues such as data and code availability and research ethics. The reporting summary document is published alongside the associated paper and, to enable reuse by other journals and institutions, is made available under an open-access licence (Announcement 2017) (Fig. 2).

4.2   Editorial and Peer-Review Procedures to Support Transparency and Reproducibility

Peer review is important for assessing and improving the quality of published research (even if evidence of its effectiveness is often questioned (Smith 2010)). Most journals have been, understandably, reluctant to give additional mandatory tasks to peer reviewers – who may already be overburdened with the continuing increase in volume of publications (Kovanis et al. 2016) – to ensure journal policy compliance and assessment of research data and code. Journal policies often encourage reviewers to consider authors' compliance with data sharing policies, but formal peer review of data tends to occur only in a small number of specialist journals, such as data journals (see later in this chapter) and journals with the strictest data sharing policies. The most stringent research data policy of Springer Nature's four types of policy requires peer reviewers to access the supporting data for every publication in a journal and includes guidelines for peer reviewers of data:

> Peer reviewers should consider a manuscript's Data availability statement (DAS), where applicable. They should consider if the authors have complied with the journal's policy on the availability of research data, and whether reasonable effort has been made to make the data that support the findings of the study available for replication or reuse by other researchers.
>
> For the Data availability statement, reviewers should consider:
>
> • Has an appropriate DAS been provided?
> • Is it clear how a reader can access the data?
> • Where links are provided in the DAS, are they working/valid?
> • Where data access is restricted, are the access controls warranted and appropriate?
> • Where data are described as being included with the manuscript and/or supplementary information files, is this accurate?
>
> For the data files, where available, reviewers should consider:

nature research

Corresponding author(s):

Last updated by author(s): YYYY-MM-DD

Double-blind peer review submissions: write DBPR and your manuscript number here instead of author names.

Reporting Summary

Nature Research wishes to improve the reproducibility of the work that we publish. This form provides structure for consistency and transparency in reporting. For further information on Nature Research policies, see Authors & Referees and the Editorial Policy Checklist.

Please do not complete any field with "not applicable" or n/a. Refer to the help text for what text to use if an item is not relevant to your study.
For final submission: please carefully check your responses for accuracy; you will not be able to make changes later.

Statistics

For all statistical analyses, confirm that the following items are present in the figure legend, table legend, main text, or Methods section.

n/a	Confirmed	
☐	☐	The exact sample size (n) for each experimental group/condition, given as a discrete number and unit of measurement
☐	☐	A statement on whether measurements were taken from distinct samples or whether the same sample was measured repeatedly
☐	☐	The statistical test(s) used AND whether they are one- or two-sided *Only common tests should be described solely by name; describe more complex techniques in the Methods section.*
☐	☐	A description of all covariates tested
☐	☐	A description of any assumptions or corrections, such as tests of normality and adjustment for multiple comparisons
☐	☐	A full description of the statistical parameters including central tendency (e.g. means) or other basic estimates (e.g. regression coefficient) AND variation (e.g. standard deviation) or associated estimates of uncertainty (e.g. confidence intervals)
☐	☐	For null hypothesis testing, the test statistic (e.g. F, t, r) with confidence intervals, effect sizes, degrees of freedom and P value noted *Give P values as exact values whenever suitable.*
☐	☐	For Bayesian analysis, information on the choice of priors and Markov chain Monte Carlo settings
☐	☐	For hierarchical and complex designs, identification of the appropriate level for tests and full reporting of outcomes
☐	☐	Estimates of effect sizes (e.g. Cohen's d, Pearson's r), indicating how they were calculated

Our web collection on statistics for biologists contains articles on many of the points above.

Software and code

Policy information about availability of computer code

Data collection	*Provide a description of all commercial, open source and custom code used to collect the data in this study, specifying the version used OR state that no software was used.*
Data analysis	*Provide a description of all commercial, open source and custom code used to analyse the data in this study, specifying the version used OR state that no software was used.*

For manuscripts utilizing custom algorithms or software that are central to the research but not yet described in published literature, software must be made available to editors/reviewers. We strongly encourage code deposition in a community repository (e.g. GitHub). See the Nature Research guidelines for submitting code & software for further information.

Data

Policy information about availability of data

All manuscripts must include a data availability statement. This statement should provide the following information, where applicable:
- Accession codes, unique identifiers, or web links for publicly available datasets
- A list of figures that have associated raw data
- A description of any restrictions on data availability

Provide your data availability statement here.

Field-specific reporting

Please select the one below that is the best fit for your research. If you are not sure, read the appropriate sections before making your selection.

☐ Life sciences ☐ Behavioural & social sciences ☐ Ecological, evolutionary & environmental sciences

nature research | reporting summary

October 2018

1

Fig. 2 The Nature Research Reporting Summary Checklist, available under a Creative Commons attribution licence from https://www.nature.com/authors/policies/ReportingSummary.pdf

- Are the data in the most appropriate repository?
- Were the data produced in a rigorous and methodologically sound manner?
- Are data and any metadata consistent with file format and reporting standards of the research community?
- Are the data files deposited by the authors complete and do they match the descriptions in the manuscript?
- Do they contain personally identifiable, sensitive or inappropriate information?

However, as of 2019 fewer than ten journals have implemented this policy of formal data peer review as a mandatory requirement, and journal policies on data sharing and reproducibility tend to focus on transparent reporting, such as including links to data sources. This enables a motivated peer reviewer to assess aspects of a study, such as data and code, more deeply, but this is not routinely expected.

In specific disciplines, journals and study designs, additional editorial assessment and statistical review are routinely employed. Some medical journals, such as *The Lancet*, consistently invite statistical review of clinical trials, and statistical reviewers have been found to increase the quality of reporting of biomedical articles (Cobo et al. 2007). Not all journals, such as those without full-time editorial staff, have sufficient resources to statistically review all research papers. Instead, journals may rely on editors and nonstatistical peer reviewers identifying if statistical review is warranted and inviting statistical review case-by-case.

Some journals have taken procedures on assessing reproducibility and transparency even further. The journal *Biostatistics* employs an Associate Editor for reproducibility, who awards articles "kite marks" for reproducibility, which are determined by the availability of code and data and if the Associate Editor for reproducibility is able to reproduce the results in the paper (Peng 2009). Another journal, *npj Breast Cancer*, has involved an additional editor, a Research Data Editor (a professional data curator), to assess every accepted article and give authors editorial support to describe and share link to the datasets that support their articles (Kirk and Norton 2019).

4.3   Image Manipulation and Plagiarism Detection

Plagiarism and self-plagiarism are common forms of misconduct and common reasons for papers being retracted (Fang et al. 2012). In the last decade, many publishers have adopted plagiarism detection software, and some apply this systematically to all submissions. Plagiarism detection software, such as iThenticate, works by comparing manuscripts against a database of billions of web pages and 155 million content items, including 49 million works from 800 scholarly publishers that participate in CrossRef Similarity Check (https://www.crossref.org/services/similarity-check/). Plagiarism detection is an important mechanism for publishers, editors and peer reviewers to maintain quality and integrity in the scholarly record. Although less systematically utilised in the editorial process, software for automated detection of image manipulation – a factor in about 40% of retractions in the biomedical literature and thought to affect 6% of published papers – is also available to journals (Bucci 2018).

5   Better Scholarly Communication Infrastructure and Innovation

Publishers provide and utilise scholarly communication infrastructure, which can be both an enabler and a barrier to reproducibility. In this chapter, "scholarly communication infrastructure" means journals, article types, data repositories, publication platforms and websites, content production and delivery systems and manuscript submission and peer-review systems.

5.1   Tackling Publication (Reporting) Bias

Publication bias, also known as reporting bias, is the phenomenon in which only some of the results of research are published and therefore made available to inform evidence-based decision-making. Papers that report "positive results", such as positive effects of drugs on a condition or disease, are more likely to be published, are more likely to be published quickly and are likely to be viewed more favourably by peer reviewers (McGauran et al. 2010; Emerson et al. 2010). In healthcare-related research, this is a pernicious problem, and widely used healthcare interventions, such as the antidepressant reboxetine (Eyding et al. 2010), have been found to be ineffective or potentially harmful, when unpublished results and data are obtained and combined with published results in meta-analyses.

5.1.1   Journals
Providing a sufficient range of journals is a means to tackle publication bias. Some journals have dedicated themselves exclusively to the publication of "negative" results, although have remained niche publications and many have been discontinued (Teixeira da Silva 2015). But there are many journals that encourage publication of all methodologically sound research, regardless of the outcome. The BioMed Central (BMC) journals launched in 2000 with this mission to assess scientific accuracy rather than impact or importance and to promote publication of negative results and single experiments (Butler 2000). Many more "sounds science" journals – often multidisciplinary "mega journals" including *PLOS One*, *Scientific Reports* and *PeerJ* – have since emerged, almost entirely based on an online-only open-access publishing model (Björk 2015). There is no shortage of journals to publish scientifically sound research, yet publication bias persists. More than half of clinical trial results remain unpublished (Goldacre et al. 2018).

5.1.2   Preregistration of Research

Preregistration of studies and study protocols, in dedicated databases, before data are collected or patients recruited, is another means to reduce publication bias. Registration is well established – and mandatory – for clinical trials, using databases such as ClinicalTrials.gov and the ISRCTN register. The prospective registration of clinical trials helps ensure the data analysis plans, participant inclusion and exclusion criteria and other details of a planned study are publicly available before publication of results. Where this information is already in the public domain, it reduces the potential for outcome switching or other sources of bias to occur in the reported results of the study (Chan et al. 2017). Clinical trial registration has been common since 2005, when the ICMJE introduced a requirement for prospective registration of trials as a condition for publication in its member journals. Publishers and editors have been important in implementing this requirement to journals.

Preregistration has been adopted by other areas of research, and databases are now available for preregistration of systematic reviews (in the PROSPERO database) and for all other types of research, with the Open Science Framework (OSF) and the Registry for International Development Impact Evaluations (RIDIE).

Registered Reports

A more recent development for preregistration is a new type of research article, known as a registered report. Registered reports are a type of journal article where the research methods and analysis plans are both pre-registered and submitted to a journal for peer review before the results are known (Robertson 2017). Extraordinary results can make referees less critical of experiments, and with registered reports, studies can be given in principle acceptance decisions by journals before the results are known, avoiding unconscious biases that may occur in the traditional peer-review process. The first stage of the peer-review process used for registered reports assesses a study's hypothesis, methods and design, and the second stage considers how well experiments followed the protocol and if the conclusions are justified by the data (Nosek and Lakens 2014). Since 2017, registered reports began to be accepted by a number of journals from multiple publishers including Springer Nature, Elsevier, PLOS and the BMJ Group.

Protocol Publication and Preprint Sharing

Predating registered reports, in clinical trials in particular, it has been common since the mid-2000s for researchers to publish their full study protocols as peer-reviewed articles in journals such as *Trials* (Li et al. 2016). Another form of early sharing of research results, before peer review has taken place or they are submitted to a journal, is preprint sharing. Sharing of preprints has been common in physical sciences for a quarter of century or more, but since the 2010s preprint servers for biosciences (biorxiv.org), and other disciplines, have emerged and are growing rapidly (Lin 2018). Journals and publishers are, increasingly, encouraging their use (Luther 2017).

5.2   Research Data Repositories

There are more than 2000 data repositories listed in re3data (https://www.re3data.org/), the registry of research data repositories (and more than 1,100 databases in the curated FAIRsharing resource on data standards, policies and databases https://fairsharing.org/). Publishers' research data policies generally preference the use of third party or research community data repositories, rather than journals hosting raw data themselves. Publishers can help enable repositories to be used, more visible, and valued, in scholarly communication. This is beneficial for researchers and publishers, as connecting research papers with their underlying data has been associated with increased citations to papers (Dorch et al. 2015; Colavizza et al. 2019).

Publishers often provide lists of recommended or trusted data repositories in their data policies, to guide researchers to appropriate repositories (Callaghan et al. 2014) as well as linking to freely available repository selection tools (such as https://repositoryfinder.datacite.org/). Some publishers – such as Springer Nature via its Research Data Support helpdesk (Astell et al. 2018) – offer free advice to researchers to find appropriate repositories. The journal *Scientific Data* has defined criteria for trusted data repositories in creating and managing its list of data repositories and makes its recommended repository list available for reuse (Scientific Data 2019).

Where they are available, publishers generally promote the use of community data repositories – discipline-specific, data repositories and databases that are focused on a particular type or format of data such as GenBank for genetic sequence data. However, much research data – sometimes called the "long tail" of research data (Ferguson et al. 2014) – do not have common databases, and, for these data, general-purpose repositories such as figshare, Dryad, Dataverse and Zenodo are important to enable all research data to be shared permanently and persistently.

Publishers, and the content submission and publication platforms they use, can be integrated with research data repositories – in particular these general repositories – to promote sharing of research data that support publications. The Dryad repository is integrated to varying extents with a variety of common manuscript submission systems such as Editorial Manager and Scholar One. Integration with repositories makes it easier and more efficient for authors to share data supporting their papers. The journal *Scientific Data* enables authors to deposit data into figshare seamlessly during its submission process, resulting in more than a third of authors depositing data in figshare (data available via http://scientificdata.isa-explorer.org). Many publishers have invested in technology to automatically deposit small datasets shared as supplementary information files with journals articles into figshare to increase their accessibility and potential for reuse.

5.3   Research Data Tools and Services

Publishers are diversifying the products and services they provide to support researchers practice reproducible research (Inchcoombe 2017). The largest scholarly publisher Elsevier (RELX Group), for example, has acquired software used by researchers before they submit work to journals, such as the electronic lab notebook Hivebench. Better connecting scholarly communication infrastructure with researchers' workflow and research tools is recognised by publishers as a way to promote transparency and reproducibility, and publishers are increasingly working more closely with research workflow tools (Hrynaszkiewicz et al. 2014).

While "organising data in a presentable and useful way" is a key barrier to data sharing (Stuart et al. 2018), data curation as a distinct profession, skill or activity has tended to be an undervalued and under-resourced in scholarly research (Leonelli 2016). Springer Nature, in 2018, launched a Research Data Support service (https://www.springernature.com/gp/authors/research-data/research-data-support) that provides data deposition and curation support for researchers who need assistance from professional editors in sharing data supporting their publications. Use of this service has been associated with increased metadata quality (Grant et al. 2019; Smith et al. 2018). Publishers, including Springer Nature and Elsevier, provide academic training courses in research data management for research institutions. Some data repositories, such as Dryad, offer metadata curation, and researchers can also often access training and support from their institutions and other third parties such as the Digital Curation Centre.

5.4 Making Research Data Easier to Find

Publishing platforms can promote reproducibility and provenance tracking by improving the connections between research papers and data and materials in repositories. Ensuring links between journal articles and datasets are present, functional and accurate is technologically simple, but can be procedurally challenging to implement when multiple databases are involved. Connecting published articles and research data in a standardised manner across multiple publishing platforms and data repositories, in a dynamic and universally adoptable manner, is highly desirable. This is the aim of a collaborative project between publishers and other scholarly infrastructure providers such as CrossRef, DataCite and OpenAIRE. This Scholarly Link Exchange (or, "Scholix") project enables information on links between articles and data to be shared between all publishers and repositories in a unified manner (Burton et al. 2017). This approach, which publishers are important implementers of, means readers accessing articles on a publisher platform or literature database or data repository, such as Science Direct or EU PubMed Central or Dryad, will be provided with contemporaneous and dynamic information on datasets that are linked to articles in other journals or databases and vice versa.

6 Enhancing Incentives

Publications in peer-reviewed journals, and citations, are established mechanisms for assigning credit for scholarly contributions and for researchers and institutions to provide evidence for their research outputs and impact. Publishers can offer

Table 5 Examples of data, software, methods and protocol journals

Type of journal	Journal	Publisher
Data journal	*Scientific Data*	Springer Nature
Data journal	*Earth Systems Science Data*	Copernicus
Data journal	*Data in Brief*	Elsevier
Data journal	*GigaScience*	Oxford University Press/BGI
Software journal	*Journal of Open Research Software*	Ubiquity Press
Software journal	*Source Code for Biology and Medicine*	Springer Nature
Software journal	*SoftwareX*	Elsevier
Protocol journal	*Nature Protocols*	Springer Nature
Protocol journal	*Current Protocols*	Wiley
Methods journal	*Nature Methods*	Springer Nature
Methods journal	*MethodsX*	Elsevier

incentives to promote transparency by providing opportunities for additional articles and citations and new forms of incentive such as digital badges.

6.1 New Types of Journal and Journal Article

In the last 10 years, more journals, and types of journal article, have emerged that publish articles that describe specific parts of a research project. The print-biased format of traditional research articles does not always provide sufficient space to communicate all aspects of a research project. These new publications include journals that specialise in publishing articles that describe datasets or software (code), methods or protocols. Established journals have also introduced new article types that describe data, software, methods or protocols (Table 5).

Of these journals and article types, data journals and data papers are the most common. Data papers do not include a Results or Conclusion, like traditional research papers. They generally describe a publicly available dataset in sufficient detail so that another researcher can find, understand and reuse the data. Data journals generally do not publish raw data, but publish peer-reviewed papers that describe datasets (Hrynaszkiewicz and Shintani 2014). Data papers often include more detailed or technical information that may be excluded from traditional research papers, or which might only appear as supplementary files in traditional research papers. Data papers can both accompany traditional research papers and be independent articles that enable the publication of important datasets and databases that would not be considered as a traditional publication.

Papers published in data journals attract citations. While the number of articles published in data journals is steadily growing, they, however, represent a small proportion of the published literature overall (Berghmans et al. 2017).

6.2   Data and Software Citation

Research data, software and other research outputs, when published in digital repositories, can be assigned Digital Object Identifiers (DOIs), like research papers and chapters, enabling these research outputs to be individually discovered and cited and their citations measured in the same way.

Citing data and software promotes reproducibility by enabling linking and provenance tracking of research outputs. Papers can be persistently linked to the version (s) of data and code that were used or generated by the experiments they describe. Data citation can also provide more specific evidence for claims in papers, when those claims are based on published data. Citation of data and software is encouraged, and in some case required, as part of many journals' data sharing and reproducible research policies (Hrynaszkiewicz et al. 2017a). Some funding agencies, such as the National Science Foundation in the USA, encourage researcher to list datasets and software (in addition to traditional publications) as part of their bibliographic sketches (Piwowar 2013).

From the researcher's (author's) perspective, citing data and software in reference lists is the same as citing journal articles and book chapters are cited. Several datasets are cited in this chapter, such as Smith et al. (Smith et al. 2018), and software can, similarly, also be cited when it is deposited in repositories that assign DOIs. Zenodo and figshare are commonly recommended for depositing code and software so that they can be cited.

To promote data citation and to enable data citations and links to be more visible to readers, publishers have implemented changes to the structure of published content (the XML underlying the digital version of journal articles) (Cousijn et al. 2017, 2018). Publishers and other scholarly infrastructure providers, such as DataCite and CrossRef (member organisations that generate DOIs for digital research outputs), are collaborating to enable data citation to be implemented and practised consistently, regardless of where researchers publish. Data citations, in article reference lists (bibliographies), have historically appeared in a small proportion of the published literature, but data citations have been increasing year-on-year (Garza and Fenner 2018). Researchers have indicated that they value the credit they receive through data citations, in some cases equally to the credit they receive from citations to their papers (Science et al. 2017).

6.3   Digital Badges for Transparency: A New Type of Incentive

The Center for Open Science offers digital badges that are displayed on published articles to highlight, or reward, papers where the data and materials are openly available and for studies that are pre-registered. Badges signal to the reader that the content has been made available and certify its accessibility in a persistent location. More than 40 journals, in 2018, offered or were experimenting with the award of badges to promote transparency (Blohowiak 2013). The use of digital

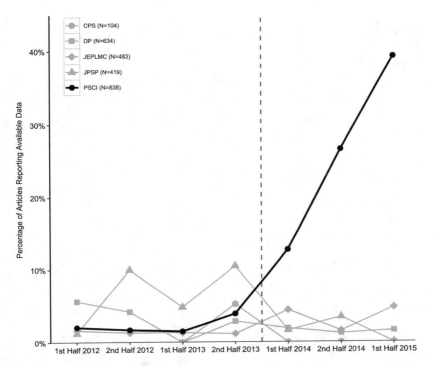

Fig. 3 Percentage of articles reporting open data by half year by journal. Darker line indicates *Psychological Science*, and dotted red line indicates when badges were introduced in *Psychological Science* and none of the comparison journals. Figure and legend reproduced from Kidwell et al. (2016)

badges is most prolific in psychology and human behavioural research journals, but they are also used in some microbiology, primatology and geoscience journals.

Awarding digital badges to authors has been associated with increased rates of data sharing by authors. When the journal *Psychological Science* (PSCI) introduced badges for articles with open data, the proportion of articles with open data increased (Fig. 3), compared to previous levels of data sharing in the journal. Data sharing also increased in the journal compared to other psychology journals (Munafò et al. 2017; Kidwell et al. 2016).

Digital badges being an effective incentive for data sharing has also been confirmed in a systematic review (Rowhani-Farid et al. 2017). The badges that are awarded by the journal *Biostatistics*' Associate Editor for reproducibility have also been associated with increased data sharing, although, in the same study, badges did not have an impact on the sharing of code (Rowhani-Farid and Barnett 2018).

Badges are usually awarded by authors self-disclosing information or they are awarded as part of the peer-review process. Another method of awarding badges adopted by *BMC Microbiology* involves the data availability statements of each paper being assessed, independently, by the publisher (Springer Nature 2018).

Box 1 Practical Recommendations for Researchers to Support the Publication of Reproducible Research

Before You Carry Out Your Research

- Check if your institution or employer, funding agency and target journals have policies on sharing and managing research data, materials and code or more broadly on reproducibility and open science.
  - Seek advice on compliance with these policies, and support including formal training, where needed.
  - Note that journal policies on data sharing are generally agnostic of whether research is industry or academically sponsored.
- Consider how you will store and manage your data and other research outputs and plan accordingly, including whether additional or specific funding is required to cover associated costs.
  - Preparing a Data Management Plan (DMP) is recommended and is often required under funding agency and institutional policies. Free tools such as https://dmponline.dcc.ac.uk/ can assist in creating DMPs.
- Determine if there are standards and expectations, and existing infrastructure such as data repositories, for sharing data in your discipline.
  - Use resources such as https://fairsharing.org/ to explore standards, policies, databases and reporting guidelines.
  - Establish if there are existing repositories for the type of data you generate. Where they exist use discipline-specific repositories for your data, and general repositories for other data types.
- Familiarise yourself with tools that enable reproducibility, and version control, particularly for computational work (Markowetz 2015).
- Where appropriate databases exist, consider preregistration of your study (for clinical trials registration in a compliant database is mandatory) as a means to reduce the potential for bias in analyses.
- For clinical studies in particular, publish your study protocol as a peer-reviewed article, or at minimum be prepared to share it with journal editors and peer reviewers.
  - Free tools such as https://www.nature.com/protocolexchange/ and https://www.protocols.io/ can be used to share methodological knowledge.
- If your target journal(s) offer them, consider preparing a registered report.

When Preparing to Submit Your Research Results to a Journal

- Register for an ORCID identifier and encourage your co-authors to do the same.
- Publish a preprint of your paper in a repository such as bioRxiv, enabling the community to give you feedback on your work and for you to assert ownership and claim credit for you work early.

(continued)

Box 1 (continued)
- Prepare your data and code for deposition in a repository, and make these available to editors and peer reviewers.
  - Use repositories rather than supplementary information files for your datasets and code.
- Consider publishing data papers, software papers or methods-focused papers to complement your traditional research papers, particularly if detailed information that enables understanding and reuse your research does not form part of your traditional papers.
- If the results of your research are inconclusive and show no difference between comparison groups ("negative results"), publish them. Many journals consider such papers.
- Always include clear statements in your publications about the availability of research data and code generated or utilised by your research.
  - If there are legitimate restrictions on the availability and reuse of your data, explain them in your data availability statements.
  - Wherever possible, include links to supporting datasets in your publications – this supports reproducibility and is associated with increased citations to papers.
- Be prepared to share with editors and peer reviewers any materials supporting your papers that might be needed to verify, replicate or reproduce the results.
  - Many repositories enable data to be shared privately before publication and in a way that protects peer reviewers' anonymity (where required).
- Cite, in your reference lists and bibliographies, any persistent, publicly available datasets that were generated or reused by your research.

After Publication of Your Research
- Be prepared to respond to reasonable requests from other scientists to reuse your data.
  - Non-compliance with data sharing policies of journals can lead to corrections, expressions of concern or retractions of papers.
- Try to view the identification of honest errors in published work – yours and others – as a positive part of the self-correcting nature of science.
- Remember working transparently and reproducibly is beneficial to your own reputation, productivity and impact as a researcher, as well as being beneficial to science and society (Markowetz 2015).

7 Making Research Publishing More Open and Accessible

The sixth and final area in which publisher can promote transparency relates to how open and accessible publishers are as organisations. This refers firstly to the content publishers distribute and secondly to the accessibility of other information and resources from publishers.

7.1 Open Access and Licencing Research for Reuse

Publishing more research open access, so that papers are freely and immediately available online, is an obvious means to increase transparency. The proportion of the scholarly literature that is published open access, each year, is increasing by 10–15%. Open access accounted for 17% of published articles in 2015 (Johnson et al. 2017), and the two largest journals in the world – *Scientific Reports* and *PLOS One* – are open-access journals.

Open-access publishing means more than access to research; it is also about promoting free reuse and redistribution of research, through permissive copyright licences (Suber 2012). Open-access journals and articles are typically published under Creative Commons attribution licences, such as CC BY, which means that the work can be copied, distributed, modified and adapted freely, by anyone, provided the original authors are attributed (the figures in this chapter are examples of this practice).

Publishing research under CC BY, or equivalent copyright licences, is important for promoting reproducibility of research because it enables published research outputs to be reused efficiently, by humans and machines. With this approach, the pace of research need not be slowed by the need to negotiate reuse rights and agreements with researchers and institutions. Meanwhile scholarly norms of acknowledging previous work (through citation) and legal requirements for attribution in copyright will ensure that researchers are credited for their contributions (Hrynaszkiewicz and Cockerill 2012).

Reuse of the research literature is essential for text and data mining research, and this kind of research can progress more efficiently with unrestricted access to and reuse of the published literature. Publishers can enable the reuse of research content published in subscription and open-access journals with text and data mining policies and agreements. Publishers typically permit academic researchers to programmatically access their publications, such as through secure content application programming interfaces (APIs), for text and data mining research (Text and Data Mining – Springer; Text and Data Mining Policy – Elsevier).

7.2 Open Publisher (Meta)Data

For other kinds of content, including research data, publishers can promote ease of access and reuse by applying and setting standards for content licences that enable reuse easily. In 2006 multiple publishers signed a joint statement agreeing not to take copyright in research data (STM, ALPSP 2006). Publishers have also promoted the use of liberal, public domain legal tools for research data and metadata. The publisher BMC introduced, in 2013, a default policy whereby any data published in their more than 250 journals would be available in the public domain, under the Creative Commons CC0 waiver (Hrynaszkiewicz et al. 2013). Publishers can also make data about their content catalogues (metadata) openly available. Springer Nature's SciGraph, for example, is a linked open data platform for the scholarly domain and collates metadata from funders, research projects, conferences, affiliations and publications (SciGraph). Many publishers also make the bibliographies (reference lists) of all their publications, subscription and open access, available openly as "open citations" (Shotton 2013; I4OC).

Beyond published articles, journals and associated metadata, publishers can share other information openly. This includes survey findings (Table 1) and the results of projects to improve transparency reproducibility – such as around data sharing policies (Hrynaszkiewicz et al. 2017a) and research data curation (Smith et al. 2018). Resources produced and curated by publishers can also be made available to the wider community (such as Scientific Data 2019).

7.3   Open for Collaboration

Publishers can promote transparency through collaboration. The biggest policy and infrastructural challenges that enable the publication of more reproducible research can only be tackled by multiple publishers collaborating as an industry and collaboration with other organisations that support the conduct and communication of research – repositories, institutions and persistent identifier providers. Progress resulting from such collaborations has been seen in data citation (Cousijn et al. 2017), data policy standardisation (Hrynaszkiewicz et al. 2017b), reporting standards to enhance reproducibility (McNutt 2014) and provenance tracking of research outputs and researchers, through persistent identification initiatives such as ORCID (https://orcid.org/organizations/publishers/best-practices). All of which, combined, help publishers and the wider research community to make practical improvements to the communication of research that support improved data quality and reproducibility.

7.3.1   The Future of Scholarly Communication?

In some respects the future of scholarly communication is already here, with dynamic, reproducible papers (Lewis et al. 2018), workflow publication, data integration and interactive data, figures and code all possible, albeit at a relatively small scale. However, these innovations remain highly unevenly distributed, and the majority of published scholarly articles remain largely static objects, with the PDF format remaining popular with many readers. Like most scientific advances, progress in scholarly communication tends not to be made through giant leaps of progress but by slow, steady, incremental improvements. However, numerous major publishers have expressed strong support for open science and are introducing practical measures to introduce and strengthen policies on transparency of all research outputs, as a prerequisite to improving reproducibility. Researchers should expect continued growth in transparency policies of journals and be prepared for demands for more transparency in the reporting of their research (see Box 1 for practical suggestions for researchers). Increasing computerisation and machine readability of papers, with integration of data and code and enhancement of metadata increasing, will promote reproducibility and new forms of research quality assessment. This will help the research community assess individual research projects more specifically than the inappropriate journal-based measure of the impact factor. Large publishers will continue to diversify the types of content they publish and diversify their businesses, evolving into service providers for researchers and institutions and including content discovery, research metrics, research tools, training and analytics in their activities alongside publishing services. Technology and services are just part of implementing reproducible research, and cultural and behavioural change – and demonstrating value and impact of reproducible research – will continue to be incentivised with policies of all stakeholders in research. Monitoring compliance with transparency and reproducibility policies remains a challenge, but increasing standardisation of policies will enable economies of scale in monitoring compliance.

Competing Interests At the time of writing this chapter, the author (IH) was employed by Springer Nature. Since July 2019 the author is employed by Public Library of Science (PLOS). Neither employer had any role in the preparation or approval of the chapter.

References

Allin K (2018) Research data: challenges and opportunities for Japanese researchers – Springer Nature survey data. https://figshare.com/articles/Research_data_challenges_and_opportunities_for_Japanese_researchers-_Springer_Nature_survey_data/6328952/1

Announcement (2017) Towards greater reproducibility for life-sciences research in Nature. Nature 546:8

Anon J (2013) Announcement: reducing our irreproducibility. Nature 496:398–398

Astell M, Hrynaszkiewicz I, Grant R, Smith G, Salter J (2018) Have questions about research data? Ask the Springer Nature helpdesk. https://figshare.com/articles/Providing_advice_and_guidance_on_research_data_a_look_at_the_Springer_Nature_Helpdesk/5890432

Baker M (2016) 1,500 scientists lift the lid on reproducibility. Nature 533:452–454

Barbui C (2016) Sharing all types of clinical data and harmonizing journal standards. BMC Med 14:63

Berghmans S et al (2017) Open Data: the researcher perspective – survey and case studies. https://data.mendeley.com/datasets/bwrnfb4bvh/1

Björk B-C (2015) Have the "mega-journals" reached the limits to growth? PeerJ 3:e981

Blohowiak BB (2013) Badges to acknowledge open practices. https://osf.io/tvyxz/files/?_ga=2.252581578.297610246.1542300800-587952028.1539080384

Bucci EM (2018) Automatic detection of image manipulations in the biomedical literature. Cell Death Dis 9:400

Burton A et al (2017) The Scholix framework for interoperability in data-literature information exchange. D-Lib Mag 23:1

Butler D (2000) BioMed central boosted by editorial board. Nature 405:384

Callaghan S et al (2014) Guidelines on recommending data repositories as partners in publishing research data. Int J Digit Curation 9:152–163

Colavizza et al (2019) The citation advantage of linking publications to research data. https://arxiv.org/abs/1907.02565

Chan A-W et al (2017) Association of trial registration with reporting of primary outcomes in protocols and publications. JAMA 318:1709–1711

Cobo E et al (2007) Statistical reviewers improve reporting in biomedical articles: a randomized trial. PLoS One 2:e332

Cousijn H et al (2017) A data citation roadmap for scientific publishers. BioRxiv. https://doi.org/10.1101/100784

Cousijn H et al (2018) A data citation roadmap for scientific publishers. Sci Data 5:180259

Dorch BF, Drachen TM, Ellegaard O (2015) The data sharing advantage in astrophysics. Proc Int Astron Union 11:172–175

Emerson GB et al (2010) Testing for the presence of positive-outcome bias in peer review: a randomized controlled trial. Arch Intern Med 170:1934–1939

Eyding D et al (2010) Reboxetine for acute treatment of major depression: systematic review and meta-analysis of published and unpublished placebo and selective serotonin reuptake inhibitor controlled trials. BMJ 341:c4737

Fang FC, Steen RG, Casadevall A (2012) Misconduct accounts for the majority of retracted scientific publications. Proc Natl Acad Sci U S A 109:17028–17033

Ferguson AR, Nielson JL, Cragin MH, Bandrowski AE, Martone ME (2014) Big data from small data: data-sharing in the "long tail" of neuroscience. Nat Neurosci 17:1442–1447

Freedman LP, Cockburn IM, Simcoe TS (2015) The economics of reproducibility in preclinical research. PLoS Biol 13:e1002165

Garza K, Fenner M (2018) Glad you asked: a snapshot of the current state of data citation. https://blog.datacite.org/citation-analysis-scholix-rda/

Giofrè D, Cumming G, Fresc L, Boedker I, Tressoldi P (2017) The influence of journal submission guidelines on authors' reporting of statistics and use of open research practices. PLoS One 12:e0175583

Goldacre B et al (2018) Compliance with requirement to report results on the EU Clinical Trials Register: cohort study and web resource. BMJ 362:k3218

Graf C (2018) How and why we're making research data more open. Wiley, Hoboken. https://www.wiley.com/network/researchers/licensing-and-open-access/how-and-why-we-re-making-research-data-more-open

Grant R, Smith G, Hrynaszkiewicz I (2019) Assessing metadata and curation quality: a case study from the development of a third-party curation service at Springer Nature. BioRxiv. https://doi.org/10.1101/530691

Grieneisen ML, Zhang M (2012) A comprehensive survey of retracted articles from the scholarly literature. PLoS One 7:e44118

Hardwicke TE et al (2018) Data availability, reusability, and analytic reproducibility: evaluating the impact of a mandatory open data policy at the journal Cognition. R Soc Open Sci 5:180448

Hrynaszkiewicz I, Cockerill MJ (2012) Open by default: a proposed copyright license and waiver agreement for open access research and data in peer-reviewed journals. BMC Res Notes 5:494

Hrynaszkiewicz I, Shintani Y (2014) Scientific Data: an open access and open data publication to facilitate reproducible research. J Inf Process Manag 57:629–640

Hrynaszkiewicz I, Busch S, Cockerill MJ (2013) Licensing the future: report on BioMed Central's public consultation on open data in peer-reviewed journals. BMC Res Notes 6:318

Hrynaszkiewicz I, Li P, Edmunds SC (2014) In: Stodden V, Leisch F, Peng RD (eds) Implementing reproducible research. CRC Press, Boca Raton

Hrynaszkiewicz I, Khodiyar V, Hufton AL, Sansone S-A (2016) Publishing descriptions of non-public clinical datasets: proposed guidance for researchers, repositories, editors and funding organisations. Res Integr Peer Rev 1:6

Hrynaszkiewicz I et al (2017a) Standardising and harmonising research data policy in scholarly publishing. Int J Digit Curation 12:65

Hrynaszkiewicz I, Simons N, Goudie S, Hussain A (2017b) Research Data Alliance Interest Group: data policy standardisation and implementation. https://www.rd-alliance.org/groups/data-pol icy-standardisation-and-implementation

Hrynaszkiewicz et al (2019) Developing a research data policy framework for all journals and publishers. https://doi.org/10.6084/m9.figshare.8223365.v1

I4OC: initiative for open citations. https://i4oc.org/

Inchcoombe S (2017) The changing role of research publishing: a case study from Springer Nature. Insights 30:13–19

Ioannidis JPA et al (2009) Repeatability of published microarray gene expression analyses. Nat Genet 41:149–155

Johnson R, Focci M, Chiarelli A, Pinfield S, Jubb M (2017) Towards a competitive and sustainable open access publishing market in Europe: a study of the Open Access Market and Policy Environment. OpenAIRE, Brussels, p 77

Kidwell MC et al (2016) Badges to acknowledge open practices: a simple, low-cost, effective method for increasing transparency. PLoS Biol 14:e1002456. https://doi.org/10.1371/journal. pbio.1002456

Kirk R, Norton L (2019) Supporting data sharing. NPJ Breast Cancer 5:8

Kovanis M, Porcher R, Ravaud P, Trinquart L (2016) The global burden of journal peer review in the biomedical literature: strong imbalance in the collective enterprise. PLoS One 11:e0166387

Leonelli S (2016) Open data: curation is under-resourced. Nature 538:41

Lewis LM et al (2018) Replication study: transcriptional amplification in tumor cells with elevated c-Myc. Elife 7

Li T et al (2016) Review and publication of protocol submissions to trials – what have we learned in 10 years? Trials 18:34

Lin J (2018) Preprints growth rate ten times higher than journal articles. https://www.crossref.org/ blog/preprints-growth-rate-ten-times-higher-than-journal-articles/

Luther J (2017) The stars are aligning for preprints – the scholarly kitchen. https://scholarlykitchen. sspnet.org/2017/04/18/stars-aligning-preprints/

Macleod MR, The NPQIP Collaborative Group (2017) Findings of a retrospective, controlled cohort study of the impact of a change in Nature journals' editorial policy for life sciences research on the completeness of reporting study design and execution. BioRxiv. https://doi.org/ 10.1101/187245

Markowetz F (2015) Five selfish reasons to work reproducibly. Genome Biol 16:274

McGauran N et al (2010) Reporting bias in medical research – a narrative review. Trials 11:37

McNutt M (2014) Journals unite for reproducibility. Science 346:679

Michener WK (2015) Ten simple rules for creating a good data management plan. PLoS Comput Biol 11:e1004525

Munafò MR et al (2017) A manifesto for reproducible science. Nat Hum Behav 1:0021

Nature (2015) Ctrl alt share. Sci Data 2:150004

Nature (2017) Extending transparency to code. Nat Neurosci 20:761

Nature (2018) Checklists work to improve science. Nature 556:273–274

Naughton L, Kernohan D (2016) Making sense of journal research data policies. Insight 29:84–89

Nosek BA, Lakens D (2014) Registered reports. Soc Psychol 45:137–141

Nosek B et al (2014) Transparency and openness promotion (TOP) guidelines. https://osf.io/ xd6gr/?_ga=2.251468229.297610246.1542300800-587952028.1539080384

Nuijten MB et al (2017) Journal data sharing policies and statistical reporting inconsistencies in psychology. Collabra Psychol 3:31

Peng RD (2009) Reproducible research and biostatistics. Biostatistics 10:405–408

Percie du Sert N et al (2018) Revision of the ARRIVE guidelines: rationale and scope. BMJ Open Sci 2:e000002

Pienta AM, Alter GC (2010) The enduring value of social science research: the use and reuse of primary research data. Russell. http://141.213.232.243/handle/2027.42/78307

Piwowar H (2013) Altmetrics: value all research products. Nature 493:159

Piwowar HA, Vision TJ (2013) Data reuse and the open data citation advantage. PeerJ 1:e175

Piwowar HA, Day RS, Fridsma DB (2007) Sharing detailed research data is associated with increased citation rate. PLoS One 2:e308

Polanin JR, Terzian M (2018) A data-sharing agreement helps to increase researchers' willingness to share primary data: results from a randomized controlled trial. J Clin Epidemiol 106:60–69

Prinz F, Schlange T, Asadullah K (2011) Believe it or not: how much can we rely on published data on potential drug targets? Nat Rev Drug Discov 10:712

Rathi V et al (2012) Clinical trial data sharing among trialists: a cross-sectional survey. BMJ 345: e7570

Robertson M (2017) Who needs registered reports? BMC Biol 15:49

Rowhani-Farid A, Barnett AG (2016) Has open data arrived at the British Medical Journal (BMJ)? An observational study. BMJ Open 6:e011784

Rowhani-Farid A, Barnett AG (2018) Badges for sharing data and code at biostatistics: an observational study. [version 2; referees: 2 approved]. F1000Res 7:90

Rowhani-Farid A, Allen M, Barnett AG (2017) What incentives increase data sharing in health and medical research? A systematic review. Res Integr Peer Rev 2:4

Savage CJ, Vickers AJ (2009) Empirical study of data sharing by authors publishing in PLoS journals. PLoS One 4:e7078

Schmidt B, Gemeinholzer B, Treloar A (2016) Open data in global environmental research: the Belmont forum's open data survey. PLoS One 11:e0146695

Science D et al (2017) The state of open data report 2017. Digital Science, London

Scientific Data (2019) Scientific Data recommended repositories. https://figshare.com/articles/Scientific_Data_recommended_repositories_June_2015/1434640

SciGraph | For Researchers | Springer Nature. https://www.springernature.com/gp/researchers/scigraph

Shamseer L, Hopewell S, Altman DG, Moher D, Schulz KF (2016) Update on the endorsement of CONSORT by high impact factor journals: a survey of journal "Instructions to Authors" in 2014. Trials 17:301

Shotton D (2013) Publishing: open citations. Nature 502:295–297

Simera I et al (2010) Transparent and accurate reporting increases reliability, utility, and impact of your research: reporting guidelines and the EQUATOR Network. BMC Med 8:24

Smith R (2010) Classical peer review: an empty gun. Breast Cancer Res 12(Suppl 4):S13

Smith G, Grant R, Hrynaszkiewicz I (2018) Quality and completeness scores for curated and non-curated datasets. https://figshare.com/articles/Quality_and_completeness_scores_for_curated_and_non-curated_datasets/6200357

Springer Nature (2018) Springer Nature launches Open data badges pilot – Research in progress blog. http://blogs.biomedcentral.com/bmcblog/2018/10/08/springer-nature-launches-open-data-badges-pilot/

STM, ALPSP (2006) Databases, data sets, and data accessibility – views and practices of scholarly publishers. https://www.stm-assoc.org/2006_06_01_STM_ALPSP_Data_Statement.pdf

Stuart D et al (2018) Whitepaper: practical challenges for researchers in data sharing. https://figshare.com/articles/Whitepaper_Practical_challenges_for_researchers_in_data_sharing/5975011

Suber P (2012) Open access. MIT Press, Cambridge. http://mitpress.mit.edu/sites/default/files/titles/content/9780262517638_sch_0001.pdf

Taichman DB et al (2017) Data sharing statements for clinical trials. BMJ 357:j2372

Teixeira da Silva JA (2015) Negative results: negative perceptions limit their potential for increasing reproducibility. J Negat Results Biomed 14:12

Text and Data Mining | Springer Nature | For Researchers | Springer Nature. https://www.springernature.com/gp/researchers/text-and-data-mining

Text and Data Mining Policy – Elsevier. https://www.elsevier.com/about/policies/text-and-data-mining

The NPQIP Collaborative Group (2019) Did a change in Nature journals' editorial policy for life sciences research improve reporting? BMJ Open Sci 3:e000035

UK Research and Innovation (2011) Common principles on data policy. https://www.ukri.org/funding/information-for-award-holders/data-policy/common-principles-on-data-policy/

Vanpaemel W, Vermorgen M, Deriemaecker L, Storms G (2015) Are we wasting a good crisis? The availability of psychological research data after the storm. Collabra 1

Vasilevsky NA, Minnier J, Haendel MA, Champieux RE (2017) Reproducible and reusable research: are journal data sharing policies meeting the mark? PeerJ 5:e3208

Vines TH et al (2013) Mandated data archiving greatly improves access to research data. FASEB J 27:1304–1308

Whitlock MC, McPeek MA, Rausher MD, Rieseberg L, Moore AJ (2010) Data archiving. Am Nat 175:145–146

Wicherts JM, Borsboom D, Kats J, Molenaar D (2006) The poor availability of psychological research data for reanalysis. Am Psychol 61:726–728

Wiley Open Science Researcher Survey (2016) https://figshare.com/articles/Wiley_Open_Science_Researcher_Survey_2016/4748332/2

Quality Governance in Biomedical Research

Anja Gilis

Contents

Abstract

Quality research data are essential for quality decision-making and thus for unlocking true innovation potential to ultimately help address unmet medical needs.

The factors influencing quality are diverse. They depend on institution type and experiment type and can be of both technical and cultural nature. A well-thought-out governance mechanism will help understand, monitor, and control research data quality in a research institution.

In this chapter we provide practical guidance for simple, effective, and sustainable quality governance, tailored to the needs of an organization performing nonregulated preclinical research and owned by all stakeholders.

GLP regulations have been developed as a managerial framework under which nonclinical safety testing of pharmaceutical and other products should be conducted. One could argue whether these regulations should be applied to all nonclinical biomedical studies. However, the extensive technical requirements of GLP may not always be fit to the wide variety of studies outside the safety arena

A. Gilis (✉)
Janssen Pharmaceutica NV, Beerse, Belgium
e-mail: agillis@its.jnj.com

© The Author(s) 2019
A. Bespalov et al. (eds.), *Good Research Practice in Non-Clinical Pharmacology and Biomedicine*, Handbook of Experimental Pharmacology 257,
https://doi.org/10.1007/164_2019_291

and may be seen as overly prescriptive and bureaucratic. In addition, GLP regulations do not take into account scientific excellence in terms of study design or adequacy of analytical methods. For these reasons and in order to allow a lean and fit for purpose approach, the content of this chapter is independent from GLP. Nevertheless, certain topics covered by GLP can be seen as valuable across biomedical research. Examples are focus on transparency and the importance of clear roles and responsibilities for different functions participating in a study.

Keywords

Change management · Fit for purpose approach · Quality governance · Research data quality · Sustainability

1 What Is Quality Governance?

The term governance is derived from the Greek verb *kubernaein [kubernáo]* (meaning to steer), and its first metaphorical use (to steer people) is attributed to Plato. Subsequently the term gave rise to the Latin verb *gubernare*, and from there it was taken over in different languages. It is only since the 1990s that the term governance is used in a broad sense encompassing different types of activities in a wide range of public and private institutions and at different levels (European Commission, http://ec.europa.eu/governance/docs/doc5_fr.pdf).

As a result of the broad application of the term governance, there are multiple definitions, among which the following is a good example (http://www.businessdictionary.com/definition/governance.html): "Establishment of policies, and continuous monitoring of their proper implementation, by the members of the governing body of an organization. It includes the mechanisms required to balance the powers of the members (with the associated accountability), and their primary duty of enhancing the prosperity and viability of the organization."

In simple terms, governance is (a) a means to monitor whether you are on a good path to achieve the intended outcomes and (b) a means to steer in the right direction so that you proactively prevent risks turning into issues. It is almost like looking in the mirror every morning and making sure you look *ok* for whatever your plans are that day, or like watching your diet and getting regular exercise in order to keep your cholesterol levels under control.

How does this translate into quality governance? Quality simply means fitness for purpose; in other words the end product of your work should be fit for the purpose it is meant to serve. In experimental pharmacology, this means that your experimental outcomes should be adequate for supporting conclusions and decisions such as on the validity of a molecular target for a novel treatment approach, on the generation of a mode of action hypothesis for a new drug, or on the safety profile of a pharmaceutical ingredient. The different activities you undertake from planning of your experiment, generating the raw data, processing these data to ultimately reporting experimental outcomes and conclusions should be free from bias and should be documented in a way that allows full reconstruction of the experiment.

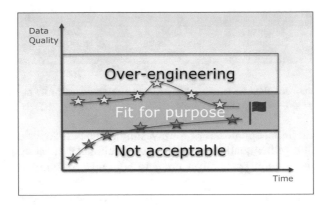

Fig. 1 A simple visualization of the purpose and concept of quality governance. Quality governance is the answer to the question: "How do I get in the green zone and stay there?" Each star represents a point in time where the organization reflects on or measures its quality level against what is considered fit for purpose. Over time, the measured outcomes will likely change, and when the outcome is in the "not acceptable" or "overengineering" zone, actions need to be taken to move back into the green fit for purpose zone. The curves represent theoretical examples of measured quality levels over time. The lower curve reflects an institution that has taken small continuous improvement steps to move data quality from not acceptable to fit for purpose level. The upper curve represents an institution that at a certain point in time finds itself "overengineering" and course-corrects to an appropriate level of quality

Quality governance in the context of this book means potential ways in which institutions can monitor research data quality over time and have a mechanism in place to detect and to deal with signals of drift. The purpose and concept of quality governance can be visualized such as in Fig. 1.

The definition of quality governance implies that, although very important, having policies or guidelines for good research practices is not sufficient. Such documents can be seen as an important building block for good quality research data. However, policies or guidelines will not reach their full impact potential and will not be sustainable over time when the monitoring component of governance is missing. Three aspects of governance are equally important: (1) there needs to be a mechanism to check whether people are applying the guidelines, (2) there needs to be a mechanism to make sure the guidelines remain adequate over time, and (3) there needs to be a mechanism to take the right actions when deviations are seen in (1) and (2). Having these three aspects of governance in place is expected to increase the likelihood of long-term sustainability and full engagement by those who are expected to apply the guidelines.

Common to all effective quality governance systems is the attention to both cultural (acceptance, engagement, sense of responsibility, etc.) and technical (guidelines, procedures, equipment, research data storage, etc.) aspects of quality. It is definitely worth investing time in effective quality governance system as it will help achieve quality research data (data fit for their purpose). Quality data lead to quality decision-making (for acceptance of the best publications or for grants to be

Fig. 2 Quality research data leads to quality decision-making which in its turn is required for unlocking optimal innovation potential

given to the best project proposals or for investing in the best active pharmaceutical ingredient). All together quality research data will ultimately lead to unlocking the best possible innovation potential in order to help address unmet medical needs (Fig. 2).

The size and type of organization (university, biotech, pharma, contract research, etc.) and the type of work that is being conducted (exploratory, confirmatory, in vivo, in vitro, etc.) will determine the more specific quality governance needs. The next parts of this chapter are meant to offer a stepwise and practical approach to install an effective and tailor-made quality governance approach.

2 Looking in the Quality Mirror (How to Define the Green Zone)

How do you know where your lab or your institution is positioned respective to the different data quality zones schematically introduced in Fig. 1? The first thing to do is to define what success means in your specific situation. In quality management terms, success is about making sure you have controlled the risks that matter to your organization. Therefore, one may start with clearly defining the risks that matter. "What is at stake?" is a good question to trigger this thinking.

2.1 What Is at Stake?

Risks can vary according to the organization type. However, the following risks are generally recognized for organizations conducting biomedical research:

(a) First and foremost comes the risk to *patients' safety*. Poor research data quality can have dramatic consequences as exemplified in 2015 in Rennes, France, where a first-in-human trail was conducted by contract research organization Biotrial on behalf of Portuguese pharmaceutical company Bial. During the trial, six healthy volunteers were hospitalized with severe neurological injuries after receiving an increased dose of the investigational compound. One patient died as a result (Regulatory Focus TM News articles 2016: https://www.raps.org/regulatory-focus%E2%84%A2/news-articles/2016/5/ema-begins-review-of-first-in-human-trial-safety-following-patient-death). The assigned investigation committee concluded that several errors and mistranslations from source documents were present in the IB that made it difficult to understand. The committee recommended enhanced data transparency and sufficiently complete preclinical safety studies (Schofield 2016).

(b) Another important risk in preclinical research is related to *animal welfare*. Animals should be treated humanely, and care should be taken that animal experiments are well designed and meticulously performed and reported to answer specific questions that cannot be answered by other means. Aspects of animal care and use can also have direct impact on research results. This topic is discussed in detail in chapter "Good Research Practice: Lessons from Animal Care and Use".

(c) There can be damage to *public trust and reputation* of the institution and/or individual scientist, for example, in case a publication is retracted. Because retraction is often perceived as an indication of wrongdoing (although this is not always the case), many researchers are understandably sensitive when their paper(s) is questioned (Brainard 2018). Improved oversight at a growing number of journals is thought to be the prime factor in the rise of annual retractions. Extra vigilance is required in case of collaborations between multiple labs or in case of delegation of study conduct, for example, to junior staff.

(d) *Business risks* may vary depending on the type of organization. The immediate negative financial consequences of using the non-robust methods (e.g., poor or failing controls) are unnecessary repeats and overdue timelines. More delayed consequences can go as far as missed collaboration opportunities. Another example can be a failure to obtain grant approvals when quality criteria of funders are not met. For biotechnological or pharmaceutical companies, there is the risk of inadequate decision-making based on poor quality data and thus investing in assets with limited value, delaying the development of truly innovative solutions for patients in need.

(e) In terms of *intellectual property*, insufficient data reconstruction can lead to refusal, unenforceability, or other loss of patent rights (Quinn 2017). In a case of a patent attack, it is essential to have comprehensive documentation of research data in order to defend your case in court, potentially many years after the tests were performed. Also, expert scientists are consulted in court and may look for potential bias in your research data.

(f) Pharmacology studies, with the exception of certain safety pharmacology studies, are not performed according to GLP regulations (OECD Principles of Good Laboratory Practices as revised in 1997 Organisation for Economic Co-operation and Development. ENV/MC/CHEM (98)17).

A large part of regulatory submission files for new drug approvals therefore consist of non-GLP studies. It is, however, important to realize that, also for non-GLP studies, the regulators do expect that necessary measures are taken to have trustworthy, unbiased outcomes and that research data are retrievable on demand. The Japanese regulators are particularly very strict when it comes to availability of data (Desai et al. 2018). *Regulators* have the authorization by law to look into your research data. They will look for data traceability and integrity. If identified, irregularities can affect the review and approval of regulatory submissions and may lead to regulatory actions that go beyond the product/ submission involved.

The above-mentioned risks are meant as examples, and the list is certainly not meant to be all inclusive.

When considering what is at stake for your specific organization, you may think just as well in terms of opportunities. As the next step, similar to developing mitigations for potential risks, you can then develop a plan to maximize the benefits or success of the opportunities you have identified, for example, enhancing the likelihood of being seen as the best collaboration partner, getting the next grant, or having the next publication accepted.

2.2   What Do You Do to Protect and Maximize Your Stakes?

When you have a clear view on the risks (and opportunities) in your organization, the next step is to start thinking about what it is that affects those stakes; in other words, what can go wrong so that your key stakes are affected. Or, looking from a positive side, what can be done to maximize the identified opportunities?

Depending on the size and complexity of your organization, the approach to this exercise may vary. In smaller organizations, you may get quick answers from a limited number of subject matter experts. However, in larger organizations, it may take more time to get the full picture. In order to get the best possible understanding about quality-related factors that influence your identified risks, it is recommended to perform a combination of interviews and reviews of research data and documents such as protocols and reports. Most often, reviews will help you find gaps, and interviews will help you define root causes.

Since you will gather a lot of information during this exercise, it is important to be well prepared, and it is advisable to define a structure for collecting and analyzing your findings.

When reviewing research data, you can group your findings in categories such as, for example:

(a) *Data storage*: Proper storage of research data can be considered as a way to safeguard your research investment. Your data may need to be accessed in the future to explain or augment subsequent research, or other researchers may wish to evaluate or use the results of your research. Some situations or practices may result in inability to retrieve data, for example, storage of electronic data in personal folders or on non-networked instrument computers that are only accessible to the scientist involved and temporary researchers coming and going without being trained on data storage practices.

(b) *Data retrieval*: It is expected that there is a way to attribute reported outcomes (in study reports and publications) and conclusions to experimental data. An easy solution is to assign to all experiments a unique identification number from the planning phase onwards. This unique number can then be used in all subsequent recordings related to data capture, processing, and reporting.

(c) *Data reconstruction*: It is expected that an independent person is able to reconstruct the presented graphs, figures, and conclusions with the information in the research data records. If a scientist leaves the institution, colleagues should be able to reconstruct his/her research outcomes. Practices that may hinder the reconstructions are, for example, missing essential information such as timepoints of blood collection, missing details on calculations, or missing batch ID's of test substances.

(d) *Risk for bias*: According to the American National Standard (T1.523-2001), the definition of bias is as follows: "Data (raw data, processed data and reported data) should be unchanged from its source and should not be accidentally or maliciously modified, altered or destroyed". Bias can occur in data collection, data analysis, interpretation, and publication which can cause false conclusions. Bias can be either intentional or unintentional (Simundic 2013). Intention to introduce bias into someone's research is immoral. Nevertheless, considering the possible consequences of a biased research, it is almost equally irresponsible to conduct and publish a biased research unintentionally. Common sources of bias for experiments (see chapter "Resolving the Tension Between Exploration and Confirmation in Preclinical Biomedical Research") are lack of upfront defined test acceptance criteria or criteria and documentation for inclusion/exclusion of data points or test replicates, use of non-robust assays, or multiple manual copy-paste steps without quality control checks. Specifically, for animal intervention studies, selection, detection, and performance bias are commonly discussed, and the SYRCLE risk of bias tool has been developed to help assess methodological quality (Hooijmans et al. 2014). It is strongly advised to involve experts for statistical analysis as early as the experimental planning phase. Obviously, every study has its confounding variables and limitations that cannot completely be avoided. However, awareness and full transparency on these known limitations is important.

(e) *Review, sign off, and IP protection*: Entry of research data in (electronic) lab notebooks and where applicable (for intellectual property reasons) witnessing of experiments is expected to occur in a timely manner (e.g., within 1 month of experimental conduct).

Likewise, when conducting interviews, you may want to consider having some general questions upfront to trigger the thinking and facilitate obtaining information that helps define route causes and thus focus areas for later actions. Some examples:

- Culture and communication
  - How is the rewarding system set up and how is it perceived? Is there a feeling that "truth seeking" behavior and good science is rewarded and seen as a success rather than positive experimental outcomes or artificial milestones? Which are the positive and negative incentives that people in this lab/institution experience?
  - How are new employees trained on the importance of quality?

- Where can employees find information on quality expectations?
- Does leadership enforce the importance of quality? Is leadership perceived to "walk the walk and talk the talk"?
- Is there a mechanism to prevent conflict of interest? Are there any undue commercial, financial, or other pressures and influences that may adversely affect the quality of research work?
- Management of resources
  - Is there sufficient personnel with the necessary competence, skills, and qualifications?
  - Are facilities and equipment suitable to deliver valid results? How is this being monitored?
  - Do computerized systems ensure integrity of data?
- Data storage
  - Is it clear to employees which research data to store?
  - Is it clear to employees where to store research data?
  - Are research data stored in a safe environment?
- Data retrieval
  - How easy is it to retrieve data from an experiment you performed last month? 1 year ago? 5 years ago?
  - How easy is it to retrieve data from an experiment conducted by your peer?
  - How easy is it to retrieve data from an experiment conducted by the postdoc who left 2 years ago?
- Data reconstruction
  - Is it clear to employees to which level of detail individual experiments should be recorded and reported?
  - Is there a process for review and approval of reports or publications?
  - During data reviews, is there attention to the ability to reconstruct experimental outcomes by following the data chain starting from protocol and raw data over processed data to reported data?
- Bias prevention
  - In study reports, is it common practice to be transparent about the number of test repeats and their outcomes regardless of whether the outcome was positive or negative?
  - In study reports, is it common practice to be transparent about statistical power, known limitations of scientific or statistical methods, etc.?
  - Are biostatistical experts consulted during study setup, data analysis, and reporting in order to ensure adequate power calculations, use of correct statistical methods, and awareness on limitations of certain methods?
  - Is there a process for review and approval of reports and publications?
  - Is there a mechanism to raise, review, and act upon any potential shortcomings in responsible conduct of research?
- Collaborations
  - Is it common practice to communicate with collaborators/partners/subcontractors on research data quality expectations?

Performing a combination of research data reviews and interviews will require some time investments but will lead to a good understanding of how your current practices affect what is at stake for your specific institution. Oftentimes, while performing this exercise, the root causes for the identified gaps will become clear. For example, people store their research data on personal drives because there are no agreements or IT solutions foreseen for central data storage, or data reconstruction is difficult because new employees are not being systematically mentored or trained on why this is important and what level of documentation is expected.

Good reading material that links in with this chapter is ICHQ9 (https://www.ich. org/fileadmin/Public_Web_Site/ICH_Products/Guidelines/Quality/Q9/Step4/Q9_ Guideline.pdf), a comprehensive guideline on quality risk management including principles, process, tools, and examples.

3 Fixing Your Quality Image (How to Get in the Green Zone)

After having looked in the quality mirror, you can ask yourself: "Do I like what I see?"

Like many things in life, quality is to be considered a journey rather than a destination and requires a continued level of attention. Therefore, it is very likely that your assessment outcome indicates that small or large adjustments are advisable.

It is important to realize that quality cannot be the responsibility of one single person or an isolated group of quality professionals in your organization. Quality is really everyone's responsibility and should be embedded in the DNA of both people and processes. The right people need to be involved in order to have true impact.

Therefore, after analyzing your quality image, it is important to articulate very clearly what you see, why improvement is needed, and who needs to get involved. Also, one should not forget to bring a well-balanced message, adapted to the target audience, and remember to emphasize what is already good.

Another common mistake is to think that research quality principles can be installed just by writing a policy, generating procedures and guidelines, and organizing training sessions, so everyone knows what to do. This may look good on paper and may even be designed to be lean and fit-for-purpose. However, building quality into the everyday activities and at all levels requires to go beyond work instructions and policies. It is the emotional connection that is the basis of success (Fig. 3) since this will make sure that quality is built into the thinking and actions of everyone involved. One can never foresee all potential situations and exceptions in a procedure, but having people's mindset right will trigger the right behavior in any situation.

Here are some hints and tips and practical examples that have been shown to be well received:

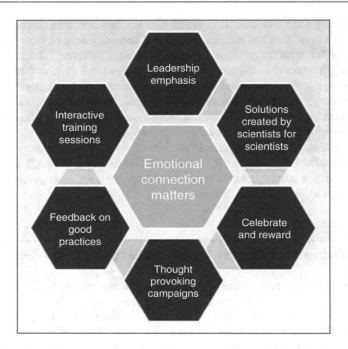

Fig. 3 Schematic representation of emotional connection elements that are relevant to consider while enhancing quality approaches. Building quality into the everyday activities and at all levels requires emotional connection. The basis of a healthy research climate is people's mindset, as the right mindset will trigger the right behavior in any situation

– Use positive language

Positive communication is key to obtain engagement. Phrases such as "How can we set ourselves up for success?" or "How can we increase our potential for innovation?" are much more effective than "We need to follow these rules for data recording."

– Use the scientists' skills

Scientists are excellent problem solvers. This skill can also do magic beyond the scientific area of expertise such as on quality-related topics. The only trigger that is needed for the scientist is the awareness that there is a situation that needs to be addressed. In this context, quality professionals are advised not to impose solutions on scientists, especially when these solutions may be perceived with even a tiny piece of increased bureaucracy. And, in the end, scientists usually know best what works well in their environment.

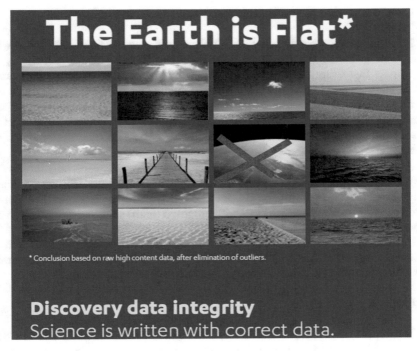

Fig. 4 Example of utilization of visuals and analogies to research data quality. This example was derived from a campaign at Janssen Pharmaceutica N.V., appealing to scientists' creativity to visualize quality. It is meant to trigger thinking on appropriateness and best practices for outlier exclusion

– Use catchy visuals or analogies

As an example, a campaign appealing to scientist's creativity can result in very engaging visuals that can be displayed in labs and corridors (example in Fig. 4) and will become the talk of the town. Positivity, creativity, and fun are key elements in building a research quality culture.

– Support from the top

The first question you will be asked by your institutions' leaders is "Why is this important?". You need a clear outline of what is at stake, what you learned during the current state analysis, and where you see gaps and a plan for change. It is best to include real examples of what can go wrong, what is currently already working well, and where improvement is still possible. Without support of the institutions' top leadership, the next steps will be extremely difficult or even impossible.

– Choose your allies – and be inclusive

It is highly likely that during your initial analysis, you have met individuals who were well aware of (parts of) the gaps and have shown interest and energy to participate in finding solutions. Chances are high that they have already started taking steps toward improvement. It is key to involve these people as you move on.

It is also likely that your initial analysis showed that, in certain parts of the organization, people have already put in place solutions for gaps that still exist in other parts of the organization. These can serve as best practice examples going forward.

Also, you may have come across some very critical people, especially those who are afraid of additional workload and bureaucracy. Although this may sound counterintuitive, it is of high value to also include this type of people in your team or to reach out to them at regular intervals with proposed (draft) solutions and ask for their input.

Finally, since the gaps you have discovered may be of diverse nature, it is important to involve experts from other disciplines such as biostatisticians, communication professionals, patent attorneys, procurement experts, IT experts, etc.

– Plan and prioritize

The gaps, best practices, and route causes defined during the initial analysis will be the starting point for change. Remember that route causes are not necessarily always of technical nature. Cultural habits or assumptions may be equally or even more important and will often take more time and effort to resolve.

It is also important to try not to change everything at once. It is better to take small incremental steps and focus on implementing some quick wins first.

Apart from prioritizing the topics that need to be tackled, it is good practice to define roles and responsibilities as well as a communication plan. For a large organization, it may be helpful to agree on a governance model for which an example is given in Fig. 5.

Change is never an easy journey, and awareness on change management principles will be helpful to achieve the ultimate goal. On the web there is a lot of useful material on change management, such as the Kotter's 8-step change model (https://www.mindtools.com/pages/article/newPPM_82.htm) that explains the hard work required to change an organization successfully. Other useful references are the RQA (Research Quality Association) booklet "Quality in research: guidelines for working in non-regulated research" that can be purchased via the RQA website (https://www.therqa.com/resources/publications/booklets/quality-in-research-book let/) and the RQA quality systems guide that can be downloaded for free (https://www.therqa.com/resources/publications/booklets/Quality_Systems_Guide/). Careful planning and building the proper foundations is key. Equally important is creating a sense of urgency and effective communication (real examples, storytelling). Having some quick wins will help to build on the momentum and increase enthusiasm.

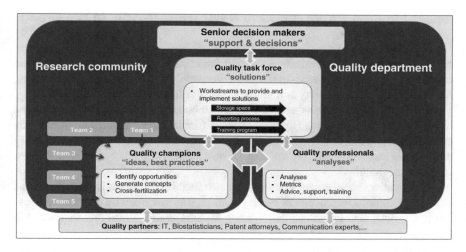

Fig. 5 Example of a governance model for the design phase of quality solutions in a large organization. There is a close collaboration between the quality department and the research community where the quality department performs analyses (such as described in Sect. 2. Looking in the quality mirror) within the different research teams. Each research team assigns a *quality champion* who works closely with the *quality professionals* and quality champions from other research teams to identify opportunities for improvement, generate concepts to fill gaps, and exchange best practices (cross-fertilization). A *quality task force* is formed consisting of both quality champions, quality professionals, and representatives from other stakeholder organizations ("*quality partners*"). This quality task force works on solutions for the entire research community, such as research data storage, research data reporting processes, or a training program according to the principles described in Sect. 3. Fixing your quality image. The quality task force provides regular updates to *senior decision-makers* whose approval is required before implementation

– Keep it simple

Try not to make guidelines too descriptive or specific and rather go with general guidance. For example, it is essential that equipment is suitable for intended use and that experimental records provide sufficient details to enable reconstruction. What exactly this means depends on the type of equipment, for what purpose it is used, and what type of experiment is conducted. General templates to document equipment maintenance or reporting templates may be helpful tools for scientists; however, it is key not to go in too much detail when specifying expectations.

– Make training fun

Interactive and fun activities during training sessions and hands-on workshops are generally well appreciated by the scientists. As an example, a quiz can be built into the training session during which scientists can work in small teams and get a small prize at the end of the session.

4   Looking Good! All Done Now? (How to Stay in the Green Zone)

After having led your organization through the change curve, you can sit back and enjoy the change you had envisioned. However, don't stay in this mode for too long. As mentioned before, quality is a journey, not a destination.

Today, it may look as if your solutions are embedded in the way people work and have become the new norm. However, 2 years from now, your initial solutions may no longer work for new technologies, newly recruited staff may not have gone through the initial roll out and training, or new needs may show up for which new solutions are required.

For a quality management system to be sustainable over time, it needs to have a built-in continuous improvement mechanism, such as described by the PDCA (plan–do–check–act), also known as the Deming cycle (Deming 1986) and visualized in Fig. 6.

First, the quality solutions need to be sustainably integrated in the way of working with clear roles and responsibilities (DO). In accordance with what has been established in the previous section (Fixing your quality image), clear quality expectations in the form of policies, best practices, or guidelines should be available to the organization, and mechanism for training of new employees and for refresher trainings of existing employees should be in place.

Ideally, all scientists advocate and apply these best practice solutions and also communicate expectations and monitor their application in their external collaborations.

Secondly, there needs to be a mechanism to monitor adherence to the green zone of Fig. 1 (CHECK). When having expectations in place, it is not sufficient to just assume that everyone will now follow them. There may be strong quantitative pressures

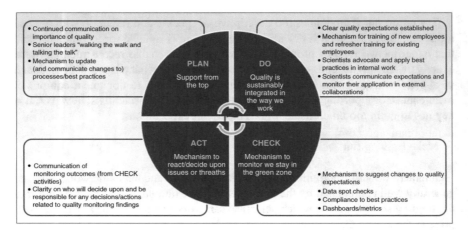

Fig. 6 Key attributes of a mature quality system represented by means of the Deming cycle (Deming 1986)

(time) that tend to make people act differently and may lead to quality being compromised. If people know that there is a certain likelihood that their work will be checked, this will provide a counterbalance for such quantitative pressures. Recent simulations support the idea that a detailed scrutiny of a relatively small proportion of research output could help to affordably maintain overall research quality in a competitive research world where quantity is highly prized (Barnett et al. 2018).

Who performs these checks may depend on the type of organization. For smaller organizations, there can be a self-assessment checklist, or a person within the lab can be assigned to do spot checks for a percentage of their time. In larger organizations, there can be an independent person or group of people (e.g., belonging to the quality department) that can perform spot checks. The benefit of a person performing these checks across different labs can be that certain practices that work well within one lab can be identified and shared with another lab that is in need of a practical solution. Such a person can also serve as go-to person for advice (e.g., on how to deal with specific situation). Whoever is performing checks, it is important to build in an element of positive reinforcement, making sure that people are recognized for applying best practices and not feel blamed when something went wrong. Also, it is important to make people "part of" the exercise and discuss any unclarities with them. This will maximize the learning opportunity as it is much more powerful to say "I am trying to reconstruct this experiment, but I am having some difficulties for which I would like to ask for your help" than to say "I noticed you forgot to document your 96-well plate layout." When performed in the right way, these checks can be a learning exercise and provide a counter-incentive for time pressure. To clarify the latter point, in contrast to quantitative metrics (such as the number of publications or the number of projects completed within business timelines), most institutions are missing metrics on data quality. When checks are performed on agreed upon quality standards (such as timely recording of data or data exclusion criteria setting before experimental conduct), the outcomes of these checks can be used as a metric for data quality. This way, the common quantitative metrics will no longer be the only drivers within a research lab, and as such, quality metrics can provide a platform to discuss route causes for poor quality data such as time pressure.

It is also worth to consider measuring the "quality culture" or "research climate" in your institution. For this purpose, the SOuRCE (Survey of Organizational Research Climate) (Baker 2015) may be useful. This is a validated 32-question survey for which responses have been shown to correlate with self-reported behavior in the conduct of research. As such, it can provide a snapshot of the research climate in an organization through the aggregated perspectives of its members.

Other applicable indicators of quality that can be considered are, for example, the degree to which (electronic) lab notebooks are used and reviewed within preset timelines or the attendance of trainings.

Besides checking whether expectations are being followed, it is equally important to make it clear to people that the quality expectations are not all carved in stone and, for good reasons, some expectations may need to be refined, altered, deleted, or added over time. Everyone should feel empowered and know how to suggest a justified change to the quality expectations. Changes to quality expectation should not only be considered to prevent drift into the "not acceptable" zone but also to avoid or course-correct when drifting into the "overengineering zone." The latter is often forgotten. Overengineering can have different causes: overinterpretation of guidelines, mitigations staying in place while the risks have disappeared (e.g., QC checks still occurring long after automation took place), solutions that have been put in place being so complex that no one is following them, etc. Conversations are the best way to detect overengineering. Such conversations can be triggered by people voicing their frustration or coming with critical questions or suggestions on certain expectations. Root-cause conversations can also be started when finding multiple times the same issues during audits or a survey can be sent out from time to time to get feedback on feasibility of quality expectations.

Third, whenever drift is seen, or when a suggestion comes up to change a certain quality expectation, there must be a mechanism to react and make decisions (ACT).

Monitoring outcomes, be it from research data spot checks or from a cultural survey, should be communicated and analyzed. If needed, follow-up actions should be defined and implemented by responsible person(s).

Last but not least, the culture of quality needs to be kept alive, and expectations need to be updated as required (PLAN). For this purpose, regular communication on the importance of quality is crucial. This can be achieved in various ways, for example, by putting posters next to copy or coffee machines, by highlighting best practices during group meetings, by e-mailing relevant publications, or by inviting external speakers. Messages coming from mentors and leaders are most impactful in this respect as well as having people recognize that these leaders themselves are "walking the walk and talking the talk."

5   Conclusion

Quality governance, when installed successfully, can be a way to provide simple and sustainable solutions that facilitate data quality and promote innovation. The basic principles of quality governance are very similar across different disciplines; however, the practical application of quality governance is dependent on multiple variables. Currently, existing guidance on quality governance for research is limited and fragmented. As a result, institutions may have policies or guidelines in place, but there is often no mechanism to monitor their application. An exception is the animal care and use aspect of research where there is legislation as well as internal and external oversight bodies (see chapter "Good Research Practice: Lessons from Animal Care and Use"). Recently, the IMI project EQIPD (European Quality in Preclinical Data, https://quality-preclinical-data.eu/) has assembled a team of both industrial and academic researchers and quality professionals to work on practical solutions to improve preclinical data quality. One of their deliverables is a tool to help institutions set up a fit-for-purpose quality system including governance aspects, aligned with the information in this chapter. Until the team has delivered their tool, we hope the guidance provided above can be of help for institutions for bringing their research data quality to the right level.

References

Baker M (2015) Metrics for ethics. Nature 520:713

Barnett AG, Zardo P, Graves N (2018) Randomly auditing research labs could be an affordable way to improve research quality: a simulation study. PLoS One 13(4):e0195613. https://doi.org/10.1371/journal.pone.0195613

Brainard J (2018) Rethinking retractions. Science 362:390–395

Deming WE (1986) Out of the crisis. Massachusetts Institute of Technology, Center for Advanced Engineering Study, Cambridge, p 88

Desai KG, Obayashi H, Colandene JD, Nesta DP (2018) Japan-specific key regulatory aspects for development of new biopharmaceutical drug products. J Pharm Sci 107:1773–1786

Hooijmans CR, Rovers MM, de Vries RB, Leenaars M, Ritskes-Hoitinga M, Langendam MW (2014) SYRCLE's risk of bias tool for animal studies. BMC Med Res Methodol 14:43

Quinn G (2017) Patent drafting: understanding the enablement requirement. http://www.ipwatchdog.com/2017/10/28/patentability-drafting-enablement-requirement/id=89721/

Schofield I (2016) Phase I trials: French Body urges transparency, says don't just follow the rules. Scrip regulatory affairs. https://pink.pharmaintelligence.informa.com/PS118626/Phase-I-Trials-French-Body-Urges-Transparency-Says-Dont-Just-Follow-The-Rules

Simundic A-M (2013) Bias in research. Biochem Med 23(1):12–15

Good Research Practice: Lessons from Animal Care and Use

Javier Guillén and Thomas Steckler

Contents

Abstract

Animal care and use play a pivotal role in the research process. Ethical concerns on the use of animals in research have promoted the creation of a legal framework in many geographical areas that researchers must comply with, and professional organizations continuously develop recommendations on specific areas of laboratory animal science. Scientific evidence demonstrates that many aspects of animal care and use which are beyond the legal requirements have direct impact

J. Guillén
AAALAC International, Pamplona, Spain

T. Steckler (✉)
Janssen Pharmaceutica NV, Beerse, Belgium
e-mail: TSTECKLE@its.jnj.com

on research results. Therefore, the review and oversight of animal care and use programs are essential to identify, define, control, and improve all of these aspects to promote the reproducibility, validity, and translatability of animal-based research outcomes. In this chapter, we summarize the ethical principles driving legislation and recommendations on animal care and use, as well as some of these laws and international recommendations. Examples of the impact of specific animal care and use aspects on research, as well as systems of internal and external oversight of animal care and use programs, are described.

Keywords
Animal care and use · Animal studies · Interplay · Preclinical data quality · Reliability

1   Ethical and Legal Framework

The use of animals for research purposes has been a subject of debate for a long time. The increase of societal concerns on this matter has been and is being reflected in the development and implementation of guidelines and strict legislation on the protection of animals used in research across the world (Guillen 2017). Although international legislation may differ between countries in some practical aspects, they are all based on the same *ethical principles*, mainly the Three Rs of Replacement, Reduction, and Refinement (Russell and Burch 1959) and, at a lesser extent, the Five Freedoms (Brambell 1965). The Three Rs are explicitly mentioned in most important international guidelines and regulations:

– *Replacement* refers to the avoidance or replacement of the use of animals in experiments where otherwise they would have been used. However, the concept of "relative replacement" based on strategies focused on reduction of animals and refinement of procedures performed on animals is also valid.
– *Reduction* refers to minimizing the number of animals needed to obtain the desired research objectives. Reduction strategies are normally based on improvements of the experimental design and/or the implementation of new techniques (e.g., imaging). An important concept is not only using less number of animals, but the right number, as using too low numbers could invalidate the research results.
– *Refinement* refers to the implementation of housing (e.g., micro- and macroenvironment), care (e.g., husbandry practices and veterinary care), and use (experimental techniques) procedures that minimize animal pain and distress.

The Five Freedoms (from hunger and thirst; from discomfort; from pain, injury, or disease; from fear and distress; and from the ability to express normal behavior), although initially proposed for the farm animal environment, are also referred to in some legislation on research animals, especially in Asia, but also have been recently considered for other related purposes, such as a recent proposal for the harm-benefit analysis to be performed as part of the ethical evaluation of research projects (Brønstad et al. 2016; Laber et al. 2016).

1.1 Recommendations for the Care and Use of Laboratory Animals

The common aspects of the ethical framework described above help with the harmonization of animal care and use in science, and of science itself through the implementation of *international guidelines and recommendations*, as well as compatible pieces of legislation (Guillen and Vergara 2017). Based on the current ethical concepts, the International Council for Laboratory Animal Science (ICLAS) collaborated with the Council for International Organizations for Medical Sciences (CIOMS) to update the International Guiding Principles for Biomedical Research Involving Animals, with the intention to guide emerging countries in developing a framework of responsibility and oversight on the use of animals in research and to serve as an international benchmark also in countries with well-developed animal-based research programs (Council for International Organizations for Medical Sciences and International Council for Laboratory Animal Science 2012). The Three Rs represent a significant aspect of this document. Also, the World Organization for Animal Health (OIE), with 180 member countries, recognizes the Five Freedoms as valuable guidance in animal welfare and more specifically describes the Three Rs and highlights their key role in the use of animals in science in Chap. 7.8 of the Terrestrial Animal Health Code (World Organization for Animal Health 2012).

One of the most widely followed sets of recommendations on animal care and use can be found in the *Guide for the Care and Use of Laboratory Animals* (*Guide*; National Research Council of the National Academy of Sciences 2011). The *Guide*, issued in the USA by the Institute for Laboratory Animal Research (ILAR) of the National Research Council (NRC), is the main nonregulatory reference in the USA, also serves as the basis for regulations and recommendations in other areas of the world (Guillen 2017), and is one of the primary standards for the accreditation of animal care and use programs across the world performed by the nonprofit organization AAALAC International (see below). The *Guide* refers to the Three Rs and to the US Government Principles for the Utilization and Care of Vertebrate Animals Used in Testing, Research, and Training (United States Government 1985), which already incorporate these same basic ethical principles. It offers recommendations on all areas of an animal care and use program, such as the institutional responsibilities (including the oversight process, the training of personnel, and the occupational health and safety program); the animal environment, housing, and management; the veterinary care; and the physical plant. The *Guide* states that it "is created by scientists and veterinarians for scientists and veterinarians to uphold the scientific rigor and integrity of biomedical research with laboratory animals as expected by their colleagues and society at large" and "establishes the minimum ethical, practice, and care standards for researchers and their institutions."

Many professional organizations publish more specific recommendations on particular areas of an animal care and use program, such as ethical review, health monitoring, or education and training. Especially noteworthy are the Federation of European Laboratory Animal Science Associations (FELASA; www.felasa.eu), the American Association for Laboratory Animal Science (AALAS; www.aalas.org), and the Canadian Council on Animal Care (CCAC; www.ccac.ca). Other organizations focus exclusively on the development and dissemination of Three Rs initiatives, like the NC3Rs in the UK (www.nc3rs.org.uk), the North American 3Rs Collaborative in the USA (http://www.na3rsc.org/home.html), or Norecopa in Norway (https://norecopa.no/).

1.2   Legislation in the USA

In the USA, legislation on the care and use of animals comes from the US Department of Agriculture (USDA) and the Public Health Service (Bradfield et al. 2017). The *Guide* details the requirements of the Public Health Service Policy (PHS Policy) on Humane Care and Use of Laboratory Animals (Public Health Service 2002) and is used by institutions to comply with the PHS Policy. Therefore, although the *Guide* is not a piece of legislation, its standards are considered as minimum requirements by the PHS. On the USDA side, the Animal Welfare Act and related Animal Welfare Regulations represent the only federal law in the USA that regulates the treatment of animals in research, with the particularity that rats, mice, and birds are not considered regulated species (United States Government 1966). However, the *Guide* standards are applied to all vertebrate species. The US system gives a lot of autonomy to the institutions and is based on the Institutional Animal Care and Use Committees (IACUCs), where participation of researchers is mandatory. The IACUC is the body responsible for the evaluation and authorization of the research protocols, and the oversight of the entire institutional animal care and use program, including the appropriate training of personnel to perform the assigned tasks.

1.3   Legislation in the European Union

Legislation in the European Union is based on Directive 2010/63/EU (European Parliament and the Council of the European Union 2010), which has been transposed into the legislation of all European member states (Guillen et al. 2017). The Directive addresses explicitly the Three Rs and distributes the main responsibilities between the public competent authorities and the users. Authorized establishments must have an Animal Welfare Body (AWB) with an advisory function on ethical matters, while the (ethical) project evaluation is assigned to the public competent authorities. However, the Directive allows member states to designate other bodies than public competent authorities for the implementation of certain tasks, and at present the project evaluation is performed in a variety of manners across the European Union, either by institutional ethics committees, external bodies, public competent authorities at regional or national level, or a combination of them (Guillen et al. 2015). Annex III of the Directive dictates the requirements for care and accommodation of animals, based on the minimum cage sizes of the Appendix A of the European Convention ETS 123, which, although not a piece of legislation, was the first pan-European document addressing in detail the protection of animals in science and was signed and ratified by a majority of the members of the Council of Europe (Council of Europe 1986, 2006). These minimum cage sizes are generally bigger than the ones recommended in the *Guide* and represent one of the most visible differences between the US and European research programs. In terms of training of personnel, the Directive requires that staff shall be adequately educated and trained before they perform certain functions including carrying out procedures on animals and designing procedures and projects and that competence is demonstrated. However, it is a competence of the member states to establish the minimum training requirements.

The European Commission has published a number of consensus documents on the implementation of the Directive, which can be found at: http://ec.europa.eu/environment/chemicals/lab_animals/interpretation_en.htm. Similar legal requirements are enforced in other European countries outside of the European Union, e.g., in Switzerland and Norway (Guillen et al. 2017; NORECOPA 2016).

1.4 Legislation in Other Countries

Other areas of the world have also developed legislation which have been extensively described elsewhere (Guillen 2017). In addition to countries such as Canada, Australia, New Zealand, or Israel having similar frameworks to those developed in the US or the European Union, many Asian countries have developed specific legislation, as well as several countries in Latin America (i.e., Brazil, Mexico, and Uruguay). Africa is the region where there is less legislation, although there are already professional associations or scientific events in some countries.

The most important aspect of the legislative initiatives is that all of them are based on the same ethical principles and try to achieve the same objectives (Guillen and Vergara 2017): Improving animal welfare standards in science is an objective per se, but this objective brings along another very important one which is the improvement of scientific quality.

Legislation that reflects international, common ethical questions is a key element in achieving these objectives. Also, legal documents normally address the same main topics: a process for the ethical evaluation (and authorization) of research protocols or projects; the need for appropriate training and competence of all personnel involved in the care and use of animals (caretakers, researchers, veterinarians, etc.); the animal environment and management (housing conditions, daily care, etc.); the need of effective veterinary care; and general requirements for facilities. Even in countries lacking specific legislation, researchers, veterinarians, and research institutions and associations work to follow these general instructions and the establishment of oversight systems, and the existence of active IACUCs or institutional ethics committees to review and improve the research protocols involving the use of animals is common also in these areas.

2 Implications for Preclinical Data Quality

2.1 Oversight Bodies Impact on Preclinical Data Quality

Thus, there is an extensive legal framework in many countries that regulates animal care and use, but how does animal care and use affect the quality of the preclinical data generated in animal experiments? First, the legal framework and its interpretation by oversight bodies (i.e., IACUCs/AWBs/ethics committees) can have a significant impact on data quality, both positive and negative.

As mentioned above, a reduction in the number of animals used according to the Three Rs is an important concept for the ethical evaluation of an animal study. But overemphasizing the need to minimize the number of animals in an experiment, without consideration of the appropriate number of animals needed to reliably answer the research question, can lead to underpowered studies with spurious results. Likewise, an uncritical refusal of study replications as unnecessary duplication of previous experiments by oversight bodies would violate the principles of good scientific methods required to gain confidence in an experimental finding (Pritt and Hammer 2017). A more balanced view by oversight bodies, on the other hand, helping with experimental design and statistical input to, e.g., determine the required sample size for the proposed studies at time of project application, can be an important step to ensure the appropriate number of animals is used and to facilitate the generation of reliable data. Clearly, this is what IACUCs/AWBs/ethical committees should strive for.

2.2 Animal Care and Use Programs Affect Preclinical Data Quality

Second, it is important to recognize that, despite the legal framework, animal facilities and their institutional animal care and use programs can differ in many aspects, even within the same country or the same organization (e.g., university). For example, there could be different barrier, hygiene, and sanitation levels to protect the health and well-being of the animals and the people working in the facility; animals could differ in microbiological status and receive different levels of veterinary care; there could be differences in the macroenvironmental (temperature, relative humidity, light intensity and duration, noise level, air circulation) and/or in the microenvironmental conditions (e.g., caging system, bedding, physical or social enrichment); there could be variations in food supplied and in water quality and also in a number of procedures, e.g., in animal acclimation, handling, transport, or surgery, to name a few, all of which could affect experimental outcome (Table 1).

Table 1 Aspects of an animal care and use program that can affect the quality of preclinical data from animal studies

• Physical plant and environmental conditions (e.g., building material, control of environmental factors, such as temperature, relative humidity, air quality)
• Training (e.g., qualifications, experience, and competence of animal technicians, researchers, veterinarians)
• Oversight (internal, by IACUC/AWB/ethics committee; external, by competent authority; or third-party accreditation, AAALAC International, CCAC)
• Housing (e.g., caging system, space, enrichment, holding room)
• Husbandry (e.g., cleaning and sanitation, food, water, bedding)
• Animal procurement (e.g., source, transport)
• Quarantine and biosecurity practices
• Health monitoring program
• Veterinary interventions
• Surgical program (techniques, asepsis, anesthetic regimens, postsurgical care)
• Pain and distress (e.g., medication, recovery)
• Euthanasia method

2.3 Health Status Influencing Preclinical Data

Fox and colleagues, for example, reported on a study designed to determine whether long-term oral supplementation with creatine, used by athletes in training, would cause histologic organ lesions in mice. Animals treated with creatine developed hepatitis but so did the control mice. Notably, *Helicobacter bilis* (*H. bilis*) was isolated from these mice and associated with hepatotoxicity seen in that study, thus confounding the experiment (Fox et al. 2004). A related *Helicobacter* species, *H. hepaticus*, has also been associated with hepatitis, inflammatory bowel disease, and cancer (Zenner 1999) and can promote drug-induced tumorigenesis in mice (Diwan et al. 1997; Nagamine et al. 2008; Stout et al. 2008). Of note, mouse *Helicobacter* infections often remain subclinical, and the animals appear healthy but can become symptomatic. The occurrence of clinical signs depends on various factors, such as strain, immunocompetency, and the gastrointestinal microbiome (Ihrig et al. 1999; Staley et al. 2009; Yang et al. 2013) and hence can lead to unexpected confounds of animal studies.

There are several other opportunistic and obligatory pathological microorganisms that can interfere with experimental outcome. It is therefore important that animals are regularly screened for the presence of these microorganisms (cf. FELASA recommendations for health monitoring in rodents, last revision: Mähler et al. 2014), either to exclude infected animals from the experiment, to initiate treatment if required (and possible), or at least to have clarity about the presence or absence of microorganisms.

However, the importance of microorganisms goes beyond agents causing clinical or subclinical disease. The gut microbiota also plays a critical role in animal and human health and disease, and its impact on animal physiology and, therefore, on how animals react in certain studies has been extensively studied in recent years

(Franklin and Ericsson 2017; Hansen et al. 2015). Both the potential impact of infectious agents and natural microbiota can be modified by routine housing and husbandry conditions.

2.4   The Impact of Housing and Husbandry

Housing conditions and husbandry can also have substantial effects on experimental rodent data, yet often researchers are not fully aware of all the environmental factors in an animal facility that can influence data quality in their experiments (Jain and Baldwin 2003; Toth 2015). These factors include cage size, positioning of the holding cage in the rack, cage material, type of bedding, ambient cage temperature, humidity, noise levels, light intensity, duration of the light/dark cycle, number of animals per cage (individual vs. social housing), food access (continuous or restricted), type of food, physical enrichment provided, cage changing practices, transporting cages with animals within a room or between rooms, and sanitation cycle of the holding room (reviewed in Castelhano-Carlos and Baumans 2009; Everitt and Foster 2004; Nevalainen 2014; Toth 2015), and this list is far from complete.

Social housing, for example, increased dopamine D2 receptor expression in dominant monkeys, but not in subordinate monkeys, when compared to individual housing, and also affected the reinforcing properties of cocaine (Morgan et al. 2002). The stability of baseline cardiovascular parameters was affected by the arrangements of pens and the social setting in dogs implanted with telemetry devices (Klumpp et al. 2006). Similarly, social enrichment has been reported to affect cardiovascular function at resting state in monkeys (Xing et al. 2015). Housing temperature affects the growth rate of tumors in mice (Hylander and Repasky 2016; Kokolus et al. 2013). These examples highlight the importance of housing conditions on preclinical data across a variety of species.

Thus, the health status of the animal, environment factors in the animal facility, daily animal care routines and experimental manipulations (e.g., recovery surgery), as well as the experience, skills, and qualifications of the people performing these activities in the animal facility (animal care staff, veterinarians, researchers) contribute to the variability of preclinical data generated in animals (Howard 2002). A reduction in the variability of experimental data generated in laboratory animals has been coupled to refinements in microbial quality monitoring and husbandry, as well as higher professional expertise (Quimby 1993), and there is additional evidence suggesting that this trend continues with additional refinements, e.g., the introduction of environmental enrichment, even though this was initially much debated (Bayne and Würbel 2014). This is important, not only for the quality of the preclinical data generated but also from an ethical perspective as high data variability requires a higher number of test animals for a study to be conclusive. Also, the scientific utility of a highly variable and non-reproducible study can be questioned, with the associated risk that animals are wasted. As has been pointed out, "laboratory animal husbandry issues are an integral but [unfortunately still] underappreciated part of investigator's experimental design" (Nevalainen 2014, p. 392). Thus, there is a plea for even more reduction of variability in further refined animal care and use programs and for more detailed reporting of animal holding conditions in publications in order to enhance transparency and consequently reproducibility of preclinical data.

With this aim in mind, guidelines have been proposed for the items to be considered when planning and reporting animal experiments (see also chapter "Minimum Information in In Vivo Research"). The PREPARE Guidelines aim to help researchers to consider all relevant items when planning animal experiments "to reduce the risk of problems, artefacts or misunderstandings arising once studies have begun" (Smith et al. 2018). The ARRIVE Guidelines were developed "to maximize the output from research using animals by optimizing the information that is provided in publications on the design, conduct, and analysis of the experiments" (Kilkenny et al. 2010). Many scientific journals have already adhered to the ARRIVE Guidelines (although their impact is debatable, cf. Hair et al. 2019) and more recently an update of the guidelines has been published (Percie du Sert et al. 2019a, b), while the impact of the PREPARE Guidelines is still to be evaluated.

3 Assessment of Animal Care and Use Programs

3.1 Internal Oversight

Internal oversight bodies, i.e., IACUCs/AWBs/ethics committees, can have a significant impact on data quality. First, they are tasked with the review of the ethical protocols, in accordance with national and international legislation and institutional policies. As part of this review process, the internal oversight body also plays an important role in advising on the experimental design of the proposed studies, including statistical considerations, and assures pain, discomfort, and distress are reduced to a minimum (Everitt and Berridge 2017; Silverman et al. 2017). As mentioned above, those factors can significantly impact on preclinical data quality. Second, the internal oversight body should conduct inspections of its own animal program and facility, at least annually or preferentially more often, also depending on legislation and policies. Besides assurance of the ethical and humane use of animals in research, this will also ensure that all aspects of an animal care and use program that can affect the quality of preclinical data from animal studies are well controlled, and it will create opportunities for further improvements of the quality of research, e.g., by assuring that surgical facilities are state of the art and that investigators conducting surgical procedures are properly trained. Third, post-approval monitoring conducted by the oversight body, primarily serving to ensure that animal use is occurring as described in the approved protocol, may also contribute to data quality. There are interdependencies between compliance, consistency, and reproducibility, and failure to reproduce an experiment has been considered as an unintended consequence of noncompliance with approved procedures (Silverman et al. 2017). Thus, the internal oversight body plays a pivotal role in the assurance of data quality in animal studies.

3.2   External Oversight

Depending on national legislation, ethical evaluation and/or authorization for animal studies may also be provided by external ethics committees, e.g. by bodies at the regional or national competent authority level, or other bodies authorized by the competent authorities to perform the ethical evaluation on their behalf. Their role in the assurance of data quality during project review is comparable to the role of an internal oversight body. In addition, many countries have a mandatory requirement for regular, announced or unannounced, inspections of animal facilities by a competent authority to monitor compliance with legal obligations. Naturally, experienced inspectors will also have an impact on preclinical data quality through assurance of a compliant animal care and use program. However, often these inspections are risk-based and may not cover all aspects of an animal care and use program that could affect preclinical data quality, and whether advice relevant to data quality is given may also depend on the profile, skills, and experience of the individual inspector. The internal oversight body is much better positioned to ensure full coverage of the aspects relevant to the quality of data from studies involving animals, to promote consistency and timely action, if required, and should take primary responsibility.

3.3   The AAALAC International Accreditation Process

AAALAC International (AAALAC) (www.aaalac.org) is a voluntary accrediting organization that enhances the quality of research, testing, and education by promoting humane and responsible research animal care and use through provision of advice and independent assessments to participating institutions and accreditation of those that meet or exceed applicable standards. More than 1,000 institutions including companies, universities, hospitals, government agencies, and other research institutions in 47 countries have earned AAALAC accreditation, demonstrating their commitment to responsible animal care and use. These institutions volunteer to participate in AAALAC's program, in addition to complying with the implementing laws that regulate animal research.

AAALAC was established in 1965 in the USA and is governed by approximately 70 scientific organizations from all around the world. The assessment and accreditation activities are performed by independent professionals with expertise in the field, who form the Council on Accreditation. The Council has three North American sections, two in the Pacific Rim, and one in Europe, each taking care of the activities in their respective geographical areas. The primary standards used by the Council are the *Guide*, the ETS 123, and the *Guide for the Care and Use of Agricultural Animals in Research and Teaching* (Federation of Animal Science Societies 2010). The Council may also use other scientific publications on different topics called Reference Resources (https://www.aaalac.org/accreditation/resources.cfm) and has to ensure that accredited programs comply with the implementing legislation in the specific location of the evaluated program. Council members are helped by ad hoc consultants/specialists, who are the same type of professionals, normally selected based on the particular expertise needed for each evaluation process.

When one institution voluntarily applies for the accreditation, it has to complete and submit the Program Description (https://www.aaalac.org/accreditation/apply.cfm), a document where all areas of the animal care and use program have to be thoroughly described. This includes the institutional responsibilities (key responsible personnel, oversight and ethical review process, competence of personnel, and occupational health and safety program), the animal environment, housing and management, the veterinary care program, and the physical plant. The Program Description is then reviewed by a Council member and the collaborating ad hoc(s), and a site visit to the institution is scheduled to evaluate the quality of the program on site. The report coming from this site visit is reviewed by and discussed with the other Council members of the same section and a decision on the accreditation status taken. Depending on the severity of the issues (if any) identified during the process, there may be mandatory issues that the institution must correct before obtaining full accreditation, and/or suggestions for improvement, which are strong recommendations for the improvement of the program that the institution can voluntarily address.

The evaluation process is based on performance standards rather than on engineering standards, which is particularly important when considering the global scope of AAALAC (Guillen 2012). While engineering standards are rigidly defined, easily measurable (e.g., minimum cage sizes), performance standards are outcome oriented, focused on goals or expected results rather than the process used to achieve the results, and have the flexibility needed in the diverse research environment. AAALAC has to make sure that institutions comply with the engineering standards which are normally part of legislation, but on top of that also apply the performance standards as described in the AAALAC Primary Standards. For example, AAALAC may accept different ethical review processes if they, in addition to be legally compliant, are effective and there is evidence of a good outcome.

The AAALAC accreditation process is compatible with quality systems like GLP or ISO. In fact, many institutions who implement GLP or ISO because they perform regulated research or have general quality systems in place (e.g., contract research organizations, pharmaceutical companies) also implement the AAALAC accreditation as this is the only global system specifically focused on animal care and use programs and carried out by independent professionals in the field. This peer-review process has been extremely successful and continues to expand in institutions around the world.

3.4   Assessments by Industry

Animal studies form an integral part of the drug development process. Those studies are either conducted within the research facilities of a company or are outsourced and performed by external service providers. To ensure external partners comply with technical requirements and ethical standards, more and more pharmaceutical companies started to formally assess the animal care and use programs of their collaborators on a regular basis, including contract research organizations (CROs), academic groups, and breeders (Mikkelsen et al. 2007; Underwood 2007). A more recent development is the joint assessment of breeders and CROs by consortia of pharmaceutical companies, which facilitates harmonization of processes across companies and enhances capacity and expertise (Interpharma 2018). In general, these animal care and use program assessments cover the aspects highlighted in Table 1, plus additional topics, such as documentation, occupational health, and safety, and often are closely oriented on the AAALAC process.

4   Conclusion

There are multiple evidences of the influence of animal care and use conditions and practices on animal-based research outcomes. The existing legislation on the use of animals in research, established upon internationally accepted ethical principles, helps creating a more common research environment that facilitates extrapolation of research results obtained in particular institutions. However, animal care and use practices may differ significantly across institutions, with potentially significant and often unknown effects on research results. Professional science-based recommendations try to complement legislation by creating standards on a number of areas, including ethical review, health monitoring, animal environment, husbandry practices, training of personnel, and others. But the implementation of the standards by research institutions still varies significantly and is very often depending on institutional or even individual commitment.

What can we learn for Good Research Practice? First and foremost, it should be clear now that the quality of animal care and use directly impacts on the quality of preclinical data. In addition, the field of animal care and use has established a framework that could be seen as a role model for Good Research Practice: Minimum requirements as defined by guidelines and legislation such as the *Guide* or the Directive 2010/63/EU set the standards for animal care and use programs in the USA and in EU member states, and both internal and external oversight bodies have been created to ensure proper implementation and adherence to these standards. The review and oversight of animal care and use programs is a key tool to not only ensure compliance with legal requirements but also to establish a well-defined research environment that considers all aspects of animal care and use that can impact research outcomes. This oversight may be internal, already mandated by legislation in many countries, and external by peers. A combination of the day-to-day internal oversight with a periodic independent external review seems to be the most efficient way to ensure the implementation of a high-quality animal care and use program where, in addition to addressing animal ethical and welfare issues, researchers can produce better quality science.

Neither the *Guide* nor the Directive 2010/63/EU is overtly prescriptive (except for a few clearly defined and nonnegotiable engineering standards), and also accrediting organizations such as AAALAC International strongly adhere to the principle of performance standards, which allows flexibility in the implementation of these standards.

These principles of minimum requirements, performance standards, and internal and external oversight could be implemented in other areas of research in a manner that is fit for the intended purpose. One may envision Good Research Practice that is guided by lean, easy to use minimal requirements defined by a quality system, based on the specific needs of the research group and working on performance standards, with day-to-day internal oversight and periodic external assessments, not to police but to improve daily research practice, possibly in combination with an accreditation process. The European Quality in Preclinical Data (EQIPD) IMI consortium (https://quality-preclinical-data.eu/) is in fact following these same principles, with the added advantage to look at animal care and use and Good Research Practice holistically, as not only does animal care and use affect the quality of preclinical data, but the need for preclinical data quality will impact on animal care and use.

References

Bayne K, Würbel H (2014) The impact of environmental enrichment on the outcome variability and scientific variability of laboratory animal studies. Rev Sci Tech 33:273–280

Bradfield JF, Bennet BT, Gillett CS (2017) Oversight of research animal welfare in the United States. In: Guillen J (ed) Laboratory animals: regulations and recommendations for the care and use of animals in research. Academic Press, Cambridge, pp 15–68

Brambell R (1965) Report of the technical committee to enquire into the welfare of animals kept underintensive livestock husbandry systems, cmd. HM Stationery Office, London, pp 1–84

Brønstad A, Newcomer CE, Decelle T, Everitt JI, Guillen J, Laber K (2016) Current concepts of harm–benefit analysis of animal experiments – report from the AALAS–FELASA working group on harm–benefit analysis – part 1. Lab Anim 50(1S):1–20

Castelhano-Carlos MJ, Baumans V (2009) The impact of light, noise, cage cleaning and in-house transport on welfare and stress of laboratory rats. Lab Anim Sci 43:311–327

Council for International Organizations for Medical Sciences and International Council for Laboratory Animal Science (2012) International guiding principles for biomedical research involving animals. www.iclas.org

Council of Europe (1986) European convention for the protection of vertebrate animals used for experimental and other scientific purposes. Eur Treat Ser 1986:123

Council of Europe (2006) Appendix A of the European convention for the protection of vertebrate animals used for experimental and other scientific purposes (ETS No. 123). Guidelines for accommodation and care of animals (Article 5 of the convention). Approved by the multilateral consultation. Cons 123:3

Diwan BA, Ward JM, Ramliak D, Anderson LM (1997) Promotion by helicobacter hepaticus-induced hepatitis of hepatic tumors initiated by n-nitrosodimethylamine in male A/JCr mice. Toxicol Pathol 25:597–605

European Parliament and the Council of the European Union (2010) Directive 2010/63/EU of the European Parliament and of the Council of 22 September 2010 on the protection of animals used for scientific purposes. Off J Eur Union 2010, L 276:33–79

Everitt JI, Berridge BR (2017) The role of the IACUC in the design and conduct of animal experiments that contribute to translational success. ILAR J 58:129–134

Everitt JI, Foster PMD (2004) Laboratory animal science issues in the design and conduct of studies with endocrine-active compounds. ILAR J 45:417–424

Federation of Animal Science Societies (2010) Guide for the care and use of agricultural animals in research and teaching, 3rd edn. Federation of Animal Science Societies, Champaign

Fox JG, Rogers AB, Whary MT, Taylor NS, Xu S, Feng Y, Keys S (2004) Helicobacter bilis-associated hepatitis in outbred mice. Comp Med 54:571–577

Franklin CL, Ericsson AC (2017) Microbiota and reproducibility of rodent models. Lab Anim 46 (4):114–122

Guillen J (2012) Accreditation of animal care and use programmes: the use of performance standards in a global environment. Anim Technol Welfare 11(2):89–94

Guillen J (2017) Laboratory animals: regulations and recommendations for the care and use of animals in research. Academic Press, Cambridge

Guillen J, Vergara P (2017) Global guiding principles: a tool for harmonization. In: Guillen J (ed) Laboratory animals: regulations and recommendations for the care and use of animals in research. Academic Press, Cambridge, pp 1–13

Guillen J, Robinson S, Decelle T, Exner C, Fentener van Vlissingen M (2015) Approaches to animal research project evaluation in Europe after implementation of directive 2010/63/EU. Lab Anim 44(1):23–31

Guillen J, Prins JB, Howard B, Degryse AD, Gyger M (2017) The European framework on research animal welfare regulations and guidelines. In: Guillen J (ed) Laboratory animals: regulations and recommendations for the care and use of animals in research. Academic Press, Cambridge, pp 117–202

Hair K, Macleod MR, Sena ES, on behalf of the IICARus Collaboration (2019) A randomised controlled trial of an intervention to improve compliance with the ARRIVE guidelines (IICARus). Res Int Peer Rev 4:1–17

Hansen AK, Krych L, Nielsen DS, Hansen CHF (2015) A review of applied aspects of dealing with gut microbiota impact on rodent models. ILAR J 56(2):250–264

Howard BR (2002) Control of variability. ILAR J 42:194–201

Hylander BL, Repasky EA (2016) Thermoneutrality, mice, and cancer: a heated opinion. Trends Cancer 2:166–175

Ihrig M, Schrenzel MD, Fox JG (1999) Differential susceptibility to hepatic inflammation and proliferation in AXB recombinant inbred mice. Am J Pathol 155:571–582

Interpharma (2018) Animal welfare report 2018. https://www.interpharma.ch/1553-animal-welfare-report-2018

Jain M, Baldwin AL (2003) Are laboratory animals stressed by their housing environment and are investigators aware that this stress can affect physiological data? Med Hypoth 60:284–289

Kilkenny C, Browne WJ, Cuthill IC, Emerson M, Altman DG (2010) Improving bioscience research reporting: the ARRIVE guidelines for reporting animal research. PLoS Biol 8(6):e1000412

Klumpp A, Trautmann T, Markert M, Guth B (2006) Optimizing the experimental environment for dog telemetry studies. J Pharmacol Toxicol Methods 54:141–149

Kokolus KM, Capitano ML, Lee CT, Eng JWL, Waight JD, Hylander BL, Sexton S, Hong CC, Gordon CJ, Abrams SI, Repasky EA (2013) Baseline tumor growth and immune control in laboratory mice are significantly influenced by subthermoneutral housing temperatura. Proc Natl Acad Sci U S A 110:20176–20181

Laber K, Newcomer CE, Decelle T, Everitt JI, Guillen J, Brønstad A (2016) Recommendations for addressing harm–benefit analysis and implementation in ethical evaluation – report from the AALAS–FELASA working group on harm–benefit analysis – part 2. Lab Anim 50(1S):21–42

Mähler M, Berard M, Feinstein R, Gallagher A, Illgen-Wilcke B, Pritchett-Corning K, Raspa M (2014) FELASA recommendations for the health monitoring of mouse, rat, hamster, Guinea pig and rabbit colonies in breeding and experimental units. Lab Anim 48:178–192

Mikkelsen LF, Hansen HN, Holst L, Ottesen JL (2007) Setting global standards for animal welfare monitoring of external contractors. AATEXX 14:731–733

Morgan D, Grant KA, Gage HD, Mach RH, Kaplan JR, Prioleau O, Nader SH, Buchheimer N, Ehrenkaufer RL, Nader MA (2002) Social dominance in monkeys: dopamine D2 receptors and cocaine self-administration. Nat Neurosci 5:169–174

Nagamine CM, Rogers AB, Fox JG, Schauer DB (2008) Helicobacter hepaticus promotes azoxymethane initiated colon tumorigenesis in BALB/c-IL10-deficient mice. Int J Cancer 122:832–838

National Research Council of the National Academy of Sciences (2011) Guide for the care and use of laboratory animals. National Academy Press, Washington

Nevalainen T (2014) Animal husbandry and experimental design. ILAR J 55:392–398

NORECOPA (2016) EUs Directive 2010/63/EU on the protection of animals used for scientific purposes. https://norecopa.no/legislation/eu-directive-201063

Percie du Sert N, Hurst V, Ahluwalia A, Alam S, Avey MT, Baker M, Browne WJ, Clark A, Cuthill IC, Dirnagl U, Emerson M, Garner P, Holgate ST, Howells DW, Karp NA, Lidster K, MacCallum CJ, Macleod M, Petersen O, Rawle F, Reynolds P, Rooney K, Sena ES, Silberberg SD, Steckler T, Würbel H (2019a) The ARRIVE guidelines 2019: updated guidelines for reporting animal research. bioRxv. https://www.biorxiv.org/content/10.1101/703181v1

Percie du Sert N, Hurst V, Ahluwalia A, Alam S, Avey MT, Baker M, Browne WJ, Clark A, Cuthill IC, Dirnagl U, Emerson M, Garner P, Holgate ST, Howells DW, Karp NA, Lidster K, MacCallum CJ, Macleod M, Petersen O, Rawle F, Reynolds P, Rooney K, Sena ES, Silberberg SD, Steckler T, Würbel H (2019b) Reporting animal research: explanation and elaboration for the ARRIVE guidelines 2019. bioRxv. https://www.biorxiv.org/content/10.1101/703355v1

Pritt SL, Hammer RE (2017) The interplay of ethics, animal welfare, and IACUC oversight on the reproducibility of animal studies. Comp Med 67:101–105

Public Health Service (2002) Public health service policy on humane care and use of laboratory animals. Office of Laboratory Animal Welfare, National Institutes of Health, Public Health Service, Bethesda

Quimby FW (1993) Twenty-five years of progress in laboratory animal science. Lab Anim 28:158–171

Russell WMS, Burch RL (1959) The principles of humane experimental technique. Universities Federation for Animal Welfare, Potters Bar

Silverman J, Macy J, Preisig PA (2017) The role of the IACUC in ensuring research reproducibility. Lab Anim 46:129–135

Smith J, Clutton RE, Lilley E, Hansen KEA, Brattelid T (2018) PREPARE: guidelines for planning animal research and testing. Lab Anim 52(2):135–141

Staley EM, Schoeb TR, Lorenz RG (2009) Differential susceptibility of P-glycoprotein deficient mice to colitis induction by environmental insults. Inflamm Bowel Dis 15:684–696

Stout MD, Kissling GE, Suárez FA, Malarkey DE, Herbert RA, Bucher JR (2008) Influence of helicobacter hepaticus infection on the chronic toxicity and carcinogenicity of triethanolamine in B6C3F1 mice. Toxicol Pathol 36:783–794

Toth LA (2015) The influence of cage environment on rodent physiology and behavior: implications for reproducibility of pre-clinical rodent research. Exp Neurol 270:72–77

Underwood W (2007) Contracting in vivo research: what are the issues? J Am Assoc Lab Anim Sci 46:16–19

United States Government (1966) Animal welfare act of 1966 (pub L. 89-544) and subsequent amendments. U.S.Code. Vol. 7, secs. 2131–2157 et seq

United States Government (1985) Principles for the utilization and care of vertebrate animals used in testing, research and training. OER Home Page – Grants Web, Office of Laboratory Animal Welfare: PHS policy on humane care and use of laboratory animals. http://grants.nih.gov/grants/olaw/references/phspol.htm#USGovPrinciples

World Organization for Animal Health (2012) Use of animals in research and education. Terrestial animal health code. https://www.oie.int/fileadmin/Home/eng/Health_standards/tahc/current/chapitre_aw_research_education.pdf

Xing G, Lu J, Hu M, Wang S, Zhai L, Schofield J, Oldman K, Adkins D, Yu H, Ren J, Skinner M (2015) Effects of group housing on ECG assessment in conscious cynomolgus monkeys. J Pharmacol Toxicol Methods 75:44–51

Yang I, Eibach D, Kops F, Brenneke B, Woltemate S, Schulze J, Bleich A, Gruber AD, Muthupalani S, Fox JG, Josenhans C, Suerbaum S (2013) Intestinal microbiota composition of interleukin-10 deficient C57BL/6J mice and susceptibility to helicobacter hepaticus-induced colitis. PLoS One 9:e70783. https://doi.org/10.1371/journal.pone.0070783

Zenner L (1999) Pathology, diagnosis and epidemiology of the rodent helicobacter infection. Comp Immunol Microbiol Infect Dis 22:41–61

Research Collaborations and Quality in Research: Foes or Friends?

Elisabetta Vaudano

Contents

Abstract

Collaboration is the cornerstone of nowadays research. Successful collaborative research and high research quality go hand in hand. Collaborative research needs to build on common and upfront expectations for the quality of its outputs. This is necessary to enable a trustful research environment where all are committed to contribute and can share the rewards. A governance and leadership are critical for this to happen as well as a policy for openness and for effective data sharing. Collaborative research is often large-scale research: to be successful it needs good research practice as an enabler. Collaborative projects are ideal vehicles to promote high research quality, among other by enabling the delivery of results of high external validity and the development and implementation of standards. Robustness of results increases when confirmed by combining different methods and tools and even more when results are obtained while sharing and learning different approaches and languages of science. When doing collaborative

E. Vaudano (✉)
The Innovative Medicines Initiative, Brussels, Belgium
e-mail: elisabetta.vaudano@imi.europa.eu

© The Author(s) 2019
A. Bespalov et al. (eds.), *Good Research Practice in Non-Clinical Pharmacology and Biomedicine*, Handbook of Experimental Pharmacology 257,
https://doi.org/10.1007/164_2019_293

research, there is the best opportunity to combine the different experience and expertise of all partners by design to create a more efficient and effective environment conductive for high-quality research. Using as example the public-private partnership type of projects created by the Innovative Medicines Initiative, the chapter covers the key aspects of the complex relationship between collaborative research and quality of research providing insights on the critical factors for delivering both a successful collaboration in research and robust high-quality research outputs.

Keywords

Data quality · Governance · Industry · Knowledge · Research output ·
Stakeholders · Trust

1   Introduction

Most modern biomedical research depends on collaboration. While research collaboration is usually thought to mean an equal partnership between two or more members who are pursuing mutually interesting and beneficial research, research collaborations come in many types and species where the relationship between the collaborators varies both in its depth and in the balance of the relationship.

At the "light" end, research collaboration might simply imply getting access to assets such as reagents, instruments, assays and data, owned by one researcher or by another researcher for its own research interests. A rather different type of collaboration is contract work where one, often private, entity enters into contractual relationships with another, often academic, institution to get a certain type of research work done. On the "heavy" end, we then have fully collaborative research where research is carried out jointly from idea generation to its implementation.

Most collaborative biomedical research occurs between two or more academic institutions. When the focus of research moves from basic to more applied research, its scope requires often involving actors beyond academia. A special type of collaboration is that between public and private institutions in public-private partnerships, such those created by the Europeans, the Innovative Medicines Initiative (IMI) (The Innovative Medicines Initiative 2009). The IMI is a joint initiative (public-private partnership) of the European Commission (EC) and the European Federation of Pharmaceutical Industries and Associations (EFPIA). The current programme, IMI2 Joint Undertaking (IMI2 JU), is part of the Horizon 2020 Framework programme. IMI fosters collaborative research among many different stakeholders, from large industries (mainly biopharmaceutical but most recently also from diagnostics, information technology and imaging industries) to small medium-sized enterprises, academia, hospitals, regulators, patient organisations, etc. These partnerships (consortia) implement a research work plan where all members work jointly towards agreed and fixed objectives and deliverables. The resources (often very significant) are provided, on one side, by funding from the EC to all eligible entities and, on the other side, through in-kind contributions from the industry members of the EFPIA and in some projects by Associated Partners to the

IMI2 JU. Importantly, neither the EFPIA members nor the IMI2 JU Associated Partners are eligible for funding. Thus, they participate at own cost and provide resources (experts' time, setting and running of assays and models, clinical studies, etc.) for the implementation of the research. The IMI has proven to be a very successful collaboration model with significant output already delivered (Faure et al. 2018). The terms of the collaboration in the IMI consortia are established by the grant agreement that all partners have to sign with the IMI and, most importantly, by the consortium agreement that all members of the consortium have to sign among themselves.

A lot has been written regarding the pros and cons of research collaboration and on the "added value" of research collaboration. Here, I would like to focus on the complex and important relation between research done collaboratively and quality of its (research) output. The aim is not to demonstrate any potential or even causal relationship but to bring to the attention of the reader the important interdependences among collaborative research and research quality, highlighting benefits and challenges.

To provide some "real-life" evidence, I will use as examples cases of collaborative research in public-private partnerships of the type created by IMI. The IMI research collaborations are highly complex, in terms of both content and stakeholders involved, and the robustness of the research outputs is affected by all factors that in isolation may be relevant for "simpler" research collaborations. At a recent event for the 10th year anniversary of the IMI, the Commissioner Moedas defined the research collaboration fostered by IMI as "radical collaboration" (The Innovative Medicines Initiative Radical Collaborations 2018).

2   Successful Collaborative Research and High Research Quality Are Interdependent

High-quality research produces results that can be confidently used as the basis for generating new knowledge or for application purposes (e.g. the development of a new drug). When choosing collaborators the trust in the quality of their research outputs is a key factor. Thus high-quality research and successful collaborative research go hand in hand. Robustness of results is increased when it can be confirmed by combining different methods and tools and even more when results are obtained while sharing and learning different approaches and languages of science. When doing collaborative research, there is the best opportunity to combine the different experiences and expertise of all partners by design to create a more efficient and effective environment conductive for high-quality research.

Most IMI consortia perform collaborative research that is highly multidisciplinary and where research teams work jointly across the public and private sector. For example, the IMI "Methods for systematic next generation oncology biomarker development", ONCOTRACK project (Methods for Systematic Next Generation Oncology Biomarker Development ONcoTRACK Project 2016), is a precompetitive research project that was created to tackle the general problem of identification and validation of clinically robust biomarkers in oncology (Schütte et al. 2017). The project team included eight EFPIA industry teams, nine universities/research institutions and four SMEs. Among the outputs of this consortium is a high-throughput screening platform for three-dimensional patient-derived colon cancer organoid cultures. The platform has been fully validated for assay robustness and reproducibility, an achievement only possible via the collaborative efforts of the multidisciplinary industry-academia team, built on a foundation of trust in the quality of the work of each partner with criteria agreed and applied from the very beginning of the partnership (Boehnke et al. 2016).

However, such complex cross-fertilisation comes with some caveats. It must not be underestimated that, for such collaborative efforts to succeed, all partners need time and good will to adjust to each other's way of working and thinking. Technical jargon can be very different; the same acronym may have very different meanings (e.g. API: "Active pharmaceutical ingredient" or "application programming interface"). The time and resources necessary for this preparatory work should be considered carefully with attention to aspects such as a good communication platform and legal support. It can be a lengthy and challenging process: timelines should be adjusted accordingly, and expectations!

Trust is the uppermost key factor of success for a research collaboration. In a collaboration, researchers depend for success on both their own results and those of their partners. Since it might be challenging and impracticable to share fully details of work going on in different laboratories, the application of high-quality standards, understood and agreed by all partners from the very start of the collaborative work, is an important enabler for a successful partnership. Once such standards are in place, there is higher motivation and opportunity for the achievement of robust high-quality results and conclusions.

Collaborative research has enabled research programmes at an unprecedented scale. Typical examples are those from genetic research. Here, large-scale collaborative genome studies have delivered huge amount of data opening new avenues for the understanding of disease biology via the use and reuse of these data by many scientists. A stringent adherence to good research practice and quality control is necessary when working at such scale. Among the many programmes is worth mentioning the "Encyclopedia of DNA Elements, ENCODE initiative funded by the National Human Genome Research Institute" (Encyclopedia of DNA Elements (ENCODE) 2019). ENCODE aims to identify all functional elements in the human and mouse genomes and make them available through the project's freely accessible database. The ENCODE project has developed standards for each experiment type to ensure high-quality, reproducible data and novel algorithms to facilitate analysis (ENCODE Consortium 2017). As a result of outreach and collaboration, enabled by such quality-driven approach, ENCODE has been highly successful, and its data are widely used to deliver high-quality publications.

While standards are needed to deliver a successful collaboration, conversely collaborative research can be a powerful tool to boost the development and implementation of standards and interoperability of results, which again significantly enhances research quality as shown by two further examples below.

Modelling and simulation (M&S), a technology providing the basis for informed, quantitative decision-making, is of high importance in modern drug development. A lack of common tools, languages and ontologies for M&S often leads to inefficient reuse of data and duplication of effort by academic, industrial and regulatory stakeholders, as well as hindering research quality. The IMI "Drug Disease Model Resources", DDMoRe (Drug Disease Model Resources DDMoRe Project 2012), consortium delivered a set of integrated tools, exchange standards and training to improve the quality and cost-effectiveness of model-informed decision-making for pharmaceutical research and development. The set of standards has been designed both for model and workflow encoding and for storage and transfer of models and associated metadata. One of the key products the project developed based on these standards is the publicly available DDMoRe model repository (DDMoRe Model Repository 2017). It provides access to more than 100 annotated and "ready to use" pharmacokinetic (PK), pharmacodynamic (PD), PK/PD, physiologically based PK (PBPK), statistical and systems biology models applied in different therapeutic areas like oncology, diabetes and neuroscience. The model repository content is quality assured by experts from the DDMoRe model review group, who provides on-demand impartial review and assesses the model's technical validity and repro-ducibility. The models that pass the review are certified and can be confidently reused by anyone either commercially or for research purposes.

Rheumatoid arthritis (RA) is a very common and debilitating condition due to many underlying disease mechanisms, thus the plethora of animal models of which the translatability and reproducibility is not well established. The IMI "Be The Cure" BTCure ("Be The Cure" BTCure Project 2012) consortium developed an infrastruc-ture to standardise procedures to generate and interpret commonly used RA animal models, as well as to generate new types of RA animal models. Their work has shown how data obtained from these models might lack quality and reproducibility, due to insufficient documentation and nomenclature, wrong presentation of results and data, as well as the selection of inappropriate models and strains. Most impor-tantly, the BTCure team proposed relevant solutions and developed training material to improve the quality of RA models (Holmdahl 2015). The achievement has been possible only by the joint collaborative work of key opinion leaders in the field of RA to build a critical mass of experts and achieve consensus via a series of dedicated workshops (e.g. BTCure Consortium 2012).

High-quality research has external validity. This does not mean its results are fully reproducible in an identical manner anywhere and anytime but that each result comes together with the awareness and understanding on the potential contextual factors that determine the variation over space or time. Collaborative research enables several partners in a consortium to replicate each other results and then share the obtained knowledge. This dramatically increases the robustness of the research outputs allowing to understand factors impinging on reproducibility of results.

For example, drug-induced liver injury (DILI) is a serious issue not only for patients and health-care professionals but also for the pharmaceutical industry and regulatory authorities. Human-specific and idiosyncratic adverse reactions are often detected only at the clinical and post-marketing stages leading to costly termination of drug development and risk for the patients with black box warnings or even withdrawal of drugs from the market. DILI remains a significant problem in drug development, suggesting that currently used in vitro models are not appropriate for effective screening. The IMI "Mechanism-Based Integrated Systems for the Prediction of Drug-Induced Liver Injury" MIP-DILI (Mechanism-Based Integrated Systems for the Prediction of Drug-Induced Liver Injury MIP-DILI Project 2012) consortium has run a comprehensive, multicentre, unbiased assessment to test this unequivocally. The consortium used a panel of compounds implicated in DILI in man, in order to determine whether any of these simple cell models per se are actually predictive of human DILI. Furthermore, by using a small panel of DILI- and non-DILI-implicated compounds and basic measures of cell health, it monitored reproducibility across different sites, thereby ensuring that data should be more definitive than any currently available (Sison-Young et al. 2017).

Multicentre collaborations can expose systematic biases and identify critical factors to be standardised. Human-induced pluripotent stem cells (iPSCs) are powerful tools for novel in vitro models in basic science and drug discovery. iPSCs need to be differentiated using lengthy complex procedures with increased possibility for variability and noise in the results. The IMI "Stem cells for biological assays of novel drugs and predictive toxicology" (STEMBANCC) project (Stem Cells for Biological Assays of Novel Drugs and Predictive Toxicology Project 2014) runs a unique assessment of the inter- and intra-laboratory reproducibility of transcriptomic and proteomic read-outs using two iPSC lines at five independent laboratories in parallel. By achieving larger sample numbers in a collaborative approach with cross-laboratory studies, the team could detect identifiable sources of variation that investigators can control. This study also strongly advocates for transparency via disclosure of identified variation-inflating confounders in published iPSC differentiation protocols (Volpato et al. 2018).

3 Quality of Research and Sustainability

Collaborative research connects researchers and institutions. The created networks are key assets that go beyond the lifetime of an individual project and can play an important function as ambassadors of the culture of quality in research. The networks can lead by example but can only be successful in time by being open to new members and by including elements of education, training and dissemination of good practice and processes.

The IMI "Combatting Bacterial Resistance in Europe" (COMBACTE) consortium ("Combatting Bacterial Resistance in Europe" COMBACTE Project 2015) has delivered a high-quality European clinical trial and laboratory network in which new antibacterial drugs can be evaluated for the treatment and prevention of infections caused by multiresistant bacteria. Over 650 laboratories and over 850 clinical trial sites all over Europe are already interconnected. Development of new antibiotics is hampered by limited market incentives but also significantly limited by suboptimal quality of trial results and lack of innovative methods robust enough to be accepted by the regulators. The COMBACTE network of statistical experts "STAT-Net" has reviewed and tested several innovative trial designs and analytical methods for randomised clinical trials, which has resulted in eight recommendations made available to the community in a white paper (de Kraker et al. 2018). The COMBACTE project has been one of the first of the IMI programme New Drugs 4 Bad Bugs (ND4BB), which today represents an unprecedented partnership between industry, academia and biotech organisations to combat antimicrobial resistance in Europe. The EUR 650 million programme comprises currently eight projects and has built the foundation for the IMI2 Accelerator programme.

Collaborative networks with a proven record of high-quality research increase their opportunities for further collaboration and funding, thus becoming sustained. A very fitting example is that of the IMI consortium "European Autism Interventions – A Multicentre Study for Developing New Medications", EU-AIMS (European Autism Interventions – A Multicentre Study for Developing New Medications Project EU-AIMS 2012), identifying markers of autism that would help in earlier and more accurate diagnosis, prognosis and the development of new therapies. The project applied the highest standard of research quality, which among others led to five "letters of support" from the European Medicines Agency (EMA) (European Medicines Agency 2015). These recognised results are the foundations of the follow-up project "Autism Innovative Medicine Studies-2-Trials", AIMS-2-Trials (Autism Innovative Medicine Studies-2-Trials AIMS-2-Trials Project 2018), which is a 115 million EUR cross-Atlantic collaborative initiative, nearly four times the resources of EU-AIMS.

4   The Importance of Effective Governance in Collaborative Research

Not all collaborations are successful in delivering high-quality research. For having a positive impact on quality of research, a collaboration must be based on transparency among all involved stakeholders, meaning clearly established governance rules for the implementation of the research work plan and sharing of knowledge thereof. Rules have to be in place from the very start of the collaboration, be communicated to all collaborators and be agile and not overly complicated, to avoid being seen as a burden instead of an enabler. An efficient governance is necessary to ensure good working relationships in the collaboration, a sense of high degree of accountability in all partners and spirit of joint responsibility while also keeping expectations realistic. Governance has to provide leadership for the collaborator's team and create a "trust environment" which in the case of the most complex partnerships might need a third party to act as a neutral broker, like in the case of the IMI Programme Office (Goldman 2012).

The collaboration governance must guarantee respect for the rights of all collaborators and that the collaboration will be rewarding for all involved. This latter can be seen as a challenge when trying to balance, on one side, the need for open innovation and knowledge sharing and, on the other, the need for intellectual property rights (IPRs) to be protected. A one-size-fits-all solution does not exist for all types of research collaborations, but some principles can be generally applicable. First, there needs to be a built-in degree of flexibility when establishing the rules for IPRs, with opportunity for adaptation with the development of the research programme. Secondly, IPR agreements must be agreed before the collaboration starts, providing consortia with less legal uncertainty and avoiding useless a posteriori discussions. Most importantly legal agreements cannot and must not substitute a well-thought-through high-quality research programme. All partners have to have a clear understanding about their own knowledge/data/assets that they need to share with the other collaborators for the implementation of a high-quality research programme and achievement of its objectives. These objectives and the pathway to achieve them have to be clear and sound.

To achieve impact and promote sustainability, all partners have to agree on the sharing of the research outputs among themselves and with third parties. In the IMI programme, all consortium members sign a consortium agreement before the start of the activities which details the collaboration rules of play and governance. A template prepared by EFPIA shows what a consortium agreement might look like (European Federation of Pharmaceutical Industries and Associations 2015). In addition, the IMI Programme Office plays a neutral role and offers impartial advice to all partners during negotiations on IPRs. This process and support ensure that the resulting agreement is in line with the IMI IPRs provisions and does not leave some project partners at a disadvantage (Laverty and Poinot 2014).

A good governance structure is important in collaborative research but is not sufficient for success. It has to be mentioned that collaborative research can still be silos research, where the collaborators share a pot of funding but otherwise work mainly independently just continuing and expanding their own internal research programme. Such "collaborations" represent a missed opportunity for cross-fertilisation and learnings and thus are not very inductive to research quality. In my own experience, these are also the consortia where most IPR-related issues arise and the signature of the consortium agreement might be delayed by lengthy, often circular, discussions.

5   On Data Sharing, Collaborative Research and Research Quality

Reluctance to data sharing has been shown to correlate with poor quality of the research results; see, for example, the paper by Wicherts et al. (Wicherts et al. 2011). Reasons for not sharing data range from lack of time and resources (Tenopir et al. 2011) to legal issues and fear for data misuse (Majumder et al. 2016; Parse Insight Consortium 2019).

Still, especially in collaborative research, data sharing is a fundamental factor for the delivery of high-quality research. It is essential for increasing reproducibility and transparency in research. This has been now recognised by publishers, by major research organisations and by funders (e.g. the Transparency and Openness Promotion guidelines (Nosek et al. 2015), a recent Nature Editorial (Nature Editorial 2017) and the OpenAIRE initiative of the European Framework Programme H2020 (European Commission 2014)).

Data sharing has to be enabled by the upfront creation of a framework that is acceptable and accepted by all partners, which is not always a simple endeavour. The IMI "Diabetes Research on patient stratification", DIRECT (Diabetes Research on Patient Stratification DIRECT Project 2013), consortium has shared its experience and learnings on how to develop a governance for data sharing in their collaboration in a recent paper published in *Life Sciences, Society and Policy* (Teare et al. 2018). What they show is that designing an internal governance structure to oversee access to a centralised database for research purposes can be a time-consuming and politically sensitive process. Partners might have different expectations and requirements for taking part. Moreover, especially for international collaborations, it might be tricky to provide a framework that is acceptable legally and ethically across borders and still is sufficiently speedy to allow research.

However, several solutions are emerging to address these concerns and enable data sharing, such as those developed by the IMI "European Medical Information Framework", EMIF, project (Trifan et al. 2018) and the Cohen Veterans Bioscience Brain Commons (Grossman 2017). Another useful resource to help the research community in the sharing of data and samples has been delivered by partners from the IMI BTCure consortium jointly with other experts from around the world. The "International Charter of principles for sharing bio-specimens and data" addresses the points necessary to enable effective and transparent data and samples sharing and even provides a general template for data and material sharing (Mascalzoni et al. 2015).

Once data sharing is enabled, it delivers significant benefits for both those sharing and the scientific community at large and is an important first step for sustainability of results. This can be exemplified by the IMI "Unrestricted leveraging of targets for research advancement and drug discovery", ULTRADD (Unrestricted Leveraging of Targets for Research Advancement and Drug Discovery ULTRADD Project 2016), project data policy. ULTRADD aims to identify and validate under-explored protein targets by profiling target-directed chemical and antibody probes in patient-cell-derived assays at the highest quality, providing biomarker and phenotypic read-outs in a more disease-relevant context. The consortium adopted a policy of full open access. The wider scientific community will therefore have ready access to much of the knowledge, data and tools generated by the project. True to its promise, the consortium has already made available to the broad community the first batch of high-quality datasets (Cell Assay Datasets 2017).

The sharing of high-quality data strongly facilitates moving research forward providing the bases for new applications and benefit to society. The IMI "Integrating bioinformatics and chemoinformatics approaches for the development of Expert systems allowing the in silico prediction of toxicities" (eTOX) consortium developed innovative strategies and novel software tools to better predict the safety and side effects of new candidate medicines for patients (Sanz et al. 2017).

The achievements were enabled by a unique shared database based on high-quality legacy information provided by EFPIA companies from their own preclinical drug toxicity studies. By the end of the project, the database has information from over 8,000 toxicity studies on almost 2,000 compounds, of which around a fifth are approved drugs. Several of the in silico algorithms developed by eTOX are now used by the pharmaceutical industry for better prediction of potential drug toxicities.

Data and knowledge in order to be shareable need to be properly managed. This can be difficult for researchers working on their own which might lack the expertise, experience, technical solutions and resources for proper data and knowledge management (DKM) during and after a project. At IMI and elsewhere, research collaborations involve more and more experts and resourcing for DKM, which is a very positive trend. The inclusion of relevant provisions for DKM in any research collaboration will enhance the quality of the research output with a cascade effect for the quality of any future project that will build on such background. A good data management and data stewardship are fundamental for data quality. Science funders, publishers and governmental agencies are beginning to require data management and stewardship plans for data generated by their grantees (see, e.g. the open access and data management for projects of IMI (Open Access and Data Management for Projects 2015)). Findable, Accessible, Interoperable and Reusable (FAIR) data (Wilkinson et al. 2016) will become more attractive for reuse and the data generators sought for new collaborations, thus automatically ensuring the sustainability of their results. Making it convenient for scientists to describe, deposit and share their data and to access data from others, plus promulgating best data practices through education and awareness, will help the future of science as well as the future of data preservation.

The awareness of this problem led to the creation of the IMI "Delivering European Translational Information & Knowledge Management Services" eTRIKS project. The eTRIKS delivered an open, sustainable research informatics and analytics platform for use by IMI (and other) projects with knowledge management needs. In addition, the project partners provide associated support, expertise and services to ensure users gain the maximum benefit from the platform. One important resource is the eTRIKS/ELIXIR-LU – IMI Data Catalogue (eTRIKS/ELIXIR-LU – IMI Data Catalogue 2015), a sustained a metadata repository linking the massive data available in a global system that can be optimally leveraged by scientists. Researchers can access the available data and are encouraged to add their data in the database to create awareness and recognition of their data contribution and demonstrate value of partner projects.

Despite the increased emphasis on the importance for DKM, often there is yet not enough awareness about the DKM needs and resources for a project. The gaps are identified too late, when activities are already up and running and all budget committed elsewhere; thus more attention is needed to this critical area in all its aspects, technical, ethical, societal and legal.

6 Enlarging the Collaborative Research Environment: Regulators and Patients as Important Partners for Research Quality

For the outputs of a biomedical translational research collaboration to impact health care, the involvement with regulators and patients is an important success factor, not the least because of the positive influence on research quality.

A partnership with regulators increases the awareness and drive of a consortium for the highest-quality results. The involvement of regulators (and in some cases also health technology assessment bodies and payers) is of high benefit to ensure that the project output meets the required standards and is of quality good enough to be taken up and used in drug development or medical practice. For example, the IMI "Understanding chronic pain and improving its treatment" EUROPAIN consortium developed rigorous processes for quality assurance of centres using in a standardised way quantitative sensory testing (QST) for measuring pain in patients. This is a novel approach for stratification of patients in clinical trials for pain medicines (Baron et al. 2017; Vollert et al. 2015). EUROPAIN engaged with the European Medicines Agency (EMA) to discuss their approach and results (The Innovative Medicines Initiative 2017a), and their approaches were considered in the updated guidelines for the development of medicines for pain treatment (European Medicines Agency 2016).

The IMI actively encourages to involve regulators as early as possible in the research work plan and has published specific guidelines for consortia (The Innovative Medicines Initiative 2017b). Many IMI projects have benefited from such interaction (Goldman et al. 2015). Similar approach is that of the Accelerating Medicines Partnership (AMP) (Accelerating Medicines Partnership AMP 2014) where the Federal Drug Administration (FDA) is a formal partner and of the Critical Path Institute (C-Path). C-Path has a large portfolio of initiatives where scientific rigour is at the basis of consensus building among participating scientists from industry and academia and the FDA to deliver new drug discovery tools (DDTs) for a given use in product development (Critical Path Institute C-Path 2005). This combination of good research practice and regulatory acceptance has allowed many of C-Path FDA-qualified DDTs to become open standards for the scientific community.

Researchers are now well aware that patients bring unique knowledge and skills to projects, which can significantly help to improve the quality of research. An exemplar collaboration where patients have actively contributed to improve research quality is the IMI "Unbiased Biomarkers for the Prediction of Respiratory Disease Outcomes" (UBIOPRED) project. Patients played a big role in the project, including participation in the scientific and the ethics boards. They helped in many aspects of the project and contributed to the fine-tuning of the research protocols. The consortium has published a very useful guide for effective engagement of patients in biomedical research (Geissler et al. 2017).

7   Why Scientists Should Consider Quality as a Key Parameter for Their Collaborative Research from the Very Start

As already stressed in the previous sections, there are multiple reasons why research quality and success in a research collaboration go hand in hand. When working as a consortium, researchers will create a collective reputation, dependencies between partners and shared ethical responsibilities. They will also have to agree in the sharing of resources and funding, which may be significant in amount and of high value from the financial and scientific point of view. It is at this very early stage that a collaborative team has to align the expectation for high research quality. To achieve impact and success in their collaborative research, scientists need to get to a common understanding and agreement on how they will deliver high-quality results ideally taking advantage of common standards and guidance when available. It is very important that this occurs upfront, before picking up the pen for writing the first line of the joint grant. In fact, it is not possible to "fix" the quality of results once they have been already delivered; this will have only a "cosmetic" effect that will be detrimental both for those producing the results and for those that will base their independent work on such evidence. Results that will not be used or will not be found useful because of poor quality will be wasted and a waste of the public or private funding. Thus, there is an important ethical value in thriving for research quality and for doing so from the very beginning.

In the world of collaborative research, reputation is of paramount importance as well as trust. If a researcher delivers results of poor quality, this will influence strongly and negatively her/his reputation. In a consortium, this might affect the reputation of everybody. Delivering poor-quality results might generate mistrust in a collaboration and thus jeopardise its implementation. In addition, other scientists will not trust either the team or the results delivered, which might affect not only the single scientist in a collaboration but also his/her institution and/or students and the other collaborators.

8   Conclusions

Collaborative research in the modern world is becoming the standard. But, as confidence in the communication media is suffering from the rise of "fake news" and the poor quality of large amount of news and data shared on the social networks, to avoid a similar fate for research results, it is vital that collaboration in research goes hand in hand with quality in research.

Luckily the awareness of the importance of fostering high research quality and rigour is emerging worldwide. In Europe, The IMI "European Quality in Preclinical Data" (EQIPD) project will deliver simple recommendations to facilitate data quality without affecting innovation in the challenging field of preclinical neuroscience research. Cornerstone of the work is agreeing across stakeholders on the key variables influencing quality and delivering a prospectively validated consensus quality management system. EQIPD will importantly foster education and training on the principles and application of quality and rigour via an online educational platform providing certified education and training (EQIPD Consortium 2018). In the United States, there is a strong attention to rigour and quality in research, and the National Institute for Health has developed a specific guidance for grantees (National Institute for Health 2018), while the National Heart, Lung and Blood Institute has published the "Study Quality Assessment Tools" resource (National Heart, Blood and Lung Institute 2017). Thus, there are several resources available (only few mentioned above), and more are emerging to help researchers deliver excellent, robust research outputs of the highest impact. However, it is important that all efforts are coordinated and aligned globally: there is still an issue of fragmentation. In addition, while the effort of the EQIPD project represents a first important European effort to foster rigour and quality in research, European researchers still lack a common guidance. To be effective such guidance must be agreed across countries and be applicable both in national and international/Europe-wide collaborations.

The upcoming Framework programme Horizon Europe has the specific objective "to support the creation and diffusion of high-quality new knowledge, skills, technologies and solutions to global challenges" (Dalli 2018). To make this objective a reality, stakeholders in research and development in Europe and beyond have to put in place an aligned and effective policy for good research practice, in a collaborative way.

References

Accelerating Medicines Partnership AMP (2014) https://www.nih.gov/research-training/accelerating-medicines-partnership-amp. Accessed 20 Feb 2019

Autism Innovative Medicine Studies-2-Trials AIMS-2-Trials Project (2018). https://www.aims-2-trials.eu. Accessed 14 Feb 2019

Baron R, Maier C, Attal N et al (2017) Peripheral neuropathic pain: a mechanism-related organizing principle based on sensory profiles. Pain 158(2):261–272

Be The Cure" BTCure Project (2012) http://btcure.eu. Accessed 12 Feb 2019

Boehnke K, Iversen PW, Schumacher D et al (2016) Assay establishment and validation of a high-throughput screening platform for three-dimensional patient-derived colon cancer organoid cultures. J Biomol Screen 21(9):931–941

BTCure Consortium (2012) WP1 standardization workshop – summary for external communication. Available via BTCure. http://btcure.eu/wp-content/uploads/2012_03_26-BTCure-WP1-workshop-summary-final_incl_images.pdf

Cell Assay Datasets (2017) UltraDD project. https://www.ultra-dd.org/tissue-platforms/cell-assay-datasets. Accessed 14 Feb 2019

Combatting Bacterial Resistance in Europe" COMBACTE Project (2015) https://www.combacte.com/about/about-combacte-net-detail/0. Accessed 12 Feb 2019

Critical Path Institute C-Path (2005) https://c-path.org/about/. Accessed 20 Feb 2019

Hubert Dalli (2018) The horizon Europe framework programme for research and innovation 2021–2027. Available via EPRS. http://www.europarl.europa.eu/RegData/etudes/BRIE/2018/627147/EPRS_BRI(2018)627147_EN.pdf. Accessed 20 Feb 2019

DDMoRe Model Repository (2017) DDMore. http://repository.ddmore.eu. Accessed 12 Feb 2019

de Kraker M, Sommer H, de Veide F et al (2018) Optimizing the design and analysis of clinical trials for Antibacterials against multidrug-resistant organisms: a white paper from COMBACTE's STAT-net. Clin Infect Dis 67(12):1922–1931

Diabetes Research on Patient Stratification DIRECT Project (2013) www.direct-diabetes.org. Accessed 12 Feb 2019

Drug Disease Model Resources DDMoRe Project (2012) www.ddmore.eu. Accessed 12 Feb 2019

ENCODE Consortium (2017) Data standards. Available via ENCODE. https://www.encodeproject. org/data-standards. Accessed 12 Feb 2019

Encyclopedia of DNA Elements (ENCODE) (2019) https://www.genome.gov/10005107/the-encode-project-encyclopedia-of-dna-elements. Accessed 12 Feb 2019

EQIPD Consortium (2018) EQIPD E-learning programme, preliminary version December 2018. Available via EQIPD https://quality-preclinical-data.eu/our-findings/training/. Accessed 20 Feb 2019

eTRIKS/ELIXIR-LU – IMI Data Catalogue (2015) https://datacatalog.elixir-luxembourg.org/ about. Accessed 16 Feb 2019

European Autism Interventions – A Multicentre Study for Developing New Medications Project EU-AIMS (2012) https://www.eu-aims.eu. Accessed 12 Feb 2019

European Commission (2014) OpenAIRE initiative of the European Framework Programme H2020. Available via OPENAIRE. https://www.openaire.eu. Accessed 14 Feb 2019

European Federation of Pharmaceutical Industries and Associations (2015) EFPIA IMI2 consortium agreement template. Available via EFPIA. https://efpia.eu/media/.../efpia-model-consortium-agreement-for-imi2-actions-2.docx. Accessed 14 Feb 2019

European Medicines Agency (2015) Letters of support EU-AIMS. Available via EMA. https:// www.ema.europa.eu/en/search/search?search_api_views_fulltext=EU-AIMS.   Accessed 20 Feb 2019

European Medicines Agency (2016) Clinical development of medicinal products intended for the treatment of pain. Available via EMA. https://www.ema.europa.eu/en/clinical-development-medicinal-products-intended-treatment-pain. Accessed 20 Feb 2019

Faure JE, Dyląg T, Norstedt I et al (2018) The European innovative medicines initiative: progress to date. Pharm Med 32:243–249

Geissler J, Ryll B, Leto di Priolo S et al (2017) Improving patient involvement in medicines research and development: a practical roadmap. Ther Innov Reg Sci 51(5):612–619

Goldman M (2012) The innovative medicines initiative: a European response to the innovation challenge. Clin Pharmacol Ther 91(3):418–425

Goldman M, Seigneuret N, Eichler HG (2015) The innovative medicines initiative: an engine for regulatory science. Nat Rev Drug Dis 14:1–2

Grossman RL (2017) An introduction to the brain commons. Available via Cohen Veterans Bioscience.   https://www.cohenveteransbioscience.org/2017/09/27/an-introduction-to-the-brain-commons. Accessed 14 Feb 2019

Holmdahl R (2015) Animal models for rheumatoid arthritis Available via HsTalks. https://hstalks. com/t/3076/animal-models-for-rheumatoid-arthritis/?biosci. Accessed 12 Feb 2019

Laverty H and Poinot M (2014) IP policy forum: intellectual property rights (IPR) in collaborative drug development in the EU: helping a European public-private partnership deliver – the need for a flexible approach to IPR. 18 Marq. Intellectual Property L Rev 31. Available via Marquette University. https://scholarship.law.marquette.edu/iplr/vol18/iss1/19. Accessed 14 Feb 2019

Majumder MA, Cook-Deegan R, McGuire AL (2016) Beyond our borders? Public resistance to global genomic data sharing. PLoS Biol 14(11):e2000206

Mascalzoni D et al (2015) International Charter of principles for sharing bio-specimens and data. Eur J Hum Genet 23:721–728

Mechanism-Based Integrated Systems for the Prediction of Drug-Induced Liver Injury MIP-DILI Project (2012) http://www.mipdili.eu. Accessed 12 Feb 2019

Methods for Systematic Next Generation Oncology Biomarker Development ONcoTRACK Project (2016) http://www.oncotrack.eu/home/index.html. Accessed 12 Feb 2019

National Heart, Blood and Lung Institute (2017) Study quality assessment tools. Available via NHBLI. https://www.nhlbi.nih.gov/health-topics/study-quality-assessment-tools. Accessed 20 Feb 2019

National Institute for Health (2018) Guidance: rigor and reproducibility in grant applications. Available via NIH. https://grants.nih.gov/policy/reproducibility/guidance.htm. Accessed 20 Feb 2019

Nature Editorial (2017) On data availability, reproducibility and reuse. Nat Cell Biol 19:259

Nosek BA et al (2015) Promoting an open research culture. Science 348(6242):1422–1425

Open Access and Data Management for Projects (2015) Available via the innovative medicines initiative. https://www.imi.europa.eu/resources-projects/open-access-and-data-management-projects. Accessed 12 Feb 2019

Parse Insight Consortium (2019) Insight into digital preservation of research outputs in Europe. Available via Liber europe. https://libereurope.eu/wp-content/uploads/2010/01/PARSE.Insight.-Deliverable-D3.4-Survey-Report.-of-research-output-Europe-Title-of-Deliverable-Survey-Report.pdf. Accessed 14 Feb 2019

Sanz F et al (2017) Legacy data sharing to improve drug safety assessment: the eTOX project. Nat Rev Drug Discov 16:812

Schütte M, Risch T, Abdavi-Azar N et al (2017) Molecular dissection of colorectal cancer in pre-clinical models identifies biomarkers predicting sensitivity to EGFR inhibitors. Nat Commun 10(8):14262

Sison-Young RL, Lauschke VM, Johannet E et al (2017) A multicenter assessment of single-cell models aligned to standard measures of cell health for prediction of acute hepatotoxicity. Arch Toxicol 91(3):1385–1400

Stem Cells for Biological Assays of Novel Drugs and Predictive Toxicology Project (2014) STEMBANCC. https://stembancc.org. Accessed 12 Feb 2019

Teare HJA, de Masi F, Banasik K et al (2018) The governance structure for data access in the DIRECT consortium: an innovative medicines initiative (IMI) project. Life Sci Soc Policy 14:20

Tenopir C, Allard S, Douglass K et al (2011) Data sharing by scientists: practices and perceptions. PLoS One 6(6):e21101

The Innovative Medicines Initiative. (2009) https://www.imi.europa.eu. Accessed 12 Feb 2019

The Innovative Medicines Initiative (2017a) 'More successful than what we thought possible' – an interview with the Europain project coordinator. Available via IMI https://www.imi.europa.eu/projects-results/success-stories-projects/more-successful-what-we-thought-possible-interview. Accessed 21 Feb 2019

The Innovative Medicines Initiative (2017b) Guidelines on engaging with regulators. Available via IMI. https://www.imi.europa.eu/resources-projects/guidelines-engaging-regulators. Accessed 20 Feb 2019

The Innovative Medicines Initiative Radical Collaborations (2018) Available via IMI. https://www.imi.europa.eu/projects-results/success-stories-projects/radical-collaboration-shaking-pharmaceutical-industry. Accessed 12 Feb 2019

Trifan A et al (2018) A methodology for fine-grained access control in exposing biomedical data. In: Building continents of knowledge in oceans of data: the future of co-created eHealth series studies health technology and informatics, vol 247. IOS Press, Amsterdam, pp 561–565

Unrestricted Leveraging of Targets for Research Advancement and Drug Discovery ULTRADD Project (2016) https://www.ultra-dd.org. Accessed 12 Feb 2019

Vollert J, Mainka T, Baron R et al (2015) Quality assurance for quantitative sensory testing laboratories: development and validation of an automated evaluation tool for the analysis of declared healthy samples. Pain 156(12):2423–2430

Volpato V, Smith J, Sandor C et al (2018) Reproducibility of molecular phenotypes after long-term differentiation to human iPSC-derived neurons: a multi-site omics study. Stem Cell Rep 11 (4):897–911

Wicherts JM, Bakker M, Molenaar D (2011) Willingness to share research data is related to the strength of the evidence and the quality of reporting of statistical results. PLoS One 6(11): e26828

Wilkinson MD, Dumontier M, Aalbersberg IJ et al (2016) The FAIR guiding principles for scientific data management and stewardship. Sci Data 3:160018

Costs of Implementing Quality in Research Practice

O. Meagan Littrell, Claudia Stoeger, Holger Maier, Helmut Fuchs,
Martin Hrabě de Angelis, Lisa A. Cassis, Greg A. Gerhardt,
Richard Grondin, and Valérie Gailus-Durner

Contents

O. Meagan Littrell and Claudia Stoeger contributed equally to this work.

O. Meagan Littrell · G. A. Gerhardt · R. Grondin
University of Kentucky Good Research Practice Resource Center and Department of Neuroscience,
Lexington, KY, USA

C. Stoeger · H. Maier · H. Fuchs · V. Gailus-Durner (✉)
German Mouse Clinic, Institute of Experimental Genetics, Helmholtz Zentrum München, German
Research Center for Environmental Health, Neuherberg, Germany
e-mail: gailus@helmholtz-muenchen.de

M. Hrabě de Angelis
German Mouse Clinic, Institute of Experimental Genetics, Helmholtz Zentrum München, German
Research Center for Environmental Health, Neuherberg, Germany

Experimental Genetics, School of Life Science Weihenstephan, Technische Universität München,
Freising, Germany

German Center for Diabetes Research (DZD), Neuherberg, Germany

L. A. Cassis
University of Kentucky Office of the Vice President for Research and Department of Pharmacology
and Nutritional Sciences, Lexington, KY, USA

© The Author(s) 2019   399
A. Bespalov et al. (eds.), *Good Research Practice in Non-Clinical Pharmacology
and Biomedicine*, Handbook of Experimental Pharmacology 257,
https://doi.org/10.1007/164_2019_294

Abstract

Using standardized guidelines in preclinical research has received increased interest in light of recent concerns about transparency in data reporting and apparent variation in data quality, as evidenced by irreproducibility of results. Although the costs associated with supporting quality through a quality management system are often obvious line items in laboratory budgets, the treatment of the costs associated with quality failure is often overlooked and difficult to quantify. Thus, general estimations of quality costs can be misleading and inaccurate, effectively undervaluing costs recovered by reducing quality defects. Here, we provide examples of quality costs in preclinical research and describe how we have addressed misconceptions of quality management implementation as only marginally beneficial and/or unduly burdensome. We provide two examples of implementing a quality management system (QMS) in preclinical experimental (animal) research environments – one in Europe, the German Mouse Clinic, having established ISO 9001 and the other in the United States, the University of Kentucky (UK), having established Good Laboratory Practice-compliant infrastructure. We present a summary of benefits to having an effective QMS, as may be useful in guiding discussions with funders or administrators to promote interest and investment in a QMS, which ultimately supports shared, mutually beneficial outcomes.

Keywords

Automation · Cost of quality (CoQ) · Documentation · German Mouse Clinic (GMC) · Good laboratory practice (GLP) · ISO 9001 · Quality management system (QMS) · Reproducibility

Abbreviations

CoQ	Cost of quality
GLP	Good laboratory practice
GMC	German Mouse Clinic
GRP	Good research practice
HMGU	Helmholtz Zentrum München
IMPC	International Mouse Phenotyping Consortium
IT	Information technology

KPIs   Key performance indicators
LIMS   Laboratory information management system
QA   Quality assurance
QC   Quality control
QMS   Quality management system
SOPs   Standard operating procedures
UK   University of Kentucky
USD   US dollar

1   Introduction

The mention of certified or standardized laboratory requirements to other scientists is often met with apprehension, a skeptical expression, and concerned questions about taking on additional bureaucracy and time-consuming paperwork in addition to introducing what is perceived as unnecessary limits for creative and innovative scientific freedom. However, we want to dispel misconceptions that implementing quality management in research practice is only marginally beneficial or too burdensome and costly to justify. Furthermore, the treatment of quality costs is misleading and incomplete if regarding only the costs of "doing something" – in this case implementing quality management. This approach, known as "omission bias," ignores the fact that "not doing something," i.e., not implementing quality management, also comes with costs, some clear and many others hidden or absent from general estimations.

We will describe two examples of implementing quality management systems in preclinical experimental (animal) research environments – one in Europe, the German Mouse Clinic, having established ISO 9001 and the other in the United States, the University of Kentucky (UK), having established Good Laboratory Practice (GLP)-compliant infrastructure. In the end, we hope to have made a convincing case for taking a long-term approach to promoting quality, where the costs are comparatively minor and the benefits exceed initial "activation energy" and commitments needed for implementation. Finally, we present a summary of benefits to having an effective QMS, as it may be useful in guiding discussions with funders or administrators to promote interest and investment in a QMS, which ultimately supports shared, mutually beneficial outcomes.

2   German Mouse Clinic: ISO 9001

2.1   Our Mission

The German Mouse Clinic (GMC) is part of the Helmholtz Zentrum München (HMGU) and located in Munich, Germany. Understanding gene function in general, and furthermore the causation, etiology, and factors for the onset of genetic diseases, is the driving force of the GMC. Established in 2001 as a high-throughput phenotyping platform for the scientific community, we have set up different phenotyping pipelines covering various organ systems and disease areas (Gailus-Durner et al. 2005, 2009; Fuchs et al. 2017; www.mouseclinic.de).

Standardized phenotyping is designated to the areas of behavior, bone and cartilage development, neurology, clinical chemistry, eye development, immunology, allergy, steroid metabolism, energy metabolism, lung function, vision and pain perception, molecular phenotyping, cardiovascular analyses, and pathology. In a comprehensive primary screen, we can analyze 700+ parameters per mouse and collect 400+ additional metadata (Maier et al. 2015). Our collaboration partners have to provide a cohort of age-matched mutant animals from both sexes and the corresponding wildtype littermates. Expectations for high-quality and scientifically valid outcomes are very high, due to widespread financial and time resources spent collectively. Therefore, study design, performance of experiments, material and equipment, as well as training of personnel have to be well coordinated, harmonized, and advanced.

2.2   Our Team and Main Stakeholders

Our *GMC team* consists of scientists with expertise in a specific disease area (e.g., energy metabolism or pathology), technicians performing the phenotypic analysis, animal caretakers, computer scientists, a management team, and the director. In total, we have approximately 50 team members with a third of the staff working in the animal facility barrier, which represents an additional challenge to maintaining efficient communication crucial for successful output and teamwork.

Collaboration partners who send us mouse models for phenotypic analysis are scientists or clinicians from groups at the HMGU and many different academic institutions, universities, hospitals, or industry from Germany, Europe, and other countries/continents like the United States, Australia, and Asia. Since the beginning of the GMC, we have analyzed mice from 170 collaboration partners/laboratories from 20 different countries.

The GMC is a partner in a number of consortia like the International Mouse Phenotyping Consortium (IMPC, a consortium of so-called mouse clinics all over the world, http://www.mousephenotype.org, Brown and Moore 2012; Brown et al. 2018) and INFRAFRONTIER (https://www.infrafrontier.eu; Raess et al. 2016). Together with the European mouse clinics, the GMC has developed standardized phenotyping protocols (standard operating procedures (SOPs)) in the European EUMODIC program (Hrabe de Angelis et al. 2015). These SOPs have been further developed for the use in the IMPC (https://www.mousephenotype.org/impress). Starting on the European (EUMODIC) and expanding to the international level (IMPC), we had to harmonize equipment, handling of animals, and documentation of laboratories to enable the compatibility of phenotyping data from different facilities around the globe.

2.3 Needs Concerning Quality Management and Why ISO 9001

There are logistical, experimental, and analytical challenges of systemic, large-scale mouse phenotyping to ensure high-quality phenotyping data. The mutant mice we receive for analysis are generated by different technologies (e.g., gene editing, knockout, knock-in) and on various genetic backgrounds. As a unique feature, we import age-matched cohorts of mice from other animal facilities for the phenotyping screening.

Therefore, several processes need to be quality-controlled in many different areas, including *project management* (request procedure, reporting), *legal requirements* (collaboration agreement, animal welfare, and gene technology regulations), *scientific processes* (scientific question, study design, choice of phenotyping pipeline, data analysis, reproducibility), *capacity management* (many parallel projects), and *knowledge* and *information management* (sustain and transfer expertise, internal and external communication). In order to cover this complex management situation, we decided to implement a QMS within the GMC.

Being a partner in an international consortium allows benchmarking with centers in similar environments. Due to this circumstance, our *rationale* for using the ISO 9001 standard was based on the following: (1) two other benchmark institutions with similar scope had already adopted ISO 9001-based QMSs, and (2) our products, consisting of research *data*, *inference*, and *publications*, are not regulated like data of, e.g., safety, toxicity, or pharmacokinetic studies and clinical trial data used for a new drug application to the regulatory authorities. Therefore, a *process-based* system seemed to be most suitable for our needs.

An ISO 9001-based QMS is very general and serves as a framework to increase quality in many aspects. In the GMC, we have implemented measures not only to improve our processes but also to directly improve instruments and increase the quality of our research results. To this end, the implementation of quality management went hand in hand with investments in information technology (IT) structure and software development.

2.4 Challenges

Building a QMS and going through the ISO 9001 certification process was a project that required significant personnel effort and time (2 years in our case). Although efforts should not be underestimated, they should be viewed within the context of promoting benefits (see Sect. 2.6). To answer the question what resources are required and how one can get started, we describe hereafter how the process was carried out at the GMC as an example of implementing an ISO 9001 QMS in a German Research Center setting, which includes *project management and organization*, the *IT infrastructure*, as well as *social* aspects.

Project Management and Organization After an introductory quality management training by an expert consultant for all GMC staff, we formed *a project* *management team* consisting of a quality manager (lead), the two heads of the GMC, the head of the IT group, and an affiliated project manager. An initial gap analysis provided useful data about our status quo (e.g., *inventory* of existing documentation). In the beginning, the main tasks were (1) development of a *project plan* with timelines and milestones, (2) defining "quality" in our biomedical research activities, and (3) determination of the *scope* of the QMS.

GMC's current strategy and future plans were reconsidered by establishing a *quality policy* (highest possible standard of research quality) and corresponding common *quality objectives* (specific, measurable, achievable, realistic, time-based (SMART)). The relaunched version of the ISO 9001:2015 challenged us to comprehensively describe our context. The *external context* of the GMC includes political, economic, social, and technological factors like animal welfare regulations, funding, health and translational research for the society, and working with state-of-the-art technologies. Our *internal context* is represented by our technical expertise (over 15 years of experience in mouse phenotyping), knowledge and technology transfer (workshops), application and improvement of the 3 Rs (https://www.nc3rs.org.uk/), and adherence to the rules of good scientific practice (GSP) and the ARRIVE guidelines (Karp et al. 2015).

Implementation of the *process approach* was addressed by defining the GMC's key processes in a process model (Maier et al. 2015) and specifying related *key performance indicators* (KPIs) (e.g., number of publications, trainings, reported errors, measures of internal audits). Process descriptions were installed including responsibilities, interfaces, critical elements, as well as *risks and opportunities* (risk-based thinking) to ensure that processes are clearly understood by everyone, particularly new employees.

It is often claimed that the ISO 9001 *documentation management* increases bureaucracy (Alič 2013). Therefore we did not create documents just for the sake of the ISO 9001. Instead we revised existing phenotyping protocols (SOPs) by adding essential topics (like data quality control (QC)) on the one hand and implemented missing quality-related procedure instructions on the other hand (e.g., regarding *error management, data management*, calibration, etc.). All documents were transferred in a user-friendly standard format and made easily retrievable (keyword-search) using commercial wiki software. A *transparent documentation management system* was successfully put into effect by implementing these processes.

The *constant improvement* of our systemic mouse phenotyping processes is continuously ensured by using established key elements of quality management such as *error management* including corrective and preventive actions (CAPA), an *audit system*, annual *management reviews*, and highly organized *capacity and resource management* structures.

IT Infrastructure and Software Tools Although our custom-built *data management* solution "MausDB" (Maier et al. 2008) has supported science and logistics for many years, the decision to implement a QMS triggered a critical review of our entire IT infrastructure. As a result, we decided to adapt our IT structure according to our broad process model (Maier et al. 2015) to deliver more reproducible data.

To this end, MausDB has been restructured into process-specific modules. MausDB 2.0 is a state-of-the-art laboratory information management system (LIMS) for automated data collection and data analysis (Maier et al. 2015). Standardized R scripts for data visualization and statistics are custom-developed for every phenotyping test and routinely applied. Numerous QC steps are built into the LIMS, including validation of data completeness and data ranges (e.g., min/max). Additional modules cover planning of capacities and resources as well as animal welfare monitoring and a project database tool for the improvement of project management and project status tracking.

On a *data management* level, a series of SOPs regulate reproducible data handling and organization. Comprehensive data monitoring allows detection of data range shifts over time, eventually triggered by changes in methods or machinery.

On the infrastructural level, a well-defined software development process built on the Scrum methodology ensures proper IT requirements management. Thus, a continuous improvement process can be applied to our IT tools. Script-based automation of frequent tasks encompasses daily backups of data as well as software build procedures. An "IT emergency SOP" has been developed to ensure well-planned IT crisis management (e.g., in case of server failure) and provides checklists and instructions for troubleshooting.

The challenges with IT-related issues described above were primarily of two categories: resources and change management. Self-evidently, implementation of all IT improvements required years of effort. However, the resulting overall process is much more efficient and less error prone. Active change management was essential in order to convince IT and non-IT staff that changes were necessary, although these would affect daily work routines. In the end, employees have come to realize that these processes save time and produce higher data quality.

Social Aspects In a preclinical research environment, the members of a research group traditionally have a high level of freedom for work planning and execution of scientific projects. Often, only one scientist conducts one detailed research project, plans the next steps from day to day, and communicates progress to the group in regular meetings. Therefore, a research environment encourages a self-responsible, independent working structure, leaving room for innovative trial and error, and supports both a creative and a competitive mind.

When you plan to implement a QMS in a preclinical research environment, you want to preserve the positive aspects of open mindedness and combine them with more regulated processes. As the initiator, you might find yourself in the position where you see both the opportunities and possible restrictions like limitations for innovative and unrestricted science. We decided to limit the certification to the standard screening pipelines in the beginning and not to force every research project into the ISO framework. Still, we encountered expected resistance to the implementation, since people feared losing freedom and control of their work structure, as well as disruption with unnecessary additional bureaucracy. This is, in every case, a complex psychological situation. Therefore, the implementation of a QMS takes time, understanding, sympathy, and measures of change and expectation management. A good deal of stamina, patience, and commitment is indispensable.

2.5   Costs

Quality control and management of preclinical animal research is a topic of increasing importance since low reproducibility rates (Begley and Ellis 2012) have put the knowledge generated by basic research in question. Furthermore, low reproducibility rates have caused immense delays and increased costs of therapeutic drug development (Freedman et al. 2015). *NOT implementing* recommended solutions like rigorous study designs, statistics consultation, randomization and blinding of samples to reduce bias, sharing data, or transparent reporting (Kilkenny et al. 2010; Landis et al. 2012; Freedman et al. 2015, 2017) in preclinical research will keep these costs high. Practical implementation of these standards could be supported by a well-developed QMS and provide structure and ensure achievement.

However, there are many concerns about the *financial expenses* needed for implementing quality in research practice by establishing and maintaining a QMS. Initial minimum financial costs include gaining knowledge about the chosen standard, in our case through trainings by an external consultant on ISO 9001-based quality management and documentation organized for the whole staff. Designation of at least one person who coordinates the implementation of the QMS is essential, which brings about the issue of *salary costs*. We hired a quality manager for 2 years (1 FTE) and in parallel trained a project manager from our team to the standard and for becoming an auditor to take over after the first certification (0.5 FTE).

In addition, the costs for the *certification body* needs to be included in cost calculations. Different certification bodies perform the ISO 9001 certification with varying costs. The first 3-year period comprises audit fees for the initial certification and two annual surveillance audits. This is followed by a recertification in the fourth year. As an example, our certification body costs were as follows: ~6,500€ for the initial certification, ~3,000€ for an annual surveillance audit, and ~5,000€ for the recertification.

Since we implemented new IT solutions, we had additional financial costs. For transparent *documentation management*, we acquired licenses for a supporting commercial wiki software (~2,200€ per year). All other software acquirements were not directly in the context of the ISO implementation and were solely data analysis or software development related. The same is true for a permanent *statistician position* (1 FTE) to support study design and data analysis.

Costs for implementing quality management in research practice are often a deterrent as the advantages of saving this money might be more obvious than the disadvantages. However, "not investing in costs" with respect to quality lead to "silent" costs. An example is the nonconformity management: if errors and corresponding measures are not properly documented, reduction as well as detection and avoidance of recurring errors through a detailed analysis is hardly possible, and the positive effects of increased efficiency and reduced failure costs are left out. The financial gain of an effective QMS can hardly be calculated in a research environment; however, documented, reviewed, and continuously improved processes ensure identification of inefficiencies, an optimized resource management, avoidance of duplication of work, and improved management information reducing the general operating costs.

2.6   Payoffs/Benefits

Why should an institution decide to improve quality in research practice by investing in an ISO 9001 QMS? We want to list a number of fundamental arguments and provide practical examples that might open a different perspective.

Building a QMS in the GMC demanded the highest efforts in the first 2 years before the certification (in 2014). However, with increasing maturation of the QMS, the process ran more efficiently due to enhanced quality awareness and a general cultural change which both led to increased quality of output (assured by monitoring the KPIs). People started to like the environment of having a QMS, and continual improvement became a habit. Over time, the benefits associated with using a QMS will offset the efforts it took to build it in the first place. Some of the most striking benefits of having an expanded QMS are listed hereafter.

Management Reviews These are important controlling steps as they give an annual overview of the actual state of the processes including all KPIs, the content of errors and the corresponding actions, open decisions that were supposed to be closed during the year, or specific actions that are pending. To this end, this kind of review differs from the usual reporting to funding authorities. They solely serve the quality status and reinforce focus on strategic, quality-related goals that have been identified as priorities. This is particularly useful since concentrating on important issues (e.g., increased QC issues in specific tests or applying a risk-based approach) is something that is often postponed in favor of other tasks requiring frequent or immediate attention. Here you need to deal with formal numbers and can react and adapt milestones if specific problems have not been addressed adequately. Surprisingly, this kind of review enabled us to react quickly to new developments. Since the digital assembly of the KPIs is in place, the numbers can be easily reported also during the year, and fact-based decisions can be made. To this end, contrary to the belief that a QMS causes a bureaucratic burden, the QMS actually facilitates agile project management.

Audit System Internal audits are an often underestimated element of a QMS. By performing internal audits (e.g., independent phenotyping protocol reviews, complex process audits, or audits addressing current quality problems such as reduction of bias), we ensure that standardization is guaranteed, measures for improvement are defined, and prevention of undesired effects is addressed. Internal system audits as well as the third-party audits ensure the integrity and effectiveness of the QMS.

> **Box 1 First Third-Party Audit**
> BEFORE: Being part of a third-party audit was initially mentally and emotionally demanding: just before the first certification audit, personnel (afraid of the visit by the audit team) kept calling to report minor issues and ask what to do.
> AFTER: Now, members of the group are used to internal scientific method auditing and have realized that we do not run the QMS solely for the certification body but for our own benefit. Today, while presenting the systemic phenotyping methods in a third-party audit, people feel accomplished, enthusiastic, and self-confident.

Training Concept Comprehensive training of personnel is time-consuming and associated with extensive documentation. However, training ensures establishment and maintenance of knowledge. New employees complete an intensive induction training including the rules of good scientific practice (GSP), 3 Rs, awareness for working in an animal research environment, QM issues, and legal regulations. Regular QM trainings build and maintain awareness of quality issues. The *"not documented, not done" principle* is well accepted now and supports the transparency of personnel competence assessment. In addition, we are currently building an eLearning training program in order to save time for logistics in case people missed trainings.

Traceability All processes were critically assessed for traceability. On the *physical level*, temporal and spatial tracking of mice and samples (blood, tissue) is an issue. We implemented a barcoding in our LIMS to register all samples. On the *data level*, we aim to maintain full traceability of data and metadata. This means we link file-based raw data to our LIMS and capture all metadata that may influence actual data, e.g., experimenter, equipment, timestamp, and device settings. On the *process level*, all transitions between sub-processes ("waiting," "done," "cancelled") are logged. This enables us to monitor dozens of ongoing projects at any time with custom-built tools to identify and manage impediments.

Box 2 Traceability Versus Personalized Data Storage

BEFORE: Some 10 years ago, we had to ask a collaboration partner for a re-genotyping because of identity problems within a cohort of mice. Tail biopsies were sent, but the electronic list correlating the biopsy numbers to the corresponding mouse IDs was saved on a personal device unavailable for the team. Therefore, the results could not be matched and the data analysis was delayed for more than 4 weeks.

AFTER: Samples now carry a barcode label with the mouse ID and any lists are saved in a central project folder.

Reproducibility With respect to quality, reproducibility of results is of paramount importance. In addition to SOPs, which regulate how phenotyping procedures are physically carried out, we put considerable efforts in making data analysis and visualization reproducible. To this end, we seamlessly integrated R (R Core Team 2013), a free statistical computing environment and programming language with our LIMS MausDB. Upon user request in MausDB, R scripts perform customized, test-specific statistical analyses as well as data visualizations. This tool restricts user interaction to the mere selection of a data set and the respective R script ensuring that same data will always reproduce the same statistical results and the same plots.

Box 3 Taking Responsibility in Writing Up Publications

BEFORE: We always provided our collaboration partners with the raw data so that they could perform additional analyses. In the past, during drafting manuscripts, we did not verify in detail if we were able to *reproduce* their statistical analysis and figures.

AFTER: With the implementation of the QMS, we have formalized how a manuscript is processed. This process now includes a step in which our in-house statistician reproduces all analyses and figures using our data as well as additional data from the collaboration partner.

2.7 Lessons Learned/Outlook

Certification to ISO 9001 is not a requirement in nonregulated preclinical biomedical research and also does not define scientific standards, but it represents a reasonable strategy to improve data quality.

GMC's QMS: A Success Story? Our ISO 9001:2015-based QMS helps us to generate and maintain transparent and traceable data records within a broad spectrum of standardized phenotyping processes with low variability and increases collaboration partner's trust in the analysis, interpretation, and reporting of research data. This structured approach also supports compliance with manifold regulations and promotes awareness and risk-based thinking for the institutional context as well as meeting the requirements of funders, personnel, the scientific community, and the public. However, to specifically address the quality of data output, we see the need to broaden the perspective and to reach out to other parties who perform quality assessments in preclinical research.

Networking Although certification is rarely found in preclinical research, participating in a network of institutions in similar scientific research areas performing, e.g., annual internal ISO 9001 audits on a mutual basis is an opportunity to address common scientific quality problems and therefore a future goal. Positive examples are the Austrian biobanks (BBMRI.at) and a French network of technological research platforms (IQuaRe; https://www.ibisa.net) having built ISO 9001 cross-audit programs.

Limits of Automation We have learned that beyond a certain level of complexity, further automation requires increasingly and disproportionately higher efforts and is therefore limited. At the GMC, automation of data analysis and visualization works well for projects adhering to our standardized workflow. Beyond that, customization of projects adds additional complexity that is not compatible with full automation. In such projects, custom data analysis still has to be performed manually.

Innovation At the GMC, information technology has supported operative processes since 2001. In that sense, "digitalization" is not just a buzzword for us, but a continuous process that aims for measurable and sustainable improvement of our work. IT solutions implemented so far mainly cover standardized processes.

"Machine learning" is another heavily used catch phrase. In the case of the auditory brainstem response test, we currently use our vast data set to develop methods for automated detection of auditory thresholds, including deep learning by neural networks. We are sure that this will provide a more reproducible method, independent from human influences. Of course, human experts will always review and QC the results. Nevertheless, setting the scene with a QMS paves the way for future investment in modern IT technologies and digitalization.

Finally, we made the experience that you need to allow flexibility and consider not including all processes and/or details in the ISO 9001 QMS. Indeed, the ISO 9001 QMS does not require to incorporate every process, and this might be also a misconception of many principal investigators (PIs) that prevents the introduction in academic settings. Real innovation is a truly inefficient and non-directed process (Tenner 2018) that needs to reside in a protected area. As soon as innovative research projects generate either new technologies or techniques, we slowly implement these into our processes and apply quality management measures step by step. Therefore, it is also important to not allow the system to "take over," getting lost in micromanagement or obsessed by automation with new IT solutions. It is all about balancing the needs for quality and scientific freedom and keeping the expectations from all involved parties in a reasonable frame. To this end, it is necessary to allow time during the implementation process and to understand that benefits are apparent only after a longer time period.

Although efforts for implementing a QMS might be more tricky in an academic setting in a university (many PIs, high diversity of activities, rapid change of personnel) than in a mouse clinic performing highly standardized tests and procedures, the ISO 9001 standard gives a framework for introducing more quality-relevant aspects in preclinical research and helps enormously with team mindset. An ISO 9001-based QMS supports quality in manifold key process types as well as in supporting, analysis, and improvement processes like training, communication, documentation, auditing, and error management.

Nevertheless, the determination of good quality data output can only be judged by scientific peers and the respective community.

3   University of Kentucky Good Research Practice (GRP) Resource Center

The Good Research Practice (GRP) Resource Center at the University of Kentucky (UK) is a research support unit under the office of the UK Vice-President for Research. In the context of experimental life sciences, it is particularly important to note that UK is also an academic healthcare center. The UK Albert B. Chandler Medical Center is located in the health sciences campus and is comprised of six colleges of biomedical sciences (dentistry, health sciences, medicine, nursing, pharmacy, and public health) as well as the clinical facilities associated with UK HealthCare. These include the UK Chandler Hospital, Kentucky Children's Hospital, UK Good Samaritan Hospital, Markey Cancer Center, Gill Heart Institute, Kentucky Neuroscience Institute, and Kentucky Clinic, which collectively support research, education, and healthcare. The Chandler Medical Center is 1 of only 22 academic medical centers in the United States that house 3 nationally recognized federally funded centers: a National Cancer Institute-designated cancer center, an Alzheimer's Disease Center funded by the National Institute on Aging, and the Kentucky Center for Clinical and Translational Science funded as part of the NIH's Clinical and Translational Science Award Consortium.

UK is a public, land grant university established in 1865 located on a 784 acre urban campus in Lexington, Kentucky. Sixteen colleges and diverse professional schools are available including Colleges of Agriculture, Dentistry, Health Sciences, Medicine, Pharmacy, and Public Health, to name a few. Major graduate research centers offering high-quality multidisciplinary graduate training include the Graduate Centers for Nutritional Sciences and Gerontology. UK researchers have developed highly productive collaborations across diverse disciplines, in multidisciplinary and interdisciplinary research. Research and academic activities at UK span all 16 colleges, 76 multidisciplinary research centers, and 31 core research facilities. UK has over 80 national rankings for academic and research excellence and is 1 of 108 private and public universities in the country to be classified as a research university with very high research activity by the Carnegie Foundation for the Advancement of Teaching.

3.1   Our Mission

The mission of the GRP Resource Center is to assist academic scientists by providing tools to support research processes that promote data quality, integrity, and reproducibility. Our center is comprised of faculty and staff with combined experience as researchers as well as regulatory experience. Our experience includes conducting both (1) research in full compliance with US FDA Good Laboratory Practices (GLP), involving defined roles as management, study directors, and quality assurance (QA), and (2) non-GLP research, which is not required to be conducted according to GLP regulations but is carried out using many of the requirements outlined (e.g., maintaining written methods and record of method changes, rigorous data documentation and traceability, etc.), however typically without QA oversight. Using experience and training with laboratory and quality management requirements set forth by the GLP regulations, our center supports laboratories seeking either mandated requirements (i.e., GLP-compliant research) or those voluntarily seeking to strengthen current practice to conduct nonregulated research. With recent alignment of stakeholder interest in enhancing the value of preclinical research, discussed further below, we support researchers in meeting these expectations as well. This is achieved through individual consultation, group training, and other resource sharing to provide templates for building a practical and effective quality management system.

3.2   Our Stakeholder's Interests and Concerns

Most visibly, primary stakeholders in research include those with direct economic investment, which may take forms of both private and government funding mechanisms. In many governmental structures, investment can be extended to the general public. The public may contribute both direct financial investment in funding research and, more indirectly, the economic burden of supporting healthcare measures that support unfavorable health outcomes where effective therapies are delayed or do not exist. Additionally, many individuals also have a personal investment with hope for a better understanding of diseases and development of therapies for illnesses that may affect them personally or family members. Other stakeholders include publication agencies, which rely on transparent and accurate reporting of research processes and results as well as a thorough discussion of possible limitations of findings when interpreting significance of results.

Public opinion of scientific research has declined due to increasing evidence for lack of reproducibility (Begley and Ioannidis 2015; Pusztai et al. 2013). Studies conducted by Bayer included review of new drug targets in published results and found that 65% of in-house experimental data did not match published results

(Mullard 2011). Similarly, a 10-year retrospective analysis of landmark preclinical studies indicates that the percentage of irreproducible scientific findings may reach as high as 89% (Begley and Ellis 2012). In addition, studies indicate a recent surge in publication retractions. Approximately 30 retraction notices appeared annually in the early 2000s, while in 2011 the web of science indexed over 400 retractions (a 13-fold increase), despite total number of papers published rising by only 44% (Van Noorden 2011).

Although stories of scientific misconduct, particularly fraud, are often the most memorable and garner the most attention, studies support that other experimental factors related to irreproducibility are bigger contributors to the problem (Gunn 2014; Freedman et al. 2015; Collins and Tabak 2014; Nath et al. 2006). *Key contributors to irreproducibility* include biological reagents/reference materials, study design, analysis, reporting, researcher and/or publishing bias, and institutional incentives and career pressures that compromise quality (Freedman et al. 2015; Begley and Ellis 2012; Begley et al. 2015; Ioannidis 2005; Baker 2016a, b; Ioannidis et al. 2007). These factors engage the scientific community at multiple levels, highlighting a need for a cultural shift and prioritization on quality, which includes the primary scientist, but also call into action funding agencies/institutes, publishers, and institutional policy makers who each have responsibilities in directing the research process and setting benchmarks for success. In the United States, a major stakeholder in research quality, the National Institutes of Health (NIH), has recognized the combination and interaction of factors that contribute to irreproducibility in preclinical research and acknowledged that there is a community responsibility to the reproducibility mission. Several NIH institutes and centers are testing measures to better train researchers and better evaluate grant applications to enhance data reproducibility (Begley and Ellis 2012; Collins and Tabak 2014), and major publishers have committed to taking steps to increase the transparency of published results (McNutt 2014; Announcement 2013).

3.3 How to Address Data Irreproducibility

With the abovementioned contributors to irreproducibility in mind, several research processes may be isolated that would benefit from implementing a quality management system. At the forefront are standardized and comprehensive documentation procedures that allow adequate study reconstructability. In research environments with frequent personnel turnover, which is particularly the case in academic settings, accessing traceable records is a cornerstone to further isolation of specific sources of irreproducibility as well as accurate and transparent reporting of experimental methods and results. For example, documentation of the key reagent batch/lot information may be critical in troubleshooting discrepancies in bioanalytical results. A secure, indexed archiving system that protects the integrity of materials is also necessary to facilitate expedient access to study materials over time. Standardization of experimental design elements is also needed with agreement among and within laboratories for critical aspects, which include, but are not limited to, key chemical/

biological resources, personnel training, equipment testing, statistical methods, and reporting standards (Nath et al. 2006; Kilkenny et al. 2010; Landis et al. 2012). Standardization and best practice guidelines for particular technologies and fields of study are continuing to be developed, debated, and refined (Taussig et al. 2018; Almeida et al. 2016); however, preclinical research would also benefit from reaching an agreement on quality management expectations common to many research applications.

3.4   Why Build a GLP-Compliant Quality Management System in Academia

Our decision to become GLP-compliant resulted from requests from industry partners who were seeking continuity through early discovery work to GLP-compliant safety studies. This circumstance arose out of a combination of experience and capabilities unique to our team with magnetic resonance imaging-targeted intracranial drug delivery in large animal models.

The GLP regulations describe the minimum requirements for conducting non-clinical studies that support or are intended to support research or marketing permits for products regulated by the FDA and Environmental Protection Agency (Pesticide Programs: Good Laboratory Practice Standards 1983; Nonclinical Laboratory Studies: Good Laboratory Practice Regulations 1978). The GLP regulations were formalized as laws in 1978 following evidence for fraudulent and careless research at major toxicology laboratories, namely, Industrial Bio-Test Laboratories and Searle. In one example of a study that was executed poorly, validation of methods used to generate test article mixtures had not been carried out, leading to nonhomogeneous mixtures and resulting in uncontrolled dosing. In other cases, study records were poorly maintained and reports appeared falsified: animals that were documented as deceased were later reported alive in study reports. Investigations culminated in the conclusion that toxicology data could not be considered valid for critical decision-making by the agency, putting public health at a tremendous risk (Baldeshwiler 2003; Bressler 1977).

With the intent to assure the quality, integrity, and reproducibility of data, GLP regulations direct conditions under which studies are initiated, planned, performed, monitored, and reported. To name select general examples, meeting GLP regulatory requirements involves:

- Characterization of key reagents (e.g., test/control articles)
- Traceable, accurate, and archived study records
- Established personnel organization
- Training of all personnel
- A written study protocol and other written methods for facility processes with record of any changes

- An independent monitoring entity, quality assurance unit (QAU), which performs inspections and audits to assure facility and regulatory requirements were met and reports accurately reflect the raw data

These elements, although not all-inclusive, are key components that address the quality management needs discussed above. Since achieving data quality is also a priority in nonregulated research, as it is a basis for drug development, adapting GLP elements and applying those as voluntary quality practices is one option to meet quality needs. At the University of Kentucky, the GRP Resource Center has taken a lead role in developing a model to guide individual researchers and facility directors who are seeking to strengthen current practices but are not conducting studies where GLP compliance is required. Our focus to date has involved initial and ongoing assessment of centralized, shared use research core facilities and instruction of research trainees. Our quality consultation services are also available by researcher request for both voluntary quality management consultation and GLP needs assessment, as facility needs dictate. Select GLP elements are listed as applicable to nonregulated preclinical laboratory research and achieving quality (see Table 1).

3.5 Challenges

Our discussion will focus on GLP suitability and limitations for quality achievement needs and implementation in preclinical research. Implementing an effective QMS, particularly those outlined in GLP regulations, is met with specific needs such that increased resource allocation should be expected including time, personnel, and associated fiscal investment. For example, few preclinical laboratories have the resources necessary to characterize test/control articles to GLP standards. Additionally, GLP compliance presents a greater challenge in the university setting where nonregulated operational units may interact with the GLP infrastructure. Also, complex administrative relationships may result in difficulty defining GLP organizational structure. Furthermore, personnel concerns include high turnover, inherent to the training environment of academia, as well as obtaining support for independent personnel to carry out QA functions, particularly where GLP studies may not be performed on a continual basis. However, compliance challenges can be lessened when executive administrators value quality and can offer support to address these needs. These unique GLP challenges and options for supporting a GLP program are discussed further in the literature (Hancock 2002; Adamo et al. 2012, 2014). Understandably, studies that require GLP compliance do not present substantial flexibility in quality management components. Nonetheless, using GLP elements as a template for nonregulated preclinical studies, where GLP is not mandated, is a viable option and may substantially lower variation and irreproducibility by focusing on top contributors to bad quality (Freedman et al. 2015). It is important to note that although these self-imposed standards are non-GLP, they still add value. Thinking in terms of a quality spectrum rather than all-or-nothing may be a more productive model for addressing reproducibility issues in a resource-limited setting.

Table 1 Total CoQ as represented by cost of failure + cost of achieving

CoQ failure	CoQ achieving	Reduced total CoQ

• Erroneous results → tentative decisions
• Inconsistent record keeping
• Repeating experiments → misutilized resources
• Compromised ethics (use of animals, clinical specimens, etc.)
• Lack of stakeholder confidence
• Missed funding opportunities (government/industry)
• Therapeutic delay
• Embarrassment
	• Investment (time + resources) in preventing failure:
	  – Standardizing documentation, e.g., developing SOP(s) and appropriate data forms
	  – Standardizing technical and administrative procedures, e.g., developing SOP for sample processing, record retention/archival, etc.
	  – Training
	  – Key resource authentication – e.g., standardized reference material, characterization of test/control articles, test systems, etc.
	  – Facility +process suitability assessment
	  – Defining personnel organizational structure w/ quality management, e.g., management, study director, quality assurance
	  – Instrument/equipment preventive maintenance
	  – Record process deviations
	  – Report standardization
	• Investment (time + resources) in quality appraisal:
	  – Audits/inspection mechanism
	  – Equipment testing
	  – Proficiency assessment
	  – Review process deviations
		• Reliable data → increased confidence in decisions
		• Transparent records + accurate and efficient reporting process
		• Greater resources availability for biological inquiry + extending findings
		• Restored stakeholder confidence
		• Maintain and expand funding opportunities
		• More efficient therapeutic development

For example, without support for an independent QAU, a periodic peer-review process among laboratory staff can be implemented to verify that critical record keeping, for example, legibility and traceability, is present in study records (Baker 2016b; Adamo et al. 2014).

3.6   Costs

Although definitions and models vary, the cost of quality (CoQ) has been described as the sum of those costs associated with both failure and achieving the desired quality (Wood 2007). Treatment of CoQ in the clinical laboratory is helpful as a template to evaluate the nonclinical laboratory concerning both quality failure and quality achievement costs (Berte 2012a; Wood 2007; Carlson et al. 2012; Feigenbaum 1991). Models subdivide quality costs that are relevant to the assessment of research laboratory costs and include prevention, appraisal, internal failure, and external failure. As examples, prevention quality costs include quality management planning, process validation, training, preventive maintenance, and process improvement measures, and appraisal costs include those associated with verification that desired quality is being achieved like periodic proficiency testing, instrument calibration, quality control, and internal inspections/audits. Both prevention and appraisal costs are part of the cost of achieving quality, while internal and external costs are associated with failure, occurring either before or after delivery of a product/service, respectively. For example, obtaining a contaminated sample that must be collected again is an internal cost, while recalling released results is an external cost (Berte 2012a).

In calculating quality failure costs in preclinical research, using reproducibility as an indicator for variation/quality and a modest estimation of 50% irreproducibility rate, this cost is estimated to exceed 28 billion dollars in the United States annually (Freedman et al. 2015). Of note, lack of agreement defining reproducibility was noted in analyses and includes both variation in results and ability to carry out replication using the available methods, and this also does not imply that irreproducible studies were of zero value (Freedman et al. 2015; Begley and Ellis 2012; Mullard 2011; Ioannidis et al. 2009). Additional failure costs are publication retractions and time/resources consumed investigating confounding results and repeating experiments, which are resources consequently unavailable for consideration of biological questions and extending findings (Baker 2016b), ultimately contributing to delayed development of effective therapies (Freedman and Mullane 2017). Cost of quality models may also include "intangible" costs associated with quality failure such as lost opportunities (Schiffauerova and Thomson 2006). As has been described, one such cost is the lack of confidence among stakeholders in biomedical research and conceivably lost opportunities due to unreliable results. Quality deficiencies may also cause undue damage to workplace morale, for example, if deficiencies result in blame placed on personnel when a lack of quality management processes or ineffective processes are actually at fault (Berte 2012a). In implementing measures to achieve quality, intangible costs may also be incurred if quality achievement processes are perceived to result from distrust in personnel competency.

Estimations from clinical laboratories indicate that approximately 35% of total operating costs are associated with CoQ, with CoQ failure and achievement accounting for 25% and 10% of the total, respectively (Menichino 1992). Although estimates specific to preclinical research are lacking, in this laboratory example, the cost of failure exceeds the cost of achieving. This makes investment in processes that promote quality (prevention, appraisal) particularly valuable since these typically require less investment than quality failures and, if implemented effectively, reduce the cost of failure by reducing the total CoQ. Furthermore, costs associated with maintaining an effective quality management system would likely decrease over time as corrective and preventive actions are implemented (Berte 2012a, b). We estimate that the initial implementation costs to meet GLP requirements totaled

$623,000 USD over a 3-year period. These costs related to achieving quality can be divided broadly into three categories: (1) internal staff/administrative costs related to establishing written methods for standardizing laboratory and other facility operations (52% of the costs); (2) appraisal costs, which includes gap analyses/external consulting and training of internal QA associates (47% of the costs); and (3) equipment maintenance and testing (1% of the costs). Of note, these categories were difficult to strictly partition among appraisal, preventing quality failure, and equipment. For example, standardizing reagents were used to both prevent failure, when used for equipment maintenance activities, and appraisal activities, if used to test equipment. Personnel effort (as accounted for as internal staff/administrative costs) also crosses categories since effort includes both time spent standardizing equipment testing criteria (to prevent failure) and time spent evaluating conformance to those criteria (appraisal). Thus, there may be some flexibility in the estimated contribution of each category to overall cost.

Although implementation costs were high, these were met with gained opportunities. Research funding by industry contracts during and after the establishment of our GLP-compliant infrastructure increased 109%, from $3.5 million USD in total support in the 6 years preceding its existence to $7.3 million USD in total support over the following 6-year period. While difficult to directly quantify, our facility collaborative opportunities were also broadened through increased industry recognition due to having experience using a QM framework familiar to the pharmaceutical/device industry. As would be predicted from other models (Berte 2012a, b), we experienced a sharp decline in costs after having established our QM framework. For example, our total annual maintenance cost represents, on average, only about 9% of the total implementation cost. Maintenance costs were reduced across all three categories, however most dramatically in costs related to appraisal/external consulting and establishing a QAU, which represents approximately 1% of the initial appraisal costs. Of note, our initial appraisal costs, which were directed toward full compliance with GLP regulations, involved identifying and working with specialized external consultants who have regulatory expertise. In research settings where quality needs do not engage GLP regulations, other options may be available including utilizing internal personnel for appraisal activities.

Costs incurred due to quality failures are difficult to quantify or even estimate; however, the further downstream in the research process quality failures is exposed, the greater the cost (Campanella 1999). For example, in the case that quality failure (as represented by variation or irreproducibility in the research results in preclinical research) is realized after publication, the cost of failure is the highest, requiring supporting all of the research costs again (Berte 2012a). Potentially, additional intangible costs are incurred if reported results were used as the scientific premise for additional research, an area that has been described clearly by NIH (Collins and Tabak 2014). Therefore, allocating resources to support the costs of attaining quality early in the research process (prevention and appraisal) is a worthwhile investment that, overtime, reduces the CoQ by reducing the cost of failure. Resources spent on services/reagents to perform preventive maintenance/testing of equipment are examples of such quality attainment costs. As outlined in Table 1, GLP regulations describe several requirements that support achieving quality which are included in a consideration of CoQ (Nonclinical Laboratory Studies: Good Laboratory Practice Regulations 1978; Berte 2012a).

3.7 Payoffs/Benefits

In short, major benefits to investing in quality in the research laboratory include recovering resources spent due to variation or reproducibility failures, as discussed above. Those involve both costs incurred in repeating the research process, in the case of irreproducibility, and recovering intangible costs: more certainty in extending research findings based on a solid scientific premise, regaining missed opportunities, and restoring stakeholder confidence. Prioritizing quality in preclinical research supports a more transparent research process with greater efficiency in developing therapies and furthering the understanding of health and disease. As described above, complying with research standards, like GLP regulations, is costly but may offer a competitive advantage by facilitating mutually beneficial industry partnerships earlier and more readily. Bridging this gap in translational research is of particular relevance in the academic healthcare center environment with the unique opportunity to utilize expertise and innovative technology platforms unique to academia (Stewart et al. 2016; Hayden 2014; Tuffery 2015; Cuatrecasas 2006) within a framework that can support continuity with researcher-initiated nonclinical studies through clinical trials (Adamo et al. 2014; Slusher et al. 2013; Yokley et al. 2017).

3.8 Lessons Learned/Outlook

Applying CoQ models to current estimations of the financial costs associated with irreproducible research makes a convincing case for reevaluating existing research practices. Exposure to regulatory guidelines and more general training in quality management in the academic research environment has been well received as a supplement to the research trainee curriculum, potentially broadening career opportunities. Following training workshops, we have received positive feedback, particularly regarding the benefit to research trainees, and some faculty have requested further individual consultation in order to identify quality needs and better implement voluntary quality practices in their laboratories. Researchers have also showed reluctance to implement a QMS with limited resources that can be allocated to such efforts. It is not uncommon for senior researchers to initially seek to implement only the minimum requirements to meet funding agency/journal expectations; however, we are optimistic that benefits will be realized with time.

With this objective in mind, in our consultation process, we value incremental progress in implementing quality management processes and welcome feedback and customized, researcher-initiated solutions to quality management needs. This is the approach we have taken in consultation with UK research core facilities as an ongoing process, which is currently underway.

In our experience, it is clear that higher administrative support is necessary to guide these collective efforts. Our center received early support from institutional administration (Dean's Office, College of Medicine) to supplement initial implementation costs. Additionally, ongoing administrative support from the Vice-President of Research has been indispensable in outlining quality indicators, initiating meaningful cultural change, and supporting the infrastructure necessary for continued achievement and assessment of quality. We believe our GLP experience strongly supports recent reproducibility initiatives from major funding agencies (e.g., NIH) and thus the shared interest of university leadership and researchers seeking to make meaningful contributions to the existing body of knowledge through their research efforts.

4 Conclusions: Investing in Quality

In this chapter, we have described our experience with the implementation of a QMS in research practice, using ISO 9001 and GLP as examples, and how the numerous benefits outweigh initial efforts. Existing concerns and evidence for quality collapse necessitate collaborative measures to improve the value of preclinical research. The costs associated with quality failures are considerable and often compound in the research process, leading to repeatedly encumbering research costs and/or distribution of misinformation and experiments based on unreliable results. However, investment in an effective QMS provides considerable benefits and return of costs multifold – recovering failure costs and expanding opportunities for the scientific community (see Table 2). In summary, although complete agreement and rapid adoption of quality standards in the scientific community seem hard to achieve, using general quality principles and processes that are available is not only a viable option but rather the cornerstone in promoting quality and reproducibility, effectively reducing the cost of quality.

Table 2 Major benefits of implementing quality management

- Higher efficiency, reproducibility, and quality
- Basis for new IT solutions and agile project management
- Easier communication, exchange, and comparability with other laboratories, platforms, and infrastructures with similar focus
- Openness, involvement, and transparency in the team (trust building)
- Investment in the future of high-quality expectations in a global digitalized environment

References

Adamo JE, Bauer G, Berro M et al (2012) A roadmap for academic health centers to establish good laboratory practice-compliant infrastructure. Acad Med 87(3):279–284. https://doi.org/10.1097/ACM.0b013e318244838a

Adamo JE, Hancock SK, Bens CM, Marshall M, Kleinert LB, Sarzotti-Kelsoe M (2014) Options for financially supporting GLP and GCLP quality assurance programs within an academic institution. J GXP Compliance 18(1)

Alič M (2013) ISSUES OF ISO 9001: implementation: improper praxes leading to bureaucracy. Dyn Relatsh Manag J 2(1):55–67. https://doi.org/10.17708/DRMJ.2013.v02n01a05

Almeida JL, Cole KD, Plant AL (2016) Standards for cell line authentication and beyond. PLoS Biol 14(6):e1002476. https://doi.org/10.1371/journal.pbio.1002476

Announcement (2013) Reducing our irreproducibility. Nature 496(7446):398. https://doi.org/10.1038/496398a

Baker M (2016a) 1,500 scientists lift the lid on reproducibility. Nature 533(7604):452–454. https://doi.org/10.1038/533452a

Baker M (2016b) How quality control could save your science. Nature 529(7587):456–458. https://doi.org/10.1038/529456a

Baldeshwiler AM (2003) History of FDA good laboratory practices. Qual Assur J 7:157–161. https://doi.org/10.1002/qaj.228

Begley CG, Ellis LM (2012) Drug development: raise standards for preclinical cancer research. Nature 483(7391):531–533. https://doi.org/10.1038/483531a

Begley CG, Ioannidis JP (2015) Reproducibility in science: improving the standard for basic and preclinical research. Circ Res 116(1):116–126. https://doi.org/10.1161/CIRCRESAHA.114.303819

Begley CG, Buchan AM, Dirnagl U (2015) Robust research: institutions must do their part for reproducibility. Nature 525(7567):25–27. https://doi.org/10.1038/525025a

Berte LM (2012a) The cost of quality. In: Harmening DM (ed) Laboratory management: principles and processes, 3rd edn. D.H. Publishing & Consulting, St. Petersburg, pp 339–359

Berte LM (2012b) Quality management in the medical laboratory. In: Harmening DM (ed) Laboratory management: principles and processes, 3rd edn. D.H. Publishing & Consulting, St. Petersburg, pp 3–34

Bressler J (1977) The Bressler report

Brown SD, Moore MW (2012) Towards an encyclopaedia of mammalian gene function: the international mouse phenotyping consortium. Dis Model Mech 5(3):289–292. https://doi.org/10.1242/dmm.009878

Brown SDM, Holmes CC, Mallon AM et al (2018) High-throughput mouse phenomics for characterizing mammalian gene function. Nat Rev Genet 19(6):357–370. https://doi.org/10.1038/s41576-018-0005-2

Campanella J (1999) Principles of quality costs, 3rd edn. American Society for Quality Press, Milwaukee

Carlson RO, Amirahmadi F, Hernandez JS (2012) A primer on the cost of quality for improvement of laboratory and pathology specimen processes. Am J Clin Pathol 138(3):347–354. https://doi.org/10.1309/AJCPSMQYAF6X1HUT

Collins FS, Tabak LA (2014) Policy: NIH plans to enhance reproducibility. Nature 505(7485):612–613

Cuatrecasas P (2006) Drug discovery in jeopardy. J Clin Invest 116(11):2837–2842. https://doi.org/10.1172/JCI29999

Feigenbaum AV (1991) Total quality control, 3rd edn. McGraw-Hill, New York

Freedman S, Mullane K (2017) The academic-industrial complex: navigating the translational and cultural divide. Drug Discov Today 22(7):976–993. https://doi.org/10.1016/j.drudis.2017.03.005

Freedman LP, Cockburn IM, Simcoe TS (2015) The economics of reproducibility in preclinical research. PLoS Biol 13(6):e1002165. https://doi.org/10.1371/journal.pbio.1002165

Freedman LP, Venugopalan G, Wisman R (2017) Reproducibility2020: progress and priorities. F1000Res 6:604. https://doi.org/10.12688/f1000research.11334.1

Fuchs H, Aguilar-Pimentel JA, Amarie OV et al (2017) Understanding gene functions and disease mechanisms: phenotyping pipelines in the German mouse clinic. Behav Brain Res 352:187. https://doi.org/10.1016/j.bbr.2017.09.048

Gailus-Durner V, Fuchs H, Becker L et al (2005) Introducing the German mouse clinic: open access platform for standardized phenotyping. Nat Methods 2(6):403–404. https://doi.org/10.1038/nmeth0605-403

Gailus-Durner V, Fuchs H, Adler T et al (2009) Systemic first-line phenotyping. Methods Mol Biol 530:463–509. https://doi.org/10.1007/978-1-59745-471-1_25

Gunn W (2014) Reproducibility: fraud is not the big problem. Nature 505(7484):483. https://doi.org/10.1038/505483b

Hancock S (2002) Meeting the challenges of implementing good laboratory practices compliance in a university setting. Qual Assur J 6:15–21. https://doi.org/10.1002/qaj.165

Hayden EC (2014) Universities seek to boost industry partnerships. Nature 509(7499):146. https://doi.org/10.1038/509146a

Hrabe de Angelis M, Nicholson G, Selloum M et al (2015) Analysis of mammalian gene function through broad-based phenotypic screens across a consortium of mouse clinics. Nat Genet 47 (9):969–978. https://doi.org/10.1038/ng.3360

Ioannidis JP (2005) Why most published research findings are false. PLoS Med 2(8):e124. https://doi.org/10.1371/journal.pmed.0020124

Ioannidis JP, Polyzos NP, Trikalinos TA (2007) Selective discussion and transparency in microarray research findings for cancer outcomes. Eur J Cancer 43(13):1999–2010. https://doi.org/10.1016/j.ejca.2007.05.019

Ioannidis JP, Allison DB, Ball CA et al (2009) Repeatability of published microarray gene expression analyses. Nat Genet 41(2):149–155. https://doi.org/10.1038/ng.295

Karp NA, Meehan TF, Morgan H et al (2015) Applying the ARRIVE guidelines to an in vivo database. PLoS Biol 13(5):e1002151. https://doi.org/10.1371/journal.pbio.1002151

Kilkenny C, Browne W, Cuthill IC, Group NCRRGW et al (2010) Animal research: reporting in vivo experiments: the ARRIVE guidelines. J Gene Med 12(7):561–563. https://doi.org/10.1002/jgm.1473

Landis SC, Amara SG, Asadullah K et al (2012) A call for transparent reporting to optimize the predictive value of preclinical research. Nature 490(7419):187–191. https://doi.org/10.1038/nature11556

Maier H, Lengger C, Simic B et al (2008) MausDB: an open source application for phenotype data and mouse colony management in large-scale mouse phenotyping projects. BMC Bioinf 9:169. https://doi.org/10.1186/1471-2105-9-169

Maier H, Schutt C, Steinkamp R et al (2015) Principles and application of LIMS in mouse clinics. Mamm Genome 26(9–10):467–481. https://doi.org/10.1007/s00335-015-9586-7

McNutt M (2014) Journals unite for reproducibility. Science 346(6210):679

Menichino T (1992) A cost-of-quality model for a hospital laboratory. MLO Med Lab Obs 24 (1):47–48, 50

Mullard A (2011) Reliability of 'new drug target' claims called into question. Nat Rev Drug Discov 10(9):643–644. https://doi.org/10.1038/nrd3545

Nath SB, Marcus SC, Druss BG (2006) Retractions in the research literature: misconduct or mistakes? Med J Aust 185(3):152–154

Nonclinical Laboratory Studies: Good Laboratory Practice Regulations (1978) 21CFR 58. Federal register

Pesticide Programs: Good Laboratory Practice Standards (1983) 40 CFR 160. Federal register

Pusztai L, Hatzis C, Andre F (2013) Reproducibility of research and preclinical validation: problems and solutions. Nat Rev Clin Oncol 10(12):720–724. https://doi.org/10.1038/nrclinonc.2013.171

R Core Team (2013) R: a language and environment for statistical computing. R Foundation for Statistical Computing, Vienna

Raess M, de Castro AA, Gailus-Durner V et al (2016) INFRAFRONTIER: a European resource for studying the functional basis of human disease. Mamm Genome 27(7–8):445–450. https://doi.org/10.1007/s00335-016-9642-y

Schiffauerova A, Thomson V (2006) A review of research on cost of quality models and best practices. Int J Qual Reliab Manag 23(6):647–669. https://doi.org/10.1108/02656710610672470

Slusher BS, Conn PJ, Frye S, Glicksman M, Arkin M (2013) Bringing together the academic drug discovery community. Nat Rev Drug Discov 12(11):811–812. https://doi.org/10.1038/nrd4155

Stewart SR, Barone PW, Bellisario A et al (2016) Leveraging industry-academia collaborations in adaptive biomedical innovation. Clin Pharmacol Ther 100(6):647–653. https://doi.org/10.1002/cpt.504

Taussig MJ, Fonseca C, Trimmer JS (2018) Antibody validation: a view from the mountains. New Biotechnol 45:1. https://doi.org/10.1016/j.nbt.2018.08.002

Tenner E (2018) The efficiency paradox. Alfred a. Knopf, New York

Tuffery P (2015) Accessing external innovation in drug discovery and development. Exp Opin Drug Discov 10(6):579–589. https://doi.org/10.1517/17460441.2015.1040759

van Noorden R (2011) Science publishing: the trouble with retractions. Nature 478(7367):26–28. https://doi.org/10.1038/478026a

Wood DC (2007) The executive guide to understanding and implementing quality cost programs: reduce operating expenses and increase revenue. The ASQ quality management division economics of quality book series. ASQ Quality Press, Milwaukee

Yokley BH, Hartman M, Slusher BS (2017) Role of academic drug discovery in the quest for new CNS therapeutics. ACS Chem Neurosci 8(3):429–431. https://doi.org/10.1021/acschemneuro.7b00040

Printed in the United States
By Bookmasters